Introduction to PROBABILITY and STATISTICS
in the Life Sciences
Revised Printing

David C. Vaughan
Wilfrid Laurier University

Kendall Hunt publishing company

Photo Credits: Used under license from Shutterstock, Inc. Chapter 1, page 1, DNA © marema, 2012. Chapter 2, page 9, dice © STILLFX, 2012. Chapter 3, page 25, hurricane © Sandy MacKenzie, 2012. Chapter 4, page 51, board games © Elena Schweitzer, 2012. Chapter 5, page 73, Texas wildflowers © Rick S, 2012. Chapter 6, page 99, tropical fish © Rich Carey, 2012. Chapter 7, page 121, flower © Anastasiya Smirnova, 2012. Chapter 8, page 143, parrots © Anan Kaewkhammul, 2012. Chapter 9, page 175, suburbia in Arizona © Tim Roberts Photography, 2012. Chapter 10, page 189, seabirds on a bluff © Ewan Chesser, 2012. Chapter 11, page 209, chemical research © Alexander Raths, 2012.

Cover image © Shutterstock, Inc.

www.kendallhunt.com
Send all inquiries to:
4050 Westmark Drive
Dubuque, IA 52004-1840

Copyright © 2012 by Kendall Hunt Publishing Company

ISBN 978-1-4652-3176-5

All rights reserved. No part of this publication may be reproduced, stored in a retrieval system, or transmitted, in any form or by any means, electronic, mechanical, photocopying, recording, or otherwise, without the prior written permission of the copyright owner.

Printed in Canada
10 9 8 7 6 5 4 3

Contents

Preface vii
Acknowledgements ix

1 Current Problems in the Life Sciences — 1
    Synopsis 1
  1.1 Introduction 1
  1.2 Some Current Problems in the Life Sciences 3
      1.2.1 Genetics 3
      1.2.2 Global Warming and Prediction 4
      1.2.3 High-Dimensional Biology and Association and Correlation ... 5
      1.2.4 Comparison of Populations 6
      1.2.5 Ethics 6
  1.3 Basic Definitions 7
    Exercises 8

2 An Introduction to Probability — 9
    Synopsis 9
  2.1 Introduction 9
  2.2 Probability 10
  2.3 Basic Counting Methods 12
  2.4 Probability: Definition and Examples 14
  2.5 Conditional Probability 16
  2.6 Odds 18
  2.7 Sets and Set Notation 18
    Exercises 21

3 Distributions of Random Variables — 25
    Synopsis 25
  3.1 Introduction 25
  3.2 Distributions 27
      3.2.1 Discrete Distributions 27
          3.2.1.1 Discrete Distributions Based on Past Data 28
      3.2.2 Continuous Distributions 29
          3.2.2.1 Continuous Distributions: The Probability Density Function 30
  3.3 Bivariate and Multivariate Distributions 33
  3.4 Expectation 36
      3.4.1 Properties of Expectation 39
  3.5 Other Characteristics of Distributions 43
    Exercises 46

4 Special Discrete Random Variables and Their Distributions — 51
    Synopsis 51
  4.1 Introduction 51
  4.2 Special Discrete Distributions 52
      4.2.1 The Uniform Distribution 53
      4.2.2 The Bernoulli Distribution 54

		4.2.3	The Binomial Distribution	55
		4.2.4	The Negative Binomial Distribution	59
		4.2.5	The Geometric Distribution	61
		4.2.6	Hypergeometric Distribution	61
			4.2.6.1 The Multivariate Hypergeometric Distribution	63
		4.2.7	The Poisson Distribution	64
		4.2.8	The Multinomial Distribution	66
		Exercises		67

5 Special Continuous Random Variables — 73

Synopsis — 73
- 5.1 Introduction — 73
- 5.2 Special Continuous Distributions — 74
  - 5.2.1 The Uniform Distribution — 74
  - 5.2.2 The Gamma Distribution — 76
    - 5.2.2.1 The Exponential Distribution — 79
    - 5.2.2.2 The Chi-Square (χ^2) Distribution — 80
  - 5.2.3 The Beta Distribution — 81
- 5.3 The Normal Distribution — 82
  - 5.3.1 The Log Normal (Lognormal, Log-normal) Distribution — 86
- 5.4 Random Variables Related to the Normal — 87
  - 5.4.1 The χ^2 Distribution — 88
  - 5.4.2 Student's t Distribution — 90
  - 5.4.3 Snedecor's F Distribution — 92
- 5.5 The Bivariate Normal Distribution — 94
- 5.6 Other Important Results — 94
- 5.7 Final Observation — 95
- Exercises — 95

6 Parameter Estimation, Sampling Distributions and Inference — 99

Synopsis — 99
- 6.1 Introduction — 99
  - 6.1.1 Principles Used in Parameter Estimation — 101
- 6.2 Maximum Likelihood Estimation — 103
  - 6.2.1 Moment Estimation — 110
- 6.3 Least Squares Estimation — 111
- 6.4 Bayesian Analysis — 112
- 6.5 Sampling Distributions — 114
  - 6.5.1 Summary of Sampling Distributions — 115
  - 6.5.2 Summary of Fundamental Rules about Distributions — 117
- 6.6 Inference — 117
- Exercises — 118

7 Descriptive Statistics — 123

Synopsis — 123
- 7.1 Introduction — 123
- 7.2 Descriptive Statistics — 125
- 7.3 Measures of Central Tendency — 127
- 7.4 Measures of Variability, Dispersion, Spread or Scale — 129
- 7.5 Visual Displays — 132

		7.5.1	Visual Displays: Categorical Data	132

	7.5.2	Visual Displays: Continuous Distributions	134
7.6	Summary		140
	Exercises		140

8 Confidence Intervals and Hypothesis Testing 145

Synopsis 145
- 8.1 Introduction 145
  - 8.1.1 Interval Estimates of a Normal Mean, Variance Known 146
  - 8.1.2 Interval Estimates of a Normal Mean, Variance Unknown 149
  - 8.1.3 Interval Estimate of a Normal Variance 150
  - 8.1.4 Interval Estimate of a Proportion in Bernoulli Trials: Large Samples, Large Populations 152
    - 8.1.4.1 Interval Estimate of a Proportion in Bernoulli Trials: Exact Calculations including Small Samples 153
- 8.2 Hypothesis Testing 154
  - 8.2.1 Introduction to Hypothesis Testing 154
  - 8.2.2 Hypothesis Testing and Pivotal Statistics 157
  - 8.2.3 Structure of a Test of Hypothesis 159
- 8.3 Confidence Intervals and Tests of Hypothesis 160
- 8.4 One- and Two-Parameter Inference 161
  - 8.4.1 A Single Mean, Normal Population 161
  - 8.4.2 A Single Proportion 162
  - 8.4.3 A Single Variance 162
  - 8.4.4 Two Means 163
    - 8.4.4.1 Two (Independent) Normal Populations 163
    - 8.4.4.2 Two Dependent Normal Populations: Paired Comparisons 165
  - 8.4.5 Correlation in Dependent Populations 165
  - 8.4.6 Two Proportions from Independent Populations 166
  - 8.4.7 Two Variances 166
- 8.5 Non-Parametric Tests—An Introduction 167
- 8.6 Summary of Pivotal Statistics 167
  - Exercises 169

9 Goodness of Fit Tests and Contingency Tables 177

Synopsis 177
- 9.1 Goodness of Fit Tests and Categorical Variables 177
- 9.2 Contingency Tables 180
  - 9.2.1 2×2 Contingency Tables 180
  - 9.2.2 $r \times c$ Contingency Tables 185
  - 9.2.3 $2 \times c$ Contingency Tables 186
  - Exercises 187

10 Analysis of Variance 191

Synopsis 191
- 10.1 Introduction 191
- 10.2 The Global F Test: A Simple ANOVA 191
  - 10.2.1 ANOVA for Model 1: Fixed Effects 192
  - 10.2.2 ANOVA for Model 2: Random Effects 196
- 10.3 ANOVA in Complete Block Designs 197

10.4 ANOVA in Two-Way Classifications 199
10.5 ANOVA in Other Designs 200
10.6 Aspects of Experimental Design 200
    10.6.1 Principles in Experimental Design 202
    10.6.2 Specific Experimental Designs 206
        10.6.2.1 Completely Randomized Design 206
        10.6.2.2 Block Designs 206
        10.6.2.3 Blocking 206
        10.6.2.4 Complete Blocks 206
        10.6.2.5 Latin Squares 206
        10.6.2.6 Incomplete Blocks 207
        10.6.2.7 Factorial Designs 207
        10.6.2.8 Nested, or Hierarchical, Designs 207
Exercises 208

11 Linear Regression   211
Synopsis 211
11.1 Prediction 211
11.2 Prediction: New Values from Old 217
Exercises 218

Appendix A Notation   221
The Greek Alphabet 221
Sigma Notation 221
Common Notation and Terms 222
Discrete Distributions 222
Continuous Distributions 223

Appendix B Calculus Review   225
Improper Integration 227
Some Problems 229
Optimization 230
    Univariate Optimization 230
    Bivariate Optimization 230
Exercises 231

Appendix C Statistical Tables   233
Cumulative Distribution Function $\Phi(z)$ of the Normal, $z \geq 0.00$ 234
Quantiles of Student's t Distribution 236
The Upper $\alpha \times 100$ Percentage Points of $\chi^2(\nu)$ 237
The Upper $\alpha \times 100$ Percentage Points of $F(\nu_1, \nu_2)$: $\alpha = 0.10$ 238
The Upper $\alpha \times 100$ Percentage Points of $F(\nu_1, \nu_2)$: $\alpha = 0.05$ 240
The Upper $\alpha \times 100$ Percentage Points of $F(\nu_1, \nu_2)$: $\alpha = 0.025$ 242

References   244
Index   246

Preface

This text is designed to introduce students with at least first-year calculus to fundamental concepts of probability, as well as to a wide array of statistical methods based on those concepts. Thus, it is appropriate for any second-year university/college course introducing life science students to statistical analysis, especially in preparation for courses and projects involving data analysis. The statistical procedures concentrate on those commonly used in the life sciences to reinforce concepts by showing how the principles arise in relevant applications.

Calculus encompasses an essential array of tools and ideas for all scientists. As fundamental principles are better understood and models reflecting the complex interplay of many dynamic factors are developed, more sophisticated mathematical techniques are required to analyze these models. Many problems in the life sciences involve relationships between variables, and calculus is essential in exploring those relationships. Probability and statistics are used in quantifying interactions in a vast array of applications, and a complete understanding of the procedures arising from these disciplines makes extensive use of calculus (as well as algebra, geometry, and linear algebra). In general, students will have sufficient background in calculus to use this text if they have a good working knowledge of differential and integral calculus of functions of one variable (including limits and improper integration), and basic knowledge of partial derivatives and applications in optimizing functions of more than one variable. The multivariate material is readily introduced as an extension of univariate optimization, so it is not essential that students have had a course on multivariate calculus. A brief summary of the calculus topics mentioned here is given in Appendix B. All students in science programs (life, physical, social, formal) should be encouraged to master the fundamentals of calculus.

The term **life sciences** includes, but is not limited to: biology, chemistry, health, psychology, and kinesiology. In fact, the concepts and methods apply in any discipline in which statistical analysis plays an important role. The examples used to illustrate concepts and techniques in probability and statistics are couched in the language of problems arising in the life sciences to reinforce the value of these disciplines, but readily translate into analyzing problems in other disciplines.

One underlying theme in this text is that notation and mathematical derivations are essential elements in the development of students' skills and their understanding of probability and statistics. Many introductory texts avoid equations and formal calculations, but notation and mathematical rigor are important to ensure that students fully grasp the conditions under which procedures can be applied and the pitfalls to avoid in any statistical analysis. For example, one of the most common errors is to gather data before establishing the question(s) of interest and the necessary experimental design needed to get good information to address the questions. A driver of many procedures is the independence structure in the experimental design. For particular designs, a partitioning of sums of squares, essentially isolating terms related to specific design elements, explains the statistical analysis. Comfort with such methods not only helps with understanding of the topics introduced, but also prepares students to acquire new theories and methods, when needed, for their own experiments

The content of this book is more than a one-term course could comfortably contain. Even so, it does not cover all designs and the corresponding analyses students may encounter. It does, however, provide a sufficient background for students to acquire new methods, as necessary. Students need the direct experience of doing the calculations related to a range of designs to be able to acquire new statistical procedures and to read and understand comprehensive analyses of data in research works.

It is not enough to know how to use drop-down menus in some statistical package — users of statistics must know what the package is doing to their data and whether the chosen procedures are appropriate. Practitioners need to know what applications in statistical packages to use and how to select techniques, such as data plots, summary statistics, and test of hypotheses, from the large number of procedures available. The achievement of a high level of proficiency with statistical techniques requires a solid grasp of the fundamentals of probability and statistics.

It is valuable reinforcement to see how useful statistics is as a discipline in providing insights into subjects that interest students. The life sciences provide a rich array of standard and modern problems to showcase general and specific procedures, and are used extensively throughout this work text to reinforce concepts, techniques, and application procedures.

The first chapter provides motivation for acquiring skills in probability and statistics by posing a number of discussion problems. Basic concepts, theory, and applications of probability and statistical distributions are developed in Chapters 2 and 3. Properties of special discrete and continuous distributions, with applications to modeling outcomes of experiments, are presented in Chapters 4 and 5. Major principles used in determining estimators and their distributions are outlined in Chapter 6, and Chapter 7 discusses some of the methods in descriptive statistics to help explore the validity of assumptions. Applications of the preceding material to confidence intervals, tests of hypotheses, goodness-of-fit, contingency tables, analysis of variance, and linear regression are presented in Chapters 8–11. A brief introduction to fundamental concepts in experimental design is given at the end of Chapter 10.

Acknowledgements

This book grew out of notes developed in teaching *Statistical Methods for the Life Sciences*. This course, labeled MA241 at Wilfrid Laurier University, resulted from in-depth discussions between the Biology and Mathematics Departments in trying to improve the statistical education that life science students needed. These discussions, along with extensive examination of applications of statistics and probability, provided the background and motivation to write such a book.

I am very grateful for the support Wilfrid Laurier has provided to me over the past several months as I tried to bring this project to fruition.

Denise Gutermuth was the laboratory supervisor for this course over several years. She developed a significant number of problems that have been incorporated into this work.

This book may never have been completed without the support of my wife, MaryAnn, over my career and the production of this text. My daughter, Jennifer, provided extensive feedback on a number of chapters, and her observations clearly led to significant improvements. I am truly grateful to them both.

CHAPTER 1

CURRENT PROBLEMS IN THE LIFE SCIENCES

Synopsis

This chapter introduces a wide range of problems in the life sciences that are highly dependent on probability and statistical methods for their analysis. These examples illustrate the vast range of important problems currently under study that use probability and distributions of variables in developing models and then statistical methods for the analysis. Later chapters will show how important it is to have a solid theoretical understanding, as well as the computational mastery, of distributions and their properties. This chapter explains many of the standard terms used in scientific analyses. Exercises that develop skills in acquiring information from various sources and in thinking about the factors that should be considered in model development are provided. This is a non-technical introduction to many of the basic concepts and definitions used throughout this text.

1.1 INTRODUCTION

The Life Sciences encompass many disciplines, including Biology, Health Sciences and Health Studies, Environmental and Green Sciences, Chemistry and Pharmaceutical Development, Kinesiology and Physical Education, Biotechnology, and related areas of study. These disciplines are attacking many of the most important problems facing society, including the effects of:

- An aging population,
- climate change,
- chemical and pharmaceutical pollutants,
- the spread of disease and pathogens, and
- physical and psychological stressors.

Scientists develop models of how various factors, based on underlying principles and relationships related to the problem under study, are related. From these models, the investigators provide quantitative analyses that, for example, help make efficient use of resources. Based on this kind of analysis, effective strategies to control or eliminate factors that have a negative impact, or to enhance those that have a positive impact, can be developed. It is often the case that the same or similar principles and relationships occur in many different problems and contexts, meaning we can learn from analyzing approaches used in one problem that may apply to others.

Mathematical language, including probability and distribution theory, makes it possible to express in a succinct way these relationships by introducing equations involving variables and conditions on those variables that reflect the system under study. Mathematical and statistical methods are then used to analyze these equations and express their solutions in the language of the original problem. With a model available, scientists can explore the impact of changing parameters and conditions to determine the response of a system to such changes. Often, such models incorporate a random (or **stochastic**) element, because factors affecting the outcome of a procedure or process cannot be controlled. The effect of this randomness is quantified through assumed distributions of the random variables and the procedures that arise from these distributions. This text concentrates on problems with such random elements and the corresponding models and their analysis.

The rapid increase in computer power has made complicated problems more amenable to analysis than even a few years ago. As more and more variables or factors are taken into account (leading to greater precision and control of important systems), the greater the need for computer methods to carry out the numerical calculations. In many cases, an analytic solution to equations or systems of equations relating the variables in a particular problem is simply not possible. Given the number of variables and parameters in sophisticated models, researchers rely on **computer simulations** and **Monte Carlo** techniques to generate highly accurate (approximate) solutions. The effects of changing parameters and using random values in the stochastic inputs can be examined quickly and effectively. An exercise readers can undertake is to find out more about these techniques and how they are used in the life sciences.

It must be emphasized that in the development of a model, certain assumptions are made. Usually, problems in the life sciences are very complicated, because living organisms are not static, and many factors affect their growth and development. For example, one major problem in trying to find a cure for a particular disease is to first find out what causes it. In the last century, many became suspicious that smoking tobacco led to lung cancers and emphysema (and other ailments). However, it was difficult to separate the effects of the general environment and an individual's predisposition to such diseases from the effects of smoking. Some people who have never smoked develop these diseases, and some people who smoke a lot never develop them. How can a scientist separate the other factors from the effects of smoking in a quantitative way? This same issue is being played out today in the climate change debate: How much of an impact does specific human activity have on weather? If some changes are observed in relation to historical records, such as warmer weather, more severe weather patterns, rising ocean levels, disappearing ice packs and glaciers, changes in animal behavior, and greater incidence of particular types of melanomas, are they due to human activity negatively impacting our climate, or are they due to normal weather cycles? Clearly, these are important problems needing immediate attention and resolution.

This text covers many of the statistical principles and methods used to quantify the level of certainty we have in **inferences** drawn about a population, based on sample information. Examples illustrate these techniques, but it should be emphasized that learning a set of formulae is not sufficient. It is also important to understand why things work, and so we know when they cannot be used.

One of the first steps taken in analyzing a problem is to determine precisely the question or questions to be answered. A common mistake is to gather data and then try to determine if the information supports answers to new questions. On the other hand, there are very large data bases, full of information that may help us solve important problems, and we have no idea if there are relationships between variables, or what the nature of those relationships might be. For example, each time you purchase something on a credit or debit card, the information about the items purchased, where they were purchased, how many of each item, etc., is recorded. The intention is to determine consumer habits, but most companies do not have the expertise to examine this information in detail. A lot of work is underway related to analyzing **massive data sets**, using techniques in biostatistics and informatics. Although we will not be exploring these issues, the fundamental mathematics, statistics, and probability used in this work form the basis of such analyses.

To make full use of the power of probability and statistics, we use the question to be answered to help us **design** the experiment. For example, it is necessary to have a very clear idea of the **population** under consideration. Often, we start out wanting to examine some particular aspect of a general population, but the means at our disposal do not let us gather information from all groups in the population equally, or from specific groups at all. If we are studying the effects of acid rain on fish populations in Northern Ontario, it is likely there will be lakes and streams that are inaccessible. Further, for many types of problems we examine, there are some factors we will control and some that we either can't or choose not to control. This has an important implication: When we observe or measure an aspect of a member of the population, generally there will not be an exact value for that individual, perhaps due to the limitations of the measuring devices or from other influences. If we could repeat the experiment on that individual, small fluctuations would occur because of these uncontrolled factors. We call such experiments **random experiments**. Such values have a **random** or **stochastic** component, and in our analysis, we have to account for that randomness. This is done through probability and statistics.

In most problems, **cost** (in terms of time, resources, and ethical considerations) is a significant issue, so we have to design the procedure or **experiment** so that we have an efficient means of acquiring enough **data** of interest to reach the level of precision required. Often, an overarching goal is to quantify our **confidence** in the inference(s) we want to draw, and this requires a sufficiently large sample that is taken under very specific conditions. It is possible that we collect information that is not representative of the entire population under study or does not conform to the conditions necessary for statistical analysis, and that would mean we may draw an incorrect inference about the whole population. For example, we randomly select adult bears from a large forested area in order to study the effect of a prolonged drought on their general health. It is quite possible the method of capturing the bears might lead to selecting a sample of bears that are weaker than average or are closely related, and this would **skew** our results.

Basically, we need to measure how often outcomes (single instances or collections of possible values of the variable(s) being measured or observed) of interest should occur when a population has particular characteristics. The method used to quantify the relative rate of occurrence of an outcome or set of outcomes is based on **probability**. Probability is a measure of the relative rate at which we expect to see certain outcomes or collections of outcomes (called **events**) occur, relative to all possible outcomes. Certain types of questions and populations need particular **experimental designs**, that is, structured procedures to get data that allows for an analysis by probability. We will consider some of the populations, questions, designs, and analyses often encountered in the life sciences.

1.2 SOME CURRENT PROBLEMS IN THE LIFE SCIENCES

Biology is entering a new era. Much of biology has been observational in nature, partly due to the difficulty in controlling factors that may be relevant. However, new techniques have allowed biologists to examine in close detail elements that could not have been observed even a few years ago. By using these techniques, we can start to extract some of the underlying principles that connect the different scales of biological systems. One consequence is that many factors now can be considered. For example, it is emerging that it is not a single gene that controls a particular process, but a complex interplay between genes and other molecules, as well as various stimuli. To determine how things happen, we need to determine the major factors and eliminate extraneous factors. Statistical analysis is one of the tools used to separate the useful from the extraneous.

We will take a brief, non-technical look at some of the problems biologists and health scientists are working on, and briefly introduce some of the statistical concepts used in the analysis.

1.2.1 Genetics

Genetic analysis is one of the most fascinating areas of research and analysis. Important information now goes from the laboratory to the courtroom or to the medical profession in a very short time span. For example, methods for comparing parents' DNA with offspring DNA have made identification certain rather than guesswork. Parents are often encouraged to have their newborn children's DNA put into a database to help with future identification. Determination of potential health risks through DNA analysis is a burgeoning area of study and produces many ethical issues. It is even possible to take a human DNA sample and other measurements, and determine the likely global area of origin of an individual. The analysis of the structure of DNA has established that important characteristics of an individual are controlled (at least to some extent) by specific genes and combinations of genes. Often, there are different versions or **alleles** of specific genes, and combinations of these alleles produce variants in individuals such as hair color or resistance to specific diseases. DNA analysis is expected to continue to grow in importance in the prevention of disease and in mitigating the effects of aging, for example.

Genetics play a significant role in determining health, disease resistance, size, strength, and other characteristics. These are not controlled by a single gene, or even just by genes. The study of **genomics**, that is, the study of the entire collection of genes, is a growing area of interest because of this complex interplay between individual genes. The same is true of **proteomics**, the study of the collection of proteins. Determining which genes and proteins are actually involved in a particular process requires sifting through large databases and the use of statistical methods to find significant relationships. As our understanding of how processes are affected by, for example, protein folding, better strategies for combating certain diseases evolve. Our understanding is advanced by quantifying underlying principles, formulating a mathematical model, and then analyzing the strengths and weaknesses of that model.

> **Discussion Topic 1:** Growing evidence suggests that environment impacts physical characteristics. For example, in a generation, the average height of a Japanese soldier increased by about 15 centimeters. How is this possible? Discuss.

This example illustrates just how complex problems can be. The interplay between one's genetic makeup and the environment is a fact, so that one cannot really study just one possible factor that would impact a particular characteristic. On the other hand, creating models that include all possible factors can be a problem in that the model is so complex it loses clarity. Often, researchers resort to **Occam's Razor**, as expressed by Isaac Newton: "We are to admit no more causes of natural things than such as are both true and sufficient to explain their appearances." Another way to think of this is as parsimony in model construction: If two models make the same predictions, the simpler model is better. Often, factors are linked, sometimes weakly and sometimes strongly. If two factors are strongly linked, then the information in one is essentially the same as the information in the other; so only one of them is needed (at least, generally).

Developing a good model takes time and experimentation, along with practice in expressing principles in mathematical language. To understand the outcomes of our experiments, it is important to know where equations come from and how they are manipulated. It is also essential to know the limitations of a particular formulation: It is often assumed that random variables follow a **normal** distribution, even when the data gathered does not support such an hypothesis. This is one of the reasons many analyses go wrong — making assumptions that are not supported by the information gathered.

1.2.2 Global Warming and Prediction

Is global warming a real phenomenon, or is the Earth going through cyclical climate change? What kind of models are being used to represent the influences on Earth's atmosphere? The atmosphere is a complex system, influenced by the strength of the ozone layer, cyclical behavior of the sun, release of ash clouds from active volcanoes, the release of pollutants from factories, and many other factors. Over the past few decades, we have seen increased incidences of melanomas, marine life beachings, severe weather, and shrinking ice caps (to name a few climate-related problems). Which of the factors affect the atmosphere to a measurable extent, and how confident are we that there is a cause-and-effect relationship between these factors and the observed problems? Such questions are part of serious ongoing debates. The actual determination of the strength of these influences on life on Earth will require sophisticated mathematical models, and techniques from computer science and probability and statistics.

> **Discussion Topic 2:** If the Pacific Ocean's average temperature in the top kilometer increases by 1° Celsius, what impact will that have on marine life? How many major storms, and of what category, will we expect to see, in comparison to past years at the lower temperature? How do researchers generate their predictions? Find sources of information on these topics, and discuss the analysis scientists have presented.

Predicting the most probable outcomes associated with changes in one or more variables is essential for the development of emergency response plans, for food production, for changes in habitable regions, and many other elements of our lives. Other examples include:

We want to know the purity of a chemical produced under different temperatures, as part of a plan to optimize a process in analyzing bacteria content of samples from a meat processing plant. It is not possible to examine all temperatures in a specific range, so four specific temperatures are selected, equally spaced on a Celsius scale, including the smallest and the largest acceptable values for the process. We want to then predict the purity for a temperature not observed.

If we ensure elementary school-aged children have adequate food, will their academic performances increase? Will their social skills improve?

Will the use of growth hormones on cattle and other livestock lead to health risks for people living near farms where the hormones are being used? What is a safe level of these hormones or their by-products as they break down in the ecosystem and in the people ingesting meat from these animals?

As we gain a better understanding of our world and our relationship with it, we can improve our quality of living. As we come to grips with the multitude of factors and their interactions that impact our lives, the greater the need for mathematics and statistics to help us quantify these effects.

> **Discussion Topic 3:** A biologist is examining the impact of the introduction of a new type of mussel in the Great Lakes waterway. In particular, he wants to know the rate of growth of this population so as to determine the most probable impact of this new species. Find examples of the introduction of non-native species of plants and animals, and discuss the dangers and the reasons for these dangers associated with such intrusions.

In testing a new drug, it is important to determine the correct dosage to achieve relief of a particular symptom, while minimizing any negative side effects. Body type and sex of the patient have been shown to be significant factors, and it is not feasible to study all dosage levels with all combinations of these other factors. A medical researcher chooses certain dosage levels to administer to some of the possible combinations, to determine a grid of dosages for the different factor combinations.

> **Discussion Topic 4:** Determine some of the important ethical considerations in carrying out **clinical trials**, and discuss protocols that might be necessary in critical situations.

1.2.3 High-Dimensional Biology and Association and Correlation

Studies of the interrelationships between different factors is complicated by the sheer number of elements that may be involved. Large databases may include hundreds of variables and millions of observations. High-dimensional biology is a relatively new discipline dependent on statistical methods to sort through the many different possible relationships between a vast array of variables. It is important to isolate those variables that seem to be interrelated, and one way is through measures of **statistical association**. One particular measure is **correlation**, which is a way to quantify the strength of relationship between two variables. Genes often work in concert, rather than singly, and so attempts to determine which genes are producing which effects under specific conditions rely on determining relationships between potentially hundreds or even thousands of genetic polymorphisms, gene expressions, peptides, metabolites, proteins, and combinations of these. From this information, pathways and networks can be identified that determine how an organism will respond to stimuli.

Simpler problems may be of the form of determining if two variables are statistically related, or if having information about one has no impact on what we might see with the other. This second condition is called **independence**. On the other hand, if two variables are **significantly** correlated, then we can use one to help predict the other. Sometimes, in fact, two variables are so highly correlated that it is possible to eliminate one from the model, since the information contained in one is essentially the same as in the other.

A problem in this area is that often many tests of the presence or absence of a relationship are run, and the chances of a **false positive** or a **false negative** increase. By this, we mean that we think there is a statistical relationship when there isn't (a false positive) or think there isn't when there is (a false negative).

> **Discussion Topic 5:** One reason that the genome project is considered important is that it is expected that a complete gene map would allow researchers to develop principles that would give great predictive power in, for example, health. Given the current state of the genome project, discuss some of the issues that must resolved for it to achieve this goal.

1.2.4 Comparison of Populations

Many problems involve two or more populations, and determining if there is a distinction between them related to a particular characteristic can have important consequences. Two varieties of wheat may have different levels of resistance to disease or drought. Age may be a factor in the symptoms experienced for those who catch a particular strain of flu. Boys and girls may learn at different rates, under different teaching methods. Life expectancy for different groups with a particular disease may differ. Allergies may be more prevalent in children raised in certain environments than in others. Statistics and probability help us determine if two or more populations have the same characteristics or, in fact, there is a measurable difference.

Several independent populations may be compared at once. Agricultural areas with different characteristics, such as soil acidity, rainfall, sunlight, insects, and other factors may have different yields per acre of a strain of corn. Different levels of weekly activity may impact cardiovascular health. Forest fires may be more prevalent in forests in different climate zones. Comparisons of several subpopulations at once provide insights into the differences, if they exist, of characteristics of importance.

The examples above are assumed to involve **independent populations**. Many times, **paired comparisons** are of interest. In this case, we want to apply two treatments to the same experimental unit or subject, and then repeat this for several subjects. In human experiments, one way to do this is to use identical twins; in other cases, lab animals bred for consistent genetic makeup can be used.

> **Discussion Topic 6:** Heavy metal pollutants can have a significant impact on the food chain. If you were to develop an experiment to examine this kind of pollution in an ocean and its effects on, say, lobsters and cod, what would be some of the issues to take into account? Would it in fact be reasonable to compare these populations?

1.2.5 Ethics

There are other considerations a practicing biologist or life scientist must take into account. In clinical trials set up to test new therapies, how many people should be put at risk by using a placebo or if a drug with unknown side effects is administered? How ethical is it to use animals in experiments? How much control should we exert over the environment to the detriment of some creatures or plant life to our own benefit? Government and research institutes try to address these questions by introducing protocols scientists must adhered to (with significant penalties if they do not), requiring signatures from participants to try to ensure these protocols are followed.

There are other concerns about how we treat our data. Some researchers may "clean" their data of possible outliers, or other suspect values, without valid reasons other than they don't look right. Researchers can be affected by who is paying the bills. It is often the case that nonpositive results will not get published, whereas positive results that indicate differences will. Part of the training of researchers and workers in the biological and health sciences is in ethical behavior.

> **Discussion Topic 7:** An experimental treatment of a particular disease, run by researcher A, has a strict profile of patients who are eligible to participate. Researcher B has a patient with the disease (and will die within weeks without the new treatment and is likely to die even with the treatment), but who does not fit

Chapter 1 — Current Problems in the Life Sciences

> the profile in one key aspect. B makes use of a drug to change that characteristic, and her patient is then accepted into the program. Discuss the implications of this scenario.

1.3 BASIC DEFINITIONS

In order that information be given in a precise and concise way, a common language must be set. Some specialized terms will be used throughout this text, and the following list will prove useful in describing the many problems in the life sciences in which statistics and probability play a substantial role.

Statistical Inference: the process of drawing conclusions about a population under study from the information contained in a sample from that population.

Descriptive Statistics: the presentation of data in tables, graphs, or other displays, including summary values describing the entire set of values.

Population: the complete listing of the individuals or items under study.

Sample: a subset of the population, selected in a particular way.

Simple Random Sample: or just random sample and abbreviated r.s.: a sample of size n from the population, selected in such a way that every member of the population has the same chance of being selected for the sample.

Replacement: in a sample, when an individual is selected for the sample, and then is put back and can be drawn again.

Without Replacement: in a sample, when once an individual has been selected, that individual cannot be selected again.

Random Number Table: a table of digits [0 through 9] generated in a way so that each digit has the same chance of being selected each time a value is chosen. Such tables are often used with finite populations.

Categorical Data: a classification into classes or groups, usually a small number of such, where there is no ranking among the classes.

Pareto Diagram: a bar chart used to pictorially represent data, often about the performance of machinery or processes, in such a way that the most frequently occurring categories begin at the left. Pareto expressed the idea that usually there will be a few very frequently occurring failure types, and several relatively rare ones, and this kind of diagram helps convey such observations.

Dot Diagram: a chart of observed values, represented as dots, above a horizontal axis representing the frequency of specific observations. By choosing different symbols, information about the same quantity, but from different times or under different circumstances, can be represented and compared.

Ranked Data: a classification into classes or groups, where there is a ranking among the classes.

Metric Data: a measurement made on the individuals in the population. The values of the measurements fill out an interval of nonzero length.

Experimental Units: the individuals in the population under study.

Random Experiment: any experiment in which the outcome cannot be determined with certainty until the experiment has been completed.

Random Variable: a function defined on the outcomes of a random experiment.

Score, Outcome, or Datum: the value of the random variable as recorded for an individual from the population.

Random variables are typically denoted by X, Y, Z etc. and values of the variable by x, y, z. Many experiments have common elements. One of the goals of this course is to try to classify certain types of experiments, which is typically done by classifying the properties of the random variable. From these classifications, we can determine general properties of these variables, so that when an experiment of the appropriate type is being considered, we simply appeal to these properties.

Exercises

Note: In the following problems, you are asked to gather information from various sources. It is not enough to simply download something from the internet or to copy something from a text. One purpose of these exercises is to have you gain a wider knowledge of problems in the life sciences. But another very important aspect is for you to express what you have learned in your own words. You should use at least two sources, and you should think about the information and then write down how you understand it. To do that, you should approach the write-up as if you were going to tell someone else (who may not be an expert) what you have learned. Clear exposition is essential in scientific writing, and practicing expository skills is as much a part of a researcher's toolkit as the actual techniques used in mathematical and statistical analysis.

There are no absolutely final answers to these questions. It is also a useful exercise to consider what is not known and the reasons why some issues have not been considered fully.

1. Using resources such as the internet, discussions with faculty, textbooks, etc., find and describe two major problems in the life sciences under current analysis. Describe at least one mathematical model being used and some of the tools (without great detail) used. Use the terminology in this introduction.

2. A medical researcher wants to run a study of the effects of a new therapy on people with rheumatoid arthritis (RA). She goes to a local hospital, screens patients coming into a rehabilitation center, and chooses the first 10 patients who have RA from those coming to the center. Is this a good way to select a sample of people for such a study?

3. A fourth-year biology student is working on a project concerning the impact of heavy metal pollution in Lake Superior on fish populations. He goes to a small cove on Lake Superior and spends two days fishing. He wants to use his catch for the project. Discuss his methodology.

4. A health researcher feels that productivity of employees in the financial industry would improve if 30 minutes of physical exercise were incorporated into the work day. Discuss some of the factors that she will have to take into account to get data that would support (or refute) this assumption.

5. Using resources such as the internet, discussions with faculty, textbooks, etc., find and describe the concept of **alleles**. Describe the theoretical distribution of possible allele pairings (one from a male, one from a female) for a given gene in a population if there are two alleles and if there are three alleles. How would you go about determining if these theoretical values are reasonable for a given population? What factors could influence the distribution of such pairs in a population?

6. Using resources such as the internet, discussions with faculty, textbooks, etc., find and describe the human genome project. Is this project complete, and if it isn't, what needs to be done? Also, describe at least two potential benefits of this project and at least two potentially negative aspects.

7. Using resources such as the internet, discussions with faculty, textbooks, etc., describe some of the impacts of an aging population. In particular, consider some of the problems being analysed in the **Canadian Longitudinal Study on Aging (CLSA)**.

8. Using resources such as the internet, discussions with faculty, textbooks, etc., describe epidemiology. What is a pandemic? Describe two ways in which statistics plays a role in epidemiology.

9. In developing new drugs, there is a major problem: the number of different combinations of chemicals. Often, the sheer number of possible combinations makes it impossible to develop and test all combinations. Using standard resources, describe how probability is used to reduce the total number of combinations to a manageable number.

10. In clinical trials, protocols may include **single blind** or **double blind**. Describe these terms and why they are deemed necessary. Describe the **placebo effect**. Describe at least two ethical considerations in setting up a clinical trial on humans.

CHAPTER 2

AN INTRODUCTION TO PROBABILITY

Synopsis

The fundamental concepts and definitions that form the foundations of probability theory are introduced. Examples illustrate the need for a way to quantify any inferences drawn about a population from a sample taken from that population. Definitions of standard terms are introduced. A range of exercises is provided to help develop fluency in the language of probability and to examine basic concepts.

2.1 INTRODUCTION

In a study of relationships between aging and physical characteristics, several men age 25 were **selected at random** from a large population in Canada. A number of physical measurements were made, including height, in centimeters. One man was measured at 193 centimeters. Is this unusual? How would we be able to tell?

It is assumed that there are two alleles, say A and B, for the gene related to pea plant height, occurring in equal frequency. AA produces tall plants, AB produces medium height plants, and BB produces short plants. How would we verify that the assumption of the frequency of occurrence of each allele is valid?

The level of acidity in soil and its impact, if any, on corn yield is under study. Generally, it is believed that the higher the acidity, the lower the crop yield. How do we determine if there is such a relation, how certain can we be of this result, and how we can develop a formula that will allow us to predict corn yields for given levels of acidity?

In a parapsychology experiment, randomly selected students in Psychology form a subject group. Each is put into a soundproof and windowless booth, and, through headphones, is prompted to guess which card, of 5 cards with individual shapes, is being looked at by a control operator. This is repeated 25 times for each subject, and the number of correct guesses is recorded. One subject made no correct guesses. Is this unusual?

A clinical trial is to be established to determine the efficacy of a new drug on memory retention in Alzheimer patients. There is a current drug on the market, and approval for a new drug requires establishing that the new drug is more effective than current therapies. How does the pharmaceutical company set up the trial, and what is meant by "more effective"?

A new virus has been isolated. It can be deadly, and passes by contact from individual to individual, although some people do not exhibit symptoms regardless of the exposure. Health scientists are looking at the possible spread of this virus in the general population. How do they analyze the spread of the disease and the ultimate effect of a pandemic?

Answering questions such as these requires methods for quantifying how confident we are in the answers we give, or even how to answer the question in the first place. Rarely, if ever, will we be exactly certain: There will always be uncontrollable or unobservable factors that can influence the results, and we often have a limit on the accuracy of the measuring devices. An early problem was to measure the distance from the Earth to the moon. It was complicated by the weather, including humidity, air pressure, temperature, and similar factors. It was from this problem that scientists learned that by combining repeated observations that individually had errors in them, they could actually reduce the error by averaging observations. This concept was not universally accepted until the process of combining observations was put on a rigorous basis through the development of probability and the concept of experimental design and analysis.

2.2 PROBABILITY

In this section, we will consider some basic probability theory, rules, and methods. Probability can be viewed as the mathematics underlying statistics, as far as the material in this book is concerned. It is assumed we have a clearly defined question or set of questions to answer about some population. Typically, there will be some restrictions that must be adhered to, and one consequence is that we may have to revise the actual population to be studied. This may be a subset of an original population under study, with the subset chosen because of access, time, or other considerations. For example, we may *want* to determine the effect of a given heavy metal (mercury, e.g.) in the Great Lakes system on fish growth. However, we only have ways to get top-dwelling fish for the study, so at best the actual population is top-dwelling fish, not all fish, in the Great Lakes.

We set up an experiment not just to get data, but to get data that can be used to quantify the level of confidence we have in the answers we give to the questions. It should be noted that getting data by arbitrary means is usually not helpful: Our ideas and methods related to quantifying results rely on the adherence to rules for getting sample data. (Recall that a sample is some subset of the whole population.) If we want to use our observations from a small group chosen from the population to make inferences about the entire population, we cannot simply choose the units we want based on convenience. The most effective way is to choose a sample that forms a **simple random sample**, often denoted by **SRS**. We will discuss this and some of the methods used to generate such samples. When our sample is not of this type, we may not be able to determine a reasonable level of confidence in our answers. On the other hand, a SRS may not be the best way to get data about populations that have different subpopulations of interest (for example, male and female; old and young; diseased and disease free, etc.). **Experimental design** is used to get the best usable information to answer questions in such instances. We will consider some of these issues as we develop the various statistical procedures in this book.

The members of the population are usually called **experimental units**. We must be able to apply the experimental procedure to each member of the population, under constant conditions. We will either observe some value or attribute, or measure some value or attribute. To formalize this, we will have an outcome X of interest related to the members of the population. X may be a measurement, a count, or a placement in a list. (There are other types of outcomes, but we will consider these common types.) Theoretically, we assume we can determine in advance of performing the experiment the possible values of X. The set of all possible outcomes is the sample space, usually denoted by S. Normally, populations are too large for a census, and so we take a subset and determine X on this subset and call it a sample. The types of experiments we are interested in are called *random*, meaning that we cannot determine the outcome until we actually run the experiment. Examples of such experiments include:

Flip a coin and record the face up. We get either heads or tails, but we don't know which one until the coin has been flipped.

Apply a particular drug treatment to those with a particular disease or condition. We cannot determine in advance what the impact will be on a patient, although we can say there will be improvement, no improvement, or harm.

The purity of a yield from a chemical process under different temperatures and pressures is under study. We cannot determine the purity until the process has been run, and this could be expressed as a percentage.

Typically, in an experiment there is at least one **event**, say E, of interest. An event is simply a subset of the sample space S. Since it is often convenient to code the outcomes, any set equivalent to the complete listing of the outcomes will also be called the sample space. For example, we can code heads as 1 and tails as 0. If the original sample space is a list of non-numeric values (eye-color; win-place-show-also ran; etc.), we generally code the outcomes with some subset of integers and then replace the original sample space with this set of numbers. When we measure a particular attribute, the possible outcomes are numbers that fill out an interval (usually just one), and so we shall assume all sample spaces are sets of numbers. There are times, however, where the information in the original, non-numeric sample space is very useful, and so we switch back and forth as needed. We often call the individual outcomes in the sample space **basic** or **elementary events**.

Chapter 2 — An Introduction to Probability

There will be some events of interest, determined in advance, and what we want to know is how frequently we expect to see such events happen. One way to phrase this is: We want to know the probability that event E will happen when we perform our experiment. The usual notation is **P(E)**, read "the probability of (the event) E." This value is required to be a number between 0 and 1, and can be viewed as the relative rate of the occurrence of E, relative to all possible outcomes of the experiment.

Example 1: We toss two fair and distinguishable (say one red, one white) dice, and record the faces up in ordered pairs, with the red face first, the white face second. The sample space is {(1,1),(1,2),...,(6,6)}. In this case, the outcome X is the ordered pair resulting from one toss of the dice. There can be other ways to express the sample space, but we must always have a one-to-one correspondence between the possible outcomes for each outcome and the sample space. A *random variable* (**r.v.**) is a function on the sample space. For example, a r.v. may be Y, the sum of the faces up, with 2,3,4,...,12 a listing of the values of this random variable. There is not a one-to-one correspondence between Y's values and the outcomes in the original experiment, namely the recording of the faces up. That is not required. However, note that a value of Y corresponds to some event in the original sample space. If Y in our example is 4, then this corresponds to the event {(1, 3), (2, 2), (3, 1)}.

Recall that a function f from set A to set B is a relation with the property that for each $a \in A$, f assigns a unique element $b \in B$, and we write $f(a) = b$. More than one element in A can be assigned to the same value in B. For example, the function from the real numbers that assigns the same real number c to all reals is called a constant function. The set A is called the domain, and the set of values b in B for which there is a value $a \in A$ with $f(a) = b$ is called the range of f. If the function f has the property that for each b in its range, there is only one $a \in A$ for which $f(a) = b$, then we call f one-to-one. If the values in the range of f are real numbers, then we call it a real valued function, and generally, this is the type of function we use throughout. Thus, a random variable is a function from a sample space into the real numbers. We do not use the notation of functions for random variables in probability and statistics. Instead, random variables are denoted by uppercase letters (typically) such as X, Y, Z, and then specific values are denoted by corresponding lowercase letters. When we carry out some experiment and observe a value the random variable X, this value x is called a **realization** of X. Basically, a random variable represents the end result of applying a function to a sample space and a realization of the random variable is the actual recorded value of the random variable.

Returning to our example, consider the event E = the sum of the faces up is 7. This is the same as saying the value of the r.v. Y is 7. Because the dice are assumed fair, all combinations of the faces are going to happen with the same frequency from either theoretical considerations (the dice are fair) or from long runs (that is, a large number of repetitions) of the experiment. Thus, (1,1) occurs as often as (4,3). There are 36 possible pairings of the faces so, for example, (1,1) would occur 1 time in 36 in long runs of the experiment (that is, performing of the experiment a large number of times). This does not mean we are guaranteed that (1,1) will occur once in every 36 rolls of the dice. It means that if we carried out the experiment many times, say N, then approximately N/36 of the outcomes would be (1,1). With this understanding of what P((1,1)) = 1/36 means, we can go back to our original problem of finding P(E) = probability the sum of the faces up is 7. Clearly, we need only count the number of distinct ways a total of 7 can be rolled, and then divide that by 36. Thus, events in the sample space that produce the value 7 in the random variable Y are (1,6), (6,1), (2,5), (5,2), (4,3) and (3,4) so P(E) = 6/36. The event E would be written E = {(1,6), (6,1), (2,5), (5,2), (4,3), (3,4)}.

You should try the experiment of rolling two distinguishable dice many times (say 100 times) and recording the total of the faces showing up each time. It is not likely that in your experiment you will see the sum 7 occur one-sixth of the time (100/6 is not a whole number), but this should be approximately what you see. It is, however, possible that in 100 rolls of the dice you never observe the value $Y = 2$ in this experiment. Does this mean that your dice are not fair in some way? Not necessarily: It is possible that the dice are fair, and it just happens that $Y = 2$ was never observed for your sample of rolls of the dice. At least theoretically, we could roll the dice an unlimited number of times, so perhaps had your sample been 1,000 instead of 100, you would have observed $Y = 2$ several times. One important reason to introduce probability is to quantify just how rare (or not) a particular outcome or set of outcomes is.

This example also illustrates how important it is to keep track of all the information in the original sample space so that you can construct probabilities for related random variables on the sample space. In this case, we can argue that the elementary events in the sample space S occur with the equal probability, so that in determining probabilities of values of the random variable, we need only count the number of elementary events in S that give rise to the particular values of the random variable.

Another thing the above example illustrates is that it is necessary to be able count the number of outcomes of a particular experiment and the number of outcomes in events. Thus, certain **counting principles** are necessary. In general, if we have an experiment in which each outcome is equally likely to happen, and there are N distinct outcomes (this can be stated as the sample space has N equally likely elements), then for event E, we will take $P(E) = \#E/N$, where $\#E$ is the total number of ways that E can occur. Note in the above example, this was the situation we had. If we assign to each elementary event in such sample spaces a unique number from 1 to N, then this is a random variable, say W, and it is a one-to-one function. Another way to express this situation is: We want to assign a probability to each value of W. Since it is assumed that each outcome in S is equally likely, then $P(W = i) = 1/N$, for each $1 \leq i \leq N$, and $P(W = x) = 0$ for any other real number. We say the random variable W has a **uniform distribution** on $1, 2, \ldots, N$. The reason we can assign $P(E) = \#E/N$ for any event E that is a subset of the numbers $1, 2, \ldots, N$ will be explained below. As we go through the next few chapters, we will give names to other standard distributions of random variables. The major advantage is that we need only work out the properties of such random variables once, and then we can refer to these results whenever they arise in whatever circumstances.

We need to add to the language of sets summarized in section 2.7. In particular, two terms that we will find useful are:

> **Mutually Exclusive Events**: Two or more events are said to be mutually exclusive if their pairwise intersections are all empty.
>
> **Exhaustive Events**: Two or more events are exhaustive if their union equals the entire sample space.

We will find it useful to have a set of events that are both mutually exclusive and exhaustive. A simple example of this is a set A and its complement $\overline{A}$.

2.3 BASIC COUNTING METHODS

Two counting methods for finite sample spaces (the number of elements in the sample space is finite) are: **Permutations and Combinations**. A permutation is any arrangement of objects where order matters. A combination is any arrangement of objects where order doesn't matter, just the specific elements in the list. In effect, a permutation is a combination where we must also take into account the order of the objects in the list. For example, suppose we have a group of 10 people and we must select 3 of them to sit on a committee. If that is the only condition, then a selection that fills the committee is a combination of the 10 people. If instead the committee has a president, vice-president, and treasurer, then who is selected for which position is important. In this instance, a particular choice of three people to fill these positions is said to be a permutation of the 10 individuals, taken 3 at a time. If we want the distinct arrangements of all 10, we are finding the permutations of the 10 individuals.

A value that occurs quite often is $n!$. The notation $n!$ is read "n factorial" and is defined by: $0! = 1$, $1! = 1$, and for $n \in \mathbb{N}$ and $n \geq 2$, $n! = n(n-1)!$. Thus, for $n \geq 2$, $n!$ is the product of the natural numbers from 1 to n. This is a useful expression in discussing both permutations and combinations.

We need another important counting principle: Suppose an experiment occurs in k stages, with the number of ways of each individual stage can occur readily determined. Let $n_1, n_2, \ldots, n_k$ be the number of ways each of the k stages can occur, and assume these numbers do not depend on what happens at other stages of the experiment. Then the number of ways the entire experiment can occur is $n_1 n_2 \ldots n_k$, i.e., the product of these

Chapter 2 — An Introduction to Probability

values. Thus, if we flip a coin 10 times, at each stage (or flip), the coin lands one of two ways only. Then there are $2\ldots2 = 2^{10}$ ways the experiment can occur. If you flip a coin and then draw card, there are $2 \times 52 = 104$ different possible outcomes of this experiment. Let us call this the **multiplication principle**.

We return to the problem of finding the number of permutations and combinations. It is easier to start with permutations and then move to combinations. If we have n distinct objects, there are $n!$ ways to arrange all n. If we select only r of the n (again, assuming order matters), there are $n!/(n-r)!$ ways to select the r. To derive this formula, picture having r slots to fill. We can fill the first slot with any one of the original n items. The second slot can be filled by any one of the remaining $n-1$ items. Continuing in this fashion, we see there are $n-r+1$ items left to fill the rth slot. One way to denote the product of the natural numbers from $n-r+1$ to n is $n!/(n-r)!$. If $r = n$, that is we want to fill all n slots, this would be a permutation of all n objects, and the formula becomes $n!/(n-n+1)! = n!$.

If we have n objects to be placed in k mutually exclusive categories such that $n_1, \ldots, n_k$, respectively, are to go in these categories (these numbers are fixed, with the condition $n_1 + \cdots + n_k = n$) and the order of the objects in each individual category is irrelevant, then there are $n!/(n_1!\ldots n_k!)$ ways to arrange these n objects under the given conditions. A common notation is:

$$\frac{n!}{n_1!n_2!\ldots n_k!} = \binom{n}{n_1, n_2, \ldots, n_k}.$$

These are called the multinomial coefficients. A way to think of this is that if we were free to arrange all n objects, there would be $n!$ ways to do this. But if there are n_i indistinguishable objects, which is expressed in this problem as the order among them is irrelevant, then we have to divide $n!$ by the number of ways of permuting the n_i indistinguishable objects. We repeat this division for every set of indistinguishable objects.

As a general rule, then, if we have n objects to be arranged and certain ones are indistinguishable, then the number of distinct permutations is $n!$ divided by the number of permutations of those objects that are not distinct.

Example 2: How many distinct permutations of the letters in the word "letters" are there? Since there are 7 letters, but two t's and two e's, then there are $7!/(2!2!) = 1260$ ways to permute these letters.

Example 3: We have 12 crows to be put into three categories, each category to have 4 crows. This may be the first step to administering a drug at three different levels. The number of ways we can do this is $12!/(4!4!4!) = 34\,650$. Such considerations are used to make sure we avoid (as best we can) any systematic bias in how the birds are assigned to the different categories.

In our problem of permuting n objects r at a time, the number of such permutations is $n!/(n-r)!$. In effect, we have the total number of permutations of the n objects, divided by the number of ways the unselected individuals could have been arranged since the order of them does not matter. This naturally gives the formula for calculating numbers of combinations: If we have n objects and we select r of them, and we don't care about order within the selected group, then there are $n!/(r!(n-r)!) = nCr = \binom{n}{r}$ ways of doing this. We divide out the number of ways of permuting the selected objects because their order doesn't matter, and the number of ways of permuting the objects not selected, because their order doesn't matter. Another way to view this is that we have two mutually exclusive and exhaustive categories (selected, non-selected) with r going in the first and hence $n - r$ in the second. Obviously, $r + n - r = n$. We note that this formula is the same value that occurs in Pascal's triangle used in expanding a binomial to the nth power.

2.4 PROBABILITY: DEFINITION AND EXAMPLES

Not all of our random variables will have a finite number of equally likely outcomes. For example, if we measure people's heights, then, at least theoretically, the possible heights fill out a continuum of values, i.e., an interval. Or, as illustrated in a previous example, we may have a finite number of values of a random variable that are not equally likely — the sum of the faces on the two dice is a natural number from 2 to 12, with the probability of 2 equal to 1/36 while the probability of 7 is 1/6. How do we assign probabilities in a consistent way in situations such as this? The answer is to have a clearly defined method of determining if what we are doing is in fact a probability assignment.

Before discussing probability, we have to have a clear idea of the kind of objects we want to work with and how we are to work with them. In general, we have some set, S, say (the sample space), where a set is simply a collection of objects (usually numbers) along with a rule for deciding membership. A subset of S is another set whose members or **elements** are also elements of S. If A is a subset of S, we write $A \subseteq S$, or if S has at least one element distinct from A we write $A \subset S$ and read this as "A is a proper subset of S." We write $x \in A$ if x is an element of A. A set with no elements is called the empty set and is denoted $\emptyset = \{\}$. The empty set is a subset of every set.

The standard operations on sets are intersection, union, and complementation: If A and B are sets, then $A \cap B$ denotes the intersection of A and B and is the set of $x \in S$ such that $x \in A \cap B$ iff $x \in A$ and $x \in B$ simultaneously. We use $A \cup B$ to denote the union of A and B and is the set of $x \in S$ such that $x \in A \cup B$ iff either $x \in A$ or $x \in B$ (or both). We denote the complement of A by $\overline{A}$ (you may see the notation A^c in some books), and this is the subset of S with $x \in \overline{A}$ iff $x \notin A$. The ideas of intersection and union extend to a finite or infinite number of sets.

Often Venn diagrams are useful pictures of sets and help illustrate, but not prove, various relationships between sets. The basic Venn diagram of two sets A and B in the universal set U is:

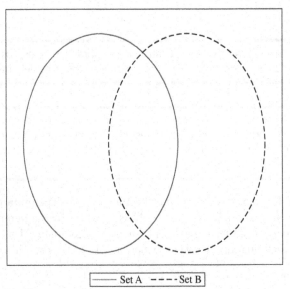

FIGURE 2-1. Venn Diagram of Two Intersecting Sets in Universe U.

An expanded introduction to sets is given in section 2.7.

Note: We will be working inside some universal set S at all times, so rather than continuing to write, for example, $A \cap B = \{x \in S | x \in A \text{ and } x \in B\}$, we will write $A \cap B = \{x | x \in A \text{ and } x \in B\}$.

Chapter 2 — An Introduction to Probability

With these basic definitions we can formally define probability:

Definition: Let S be a non-empty sample space. Let $\mathcal{F}$ be a collection of events of interest in S, including S, the empty set, unions, intersections and complements. Let P be a function from $\mathcal{F}$ to [0,1] with the following properties:

1. $P(S) = 1$.
2. $P(E) \geq 0$ for all events E in $\mathcal{F}$.
3. Let $A_1, A_2, \ldots$ be a finite or countably infinite sequence of mutually exclusive events in $\mathcal{F}$. Then $P(A_1 \cup A_2 \cup \ldots) = \sum_{i=1} P(A_i)$.

We then call P a probability (function) or probability set function defined on S.

In more advanced courses, the reason for the stated restriction on the collection of events will be explained. For finite or countable sample spaces, we assume $\mathcal{F}$ contains all subsets of S.

Special Case: Suppose for a sample space S there are N elementary events, and we know that $P(E_i) = 1/N$, for all elementary events E_i. We say this forms a uniform discrete distribution where **distribution** means how the probability is spread over the sample space. Then for any event E, $P(E) = n/N$, where $n = \#(E)$, i.e., the number of simple events in E. That this is a proper assignment of probabilities to events follows from the fact that the elementary events are mutually exclusive. As an exercise, the reader can verify that the axioms hold for this probability assignment. Note that regardless of the unions of sets involved, the end result is a set with no more than N members of the sample space.

From these basic axioms of probability, other results can be derived. For example:

Theorem 1: Let S be a sample space with probability P (or probability function P defined on S). Let A and B be subsets of S. Then:

(i) $P(\emptyset) = 0$.
(ii) $P(A) = 1 - P(\overline{A})$.
(iii) $P(A \cup B) = P(A) + P(B) - P(A \cap B)$ (the General Addition Rule). This generalizes to several sets in union.
(iv) $P(A \cup B) = P(A \cap \overline{B}) + P(\overline{A} \cap B) + P(A \cap B)$.
(v) $P(A) \leq P(B)$, whenever $A \subseteq B$.
(vi) $P(A) = P(A \cap B) + P(A \cap \overline{B})$.

It is important to learn how to use the axioms defining probability, and students should prove these results.

Example 4: Let X be a random variable. Suppose X's possible (distinct) values are $x_1, \ldots, x_n$. By observation or otherwise, we find $P(X = x_i)$ for each i. A display that lists these probabilities is called the probability distribution of X:

X	x_1	x_2	$\ldots$	x_n
$P(X = x)$	p_1	p_2	$\ldots$	p_n

It is assumed that $p_1 + \cdots + p_n = 1$. The probability distribution can also be written as a formula if there is a convenient one available, and we classify random variables by the form of the function. This will be a common situation in this book. We also define $P(X = x) = 0$ for any value x distinct from the x_i values. In this way, we can always assume our (real valued) random variable is defined on $\mathbb{R}$. This may appear artificial, but it does mean we can assume the values of our random variables are always $\mathbb{R}$ so that we do not have to deal with different ranges of values of different random variables. Thus, it unifies the development of properties of random variables.

When the random variable can take on only a finite or countably infinite number of distinct values, we call X a discrete random variable. For example, suppose we keep track of certain types of fish caught by a particular method in a large lake over a long period of time. Let A = the event of catching a trout; B = the event of catching a bass; C = the event of catching a sucker; and D = the event of catching any other fish. Using our definition of probability for this type of data, and the results of many fishing trips, we find $P(A) = 0.23$; $P(B) = 0.15$; $P(C) = 0.21$; and $P(D) = 0.41$. We assume these categories are mutually exclusive and exhaustive. It is assumed that the type of the next fish caught cannot be predicted with certainty, that is, the experiment of catching a fish in the lake is a random experiment. Let X = type of the next fish caught, and code these **categories** of fish as 1, 2, 3, 4, respectively. Then $P(X = 1) = 0.23$, or $P(X = 1 \text{ or } 3) = P(1) + P(3) = 0.23 + 0.21 = 0.44$, etc. Another calculation might be finding probabilities of complements, such as $P(\overline{C}) = 1 - 0.21 = 0.79$.

Example 5: In a particular course, there are two term tests, say Test 1 and Test 2. We will take as the population under study the set of students who wrote both tests. Suppose that over many years, records are kept and the proportion of students who passed Test 1 is 0.90, the proportion who passed Test 2 is 0.85, and the proportion who passed both tests is 0.80. Let X be the random variable such that $X = 0$ if a student fails both tests; $X = 1$ if a student passes Test 1 but not Test 2; $X = 2$ if a student passes Test 2 but not Test 1; and $X = 3$ if a student passes both tests. The given proportions can then be used to calculate probabilities of the various values of X. For example: What is the value of $P(X \geq 1)$? This would be the same as the probability a student passes at least one test, or in other words, a student passes Test 1 or Test 2. If we let A be the event a student passes Test 1 (which does not mean the student didn't pass Test 2) and B the event a student passes Test 2 (again the student may or may not have passed Test 1), then $P(A) = 0.90$, $P(B) = 0.85$ and $P(A \cap B) = 0.80$. Thus, $P(X \geq 1) = P(A \cup B) = P(A) + P(B) - P(A \cap B) = 0.90 + 0.85 - 0.80 = 0.95$. This also tells us that the probability of failing both tests is $0.05 = P(X = 0) = P(\overline{A \cup B})$. Further, $P(X = 1) = P(A \cap \overline{B}) = P(A) - P(A \cap B) = 0.90 - 0.80 = 0.10$.

2.5 CONDITIONAL PROBABILITY

In many instances, events are related, and we get some information about one event when we know that another has occurred. We quantify the relationship between events through **conditional probability**: Let A and B be events in a sample space S with probability P. Assume $P(B) \neq 0$. Then the probability of A, *conditional* on B, is defined to be $P(A \mid B) = P(A \cap B)/P(B)$. We read $P(A \mid B)$ as "the probability of A conditional on B" or "the probability of A given B." In effect we think of B as already having occurred so that B really is our sample space. The relative rate of occurrence of an event is now measured relative to B. In particular, our formula considers that part of event A that is in B. If knowing that B has occurred doesn't affect the probability that A will occur, i.e., $P(A \mid B) = P(A)$ then we say A is independent of B. More formally,

Definition 1: We say A and B are **independent** iff $P(A \cap B) = P(A)P(B)$. Note that this does not require either event A or B to have nonzero probability. In general, we say events $A_1, \ldots, A_n$ are independent iff $P(\cap A_j) = \Pi P(A_j)$ over all subcollections of the events.

Note that if one of A or B has zero probability, then A and B will be independent. If neither has zero probability, then if $P(A \mid B) = P(A)$ we will also have $P(B \mid A) = P(B)$. Thus, we simply say A and B are independent if the conditions mentioned hold.

Example 6: A tank contains 6 guppies, 4 tetras, and 5 koi. Three fish are drawn successively, with no replacement, and the type of fish and order are noted. Find the probability of getting the exact result: (guppy, tetra, koi).

Solution: First we note the **general multiplication rule** for events (not necessarily independent):
$P(A \cap B) = P(A \mid B)P(B) = P(B \mid A)P(A)$. Let G be the event that a guppy is drawn on the first draw, T the event that a tetra is drawn on the second draw, and C the event that a koi is drawn on the third draw. Then we want

Chapter 2 — An Introduction to Probability

$P(G \cap T \cap K)$. However, by the multiplication rule:

$$P(G \cap T \cap K) = P(G \cap T)P(K \mid G \cap T) = P(G)P(T \mid G)P(K \mid G \cap T) = \frac{6}{15}\frac{4}{14}\frac{5}{13} = \frac{4}{91}.$$

Once we know what happens on the first two draws, then it is simple to determine what happens on the third draw. Once we know what happens on the first draw, it is straightforward to determine what happens on the second draw. If we wanted, say, $P(T \cap K)$, we have a problem because we don't know what happened on the first draw.

An important theorem that relates some of the ideas we have discussed is:

Theorem 2: The Theorem of Total Probability. Let A be any event in a sample space with probability P. Let $E_1, \ldots, E_k$ be mutually exclusive and exhaustive events in the sample space, with $P(E_i) \neq 0$ for any i. Then $P(A) = P(A \cap E_1) + \cdots + P(A \cap E_k)$. The proof is left as an exercise. Note that this generalizes Theorem 2.1 (vi).

An immediate consequence is Bayes' Theorem (or Bayes' Rule):

Theorem 3: Bayes' Theorem. Let A and B be events with nonzero probability. Then:

$$P(B \mid A) = \frac{P(A \mid B)P(B)}{P(B)P(A \mid B) + P(\overline{B})P(A \mid \overline{B})}.$$

Proof: This is a direct consequence of the Theorem of Total Probability. Also, this extends to k mutually exclusive and exhaustive events, namely:

$$P(E_i \mid A) = \frac{P(A \mid E_i)P(E_i)}{P(E_1)P(A \mid E_1) + P(E_2)P(A \mid E_2) + \cdots + P(E_k)P(A \mid E_k)}.$$

Bayes' theorem is used to revise probabilities as more information becomes available.

Example 7: On a given multiple choice exam consisting of 100 questions, each question has 5 possible choices. It is assumed that if a student knows the answer, the student makes the correct selection. If the student does not know the correct answer, then the student randomly selects one of the 5 possible choices, i.e., each of the 5 choices is equally likely to be selected. For a student that knows 80% of the material in the course, what is the probability that if the student gives a correct answer, he or she actually knew the answer to the question. Assume students make a selection for all questions.

Solution: Let A be the event that the student answers correctly and E the event he/she knows the answer. We want $P(E \mid A)$. By Bayes' Theorem,

$$P(E \mid A) = \frac{P(A \mid E)P(E)}{P(E)P(A \mid E) + P(\overline{E})P(A \mid \overline{E})}.$$

But we know: $P(E) = 0.8$, $P(\overline{E}) = 0.2$, $P(A \mid E) = 1$ and $P(A \mid \overline{E}) = 1/5$ so that

$$P(E \mid A) = (1)(0.8)/[(1)(0.8) + (0.2)(0.2)] = 5/6.$$

Example 8: A blood test is used to determine the presence or absence of a particular steroid, and it correctly gives a positive result 96% of the time; that is, if an athlete has taken this steroid, 96 times out of 100 on average, the test so indicates. It will give a false positive on 1% of those who have not taken the steroid. It is believed that in the group of all athletes, 0.6% of them have taken the steroid. An athlete is chosen at random and tests positive. What is the probability the athlete has actually taken the steroid?

Solution: Let E be the event the athlete tests positive and C the event the athlete has actually used the steroid. Then:

$$P(C \mid E) = \frac{P(E \mid C)P(C)}{P(C)P(E \mid C) + P(\overline{C})P(E \mid \overline{C})} = (0.006)(.96)/[(.005)(.95) + (.994)(.01)] \approx .337.$$

Thus, in only about one-thrid of the cases of a positive result has the athlete actually taken the steroid. Obviously, such a test cannot be used to detect steroid usage. However, if result is negative, the probability of not having used the steroid is 0.999756.

2.6 ODDS

A term sometimes encountered is that of **odds**. This is just another way to express probability. It is the ratio of wins (or positive outcomes) to losses (negative outcomes): the odds of 6 to 5 or 6:5 in favor of an event means that in 11 times, we expect 6 wins and 5 losses. Note that we can also say 5:6 against an event, and this then means that 5 times out of 11 the event won't happen. In this context, success is the event not happening. If we have odds of a:b in favor of event E, then the probability p that E happens is $a/(a + b)$. Although a and b can be many different values and the ratio remain fixed, we take a and b to be the smallest positive integers that yield the given probability: If $p = 1/2$, then $a = 1$ and $b = 1$. If $p = 15/32$, then $a = 15$, $b = 17$, etc.

2.7 SETS AND SET NOTATION

This section may be omitted by those with a good working knowledge of sets and basic set actions (intersection, union, and complementation). Standard notation for sets used throughout this book are included, so a brief review by all may prove useful.

A **set** is any collection of objects with a rule to decide membership. This rule can be simply an exhaustive list of the objects in the set, a formula, a set of conditions, or a written statement. The objects in the set are called its **elements**. We typically use a capital (uppercase) letter to denote the set and lowercase letters to denote its elements. The notation used to express sets is set (builder) notation: "{" and "}" enclose the objects in the set; the symbol | is read "such that" (you can also simply write in these words).

Example 9: Examples of Sets: A = {1,2} is the set of the numbers 1 and 2. Note that if a set has a small number of elements, they are often written out in a list. B = {car, dragon, housefly} is a set that contains the words car, dragon, and housefly. We often have an equation, or set of equations or conditions to satisfy, and then we write the solutions in a set. We call this the **solution set**: the set of real numbers x that satisfy the equation $x^3 - x = 0$ form the elements of the solution set of the equation, and in this case, is $\{-1, 0, 1\}$. The set of continuously differentiable functions f defined on the interval (0,1) that satisfy $f'(x) = 0$ is the set functions $\{f(x) = a$ where a is a constant$\}$. We can also have a set described by a condition such as E = {points on the unit circle, centered at the origin}.

If an object x is an element of a particular set A, we write $x \in A$ and read this as "x is an element of A."

There are some sets that we encounter frequently, and we use specific notation for them. The set of **natural numbers**, that is, the set of positive whole numbers, is denoted $\mathbb{N}$, or in set notation $\mathbb{N}$ = {the set of all positive whole numbers} = $\{1, 2, 3, \ldots\}$. This last formulation cannot list all of the elements, but lists the first few with the clear implication of what else is in the set.

The set of the **integers**, or whole numbers, are denoted $\mathbb{Z}$ and sometimes written as $\{\ldots, -3, -2, -1, 0, 1, 2, 3 \ldots\}$. Thus, if we want to solve an equation inside the universal set $\mathbb{Z}$, the solutions are all whole numbers. For example, $\{x \in \mathbb{Z} \mid x^2 = 1\} = \{-1, 1\}$. Suppose instead we had $\{x \in \mathbb{Z} \mid x^2 \leq 1\}$. Then the solution set is = $\{-1, 0, 1\}$. If the problem had been to solve the equation in $\mathbb{N}$, the solution set would be $\{1\}$.

Chapter 2 — An Introduction to Probability

The real numbers are denoted $\mathbb{R}$, and perhaps the simplest way to think of them is the set of all decimal expansions. This set contains all of the rational numbers (denoted $\mathbb{Q} = \{x | x = m/n, m, n \in \mathbb{Z}, n \neq 0\}$). The rational numbers can also be thought of as all decimals that either terminate or eventually have a repeating pattern. Thus, the rationals include whole numbers, as well as numbers such as $22/7 = 3.\overline{142857}$, where the bar over the numbers at the end indicates this group repeats (in the given order) from that point on. The real numbers also contain all decimal expansions that do not have a repeating pattern. Numbers such as π from geometry and e from exponentials and logarithms are examples of numbers with non-repeating decimal expansions.

We will not need the **complex** (or **imaginary**) numbers in this course, but if i is the value such that $i^2 = -1$, then the complex numbers are denoted $\mathbb{C} = \{z = a + ib \,|\, a, b \in \mathbb{R}\}$. We mention them for completeness.

As can be seen above, there are instances where all of the elements of one set, B say, are also elements of another, A. We then say that B is a subset of A, and write either $B \subseteq A$ if B is possibly identical with A, or $B \subset A$ if there is at least one element in A that is not in B, but all elements of B are elements of A. In this last situation, we say B is a proper subset of A. Thus $\mathbb{N} \subset \mathbb{Z} \subset \mathbb{R}$. Also, $\{1, 4, 6\} \subset \{1, 2, 3, 4, 5, 6\}$.

The **cardinality** of a set is a way of expressing how many elements are in a set. For example, the set $\{2, 5, 7, 11\}$ has cardinality 4, because there are 4 (distinct) elements in it. The set $\{2, 2, 5, 7, 11, 11, 11\}$ still has cardinality 4, since we only count the number of distinct elements. What about $\mathbb{N}$? Clearly, this cannot have finite cardinality. We say that it has **countable** cardinality. Any set that can be put into one-to-one correspondence with $\mathbb{N}$ is also of countable cardinality, and this includes $\mathbb{Z}$ and $\mathbb{Q}$.

What about $\mathbb{R}$? It can be shown that there is no way to put $\mathbb{R}$ into one-to-one correspondence with $\mathbb{N}$. We will simply say that the real numbers are **uncountable**. It is in fact the case that there are a lot more real numbers than natural or rational numbers.

The real numbers are also said to form a **continuum**. Interval notation for certain subsets of the reals is quite useful: We write (a, b) to represent $\{x \in \mathbb{R} \,|\, a < x \text{ and } x < b\}$. This type of set is called an open interval in $\mathbb{R}$. Note that if $a \geq b$, then (a, b) has no elements, and we call this the empty set, $\emptyset$.

We write $[a, b]$, and call it a closed interval, to represent $\{x \in \mathbb{R} \,|\, a \leq x \text{ and } x \leq b\}$. If $a = b$, this reduces to the set with the single element a (or b). If $a > b$, then the interval has no elements, and so it is the empty set, $\emptyset$. Note: **The empty set is a subset of every set**. We call a the left endpoint, and b the right endpoint of the interval.

We also write $(a, b]$ for the set of real numbers that are greater than a and less than or equal to b, and $[a, b)$ for the set of real numbers that are greater than or equal to a and less than b. Note in interval notation, both conditions must hold for x to be in the interval. We call $b - a$ the length of an interval (where $a \leq b$ is now always assumed).

In order to extend interval notation, we introduce the symbols ∞ and $-\infty$. These are not real numbers, but are used to indicate that some set of real numbers has no upper bound, and ∞ is used here to denote this situation, or has no lower bound, and then we use $-\infty$. That means the real numbers could be written as $\mathbb{R} = (-\infty, \infty)$. This just means it is the set of all decimal expansions as usual. We can also view this as an interval, just infinite in both directions. If a is finite, then we can have intervals of the form (a, ∞), $[a, \infty)$, $(-\infty, a)$ and $(-\infty, a]$ (with definitions following the form of the above). These are also called rays. Note that it is wrong to write, e.g., $[-\infty, 3)$ because $-\infty$ is not a real number. The length of any ray is infinite.

Let A and B be sets, and assume $B \subseteq A$. The cardinality of B cannot be larger than the cardinality of A. It may not appear obvious, but the cardinality of any interval with nonzero length is the same as the cardinality of the real numbers. This can be stated as there are as many elements in, say, $(0,1)$ as there are in the reals. This fact has important implications for the development of probability for finite or countable sets versus for uncountable sets.

We also call $\mathbb{R}$, along with certain operations and conditions, one-dimensional Euclidean space. As an exercise, you should determine what exactly are the operations and conditions that make this "Euclidean space." This generalizes to $\mathbb{R}^n$, $n \in \mathbb{N}$, or n-dimensional Euclidean space (with appropriate operations and conditions) as follows: Let $(x_1, \ldots, x_n)$ denote n real numbers, $x_1, \ldots, x_n$, written in the given order. This distinguishes it from sets because the order of elements in a set does not matter. We call such a list in the round brackets an ordered n-tuple. Often, $n = 2$ and then we call it an ordered pair. In general, $\mathbb{R}^n = \{$all $(x_1, \ldots, x_n) \mid x_j \in \mathbb{R}, 1 \leq j \leq n\}$. In particular, then, $\mathbb{R}^2$ is the set of all ordered pairs of real numbers, and we often simply say "R-2". $\mathbb{R}^3$ is the set of all ordered triples of real numbers. Almost all problems in this course deal with sets of numbers or ordered n-tuples of numbers.

It should be pointed out that, even though we generally write numerical values of a set in some order, there is no order in a set. Thus, the two sets $\{1, 4, 6\}$ and $\{6, 1, 4\}$ are the same. Also, a set is simply the collection of objects that follow some rule. This means that repeated values are extraneous: $\{1, 2, 2, 3, 4, 4, 4\}$ is the same as $\{1, 2, 3, 4\}$. To avoid confusion or extra work on sets, it is best to "toss out" the repeated values.

In many problems, we work within some prescribed set, called the **universal set**, and the letter U is used in general to denote it. In probability and statistics, the universal set is most often the set of possible outcomes of some experiment, and in this case, the letter used is S. Here, S stands for **sample space**, the set of all possible outcomes of our experiment. An **event** then is just a subset of a given sample space. It is the rate of seeing the particular event, relative to all possible outcomes, that is the basis of probability. Although the outcomes of an experiment need not be numbers, we can always code the results so that we can work within number systems in all applications.

Suppose we have a universal set, S. There are three standard operations on or between sets: complementation, union, and intersection:

Complementation: Let A be a subset of S. Then the set $\{x \in S \mid x \notin A\}$ is called the **complement** of A (in S).

Union: Let A and B be subsets of S. We call $\{x \in S \mid x \in A \text{ or } x \in B\}$ the union of the sets A and B, and write it as $A \cup B$. Note that the condition to be an element of the union means that the element is either in A or in B, or in both simultaneously. The word "or" as it is used here is called the non-exclusive "or." We can extend this idea to the union of an arbitrary number of sets. For example, if $B \subseteq \mathbb{N}$, and for all $i \in B$, A_i is a subset of a sample space, then the union of this collection of sets is $\{x \mid \exists i \in B \text{ for which } x \in A_i\}$.

It is for ease of expression that we use consecutive natural numbers as the indexing set. For example, we may have a collection of sets best indexed by the integers, say B_i, $i \in \mathbb{Z}$ and then the union of this collection is as before: each element in the union must be in at least one of these sets, and we would write $\cup_{i \in \mathbb{Z}} B_i$.

Intersection: Let A and B be subsets of S. We call $\{x \in S \mid x \in A \text{ and } x \in B\}$ the intersection of A and B. Note that both conditions must hold simultaneously. As with unions, we can extend this definition to any collection of sets: If C_i is a collection of sets indexed by $i \in J$, where J is the indexing set (which is typically some subset of integers), then the intersection of all of these sets is the set of values x which are in each and every set C_i, $i \in J$. We write $\cap_{i \in J} C_i$ for the intersection.

Example 10: Let $S = \{$all outcomes of three flips of a fair coin$\}$. To write out all possible outcomes, let T denote Tails, and H denote Heads. The possible outcomes can be written as order triples: (H,H,H), (H,H,T), (H,T,H), (H,T,T), (T,H,H), (T,H,T), (T,T,H), (T,T,T). Note that it is important to write the outcomes in some systematic way to ensure all outcomes get listed. It is also useful to know in advance just how many outcomes there should be. Why should there be 8 possible outcomes?

If we let $A = \{$outcomes with at least one H$\}$, then $\overline{A}$ is $\{(T,T,T)\}$.
If we let $C = \{$there is an odd number of Heads$\}$, and $B = \{$there are two or more Tails$\}$, then:

$$C \cup B = \{(H, H, H), (H, T, T), (T, H, T), (T, T, H), (T, T, T)\} \text{ and}$$
$$C \cap B = \{(H, T, T), (T, H, T), (T, T, H)\}.$$

It may appear at times that mathematicians are preoccupied with language and often stress what seem minor points. For example, a mathematician would find writing expressions such as $1/\infty = 0$ improper. One reason for such preoccupation is, in fact, to attain clarity. If we have $A = \{2\}$ and $B = \{\{2\}\}$, are A and B the same? The answer, of course, is no. The only element of A is 2, and the only element of B is {2}. A similar distinction is made of a function, say f, and a value of that function, $f(x)$. Is the empty set the same as {0} (the set containing zero)? Again, the answer is no. The empty set has no elements, while 0 is a well-defined object in the given set. The meaning of these things is different, and when people confuse them in their writing, it probably means they really don't understand the basic concepts. Mathematicians are in fact trying to make discussions of issues simpler by using precise language that cannot be misinterpreted. Precision of language makes it possible to convey to anyone else using the same mathematical concepts and definitions ideas and results in an unambiguous way. A solid command of the fundamental mathematical language used in this book is as important as the applications using that language.

It is highly recommended that students practice writing symbols, equations and derivations, and commit to memory the definitions of the terms used throughout this book. Such practice will make it easier to write your own quantitative analyses and to understand research writings of others.

Exercises

1. Let the universal set **U** be given by: $\mathbf{U} = \{x \in \mathbb{Z} | \, |x| \leq 15\}$. Consider the following subsets: $A = \{x \in \mathbf{U} | \, x^2 \leq 9\}$, $B = \{-1, 1, 3, 8\}$, and $C = \{x \in U | x \text{ is odd }\}$. Determine the set of elements in each of the following:

    (a) $A \cap B$

    (b) $A \cup C$

    (c) $\overline{C}$

    (d) $\overline{A \cup C}$

    (e) $A \cap (B \cup C)$

    (f) $\overline{A} \cap \overline{C}$

    (g) $B \cap C$

2. List all of the subsets of $B = \{\text{king, queen, castle, knight}\}$. Before writing them out, determine how many subsets there should be, and list the reasons.

3. Consider an experiment where a fair coin is flipped three times and the order in which heads or tails appears is relevant.

    (a) How many elements are in the sample space?

    (b) Describe the sample space.

    (c) List all events that satisfy the following events:

    (i) A: All flips result in tails.

    (ii) B: The first flip results in tails.

    (iii) C: The last flip results in heads.

    (iv) D: At least one flip results in heads.

    (d) Determine P(A), P(B), P(C), P(D) from 3(c).

4. Calculate the value of each of the following by hand. Use your calculator to check your answers.

    (a) $\binom{6}{2}$ (b) $\binom{10}{7}$ (c) $\binom{20}{4}$

5. How many distinct 'words' can be created by a rearrangement of all of the letters in (a) MATH (b) STATISTICS?

6. To test the effectiveness of a new fertilizer in the presence of other factors, a researcher is designing an experiment where plants will be subject to different combinations of light, temperature and fertilizer levels. Three different light levels, five different temperature levels, and four different fertilizer levels will be used. For each combination of light level, temperature level, and fertilizer level there will be six plants grown in the environment. How many plants will the researcher need?

7. DNA strands are made up of linear chains of proteins. Typical proteins are made up of linear, unbranched chains of amino acids called polypetide chains. There are 20 standard amino acids and four bases, Adenine (A), Thymine (T), Guanine (G), and Cytosine (C) that make up amino acids in the form of three adjacent bases, (see, for example, Barrett and Elmore (1998) or Hartl and Jones (2009)). We can view each amino acid as a three-letter word made up of the letters A, T, G, and C. Note that any of the bases may appear more than once in the code for an amino acid.

   (a) How many different expressions made up of the three bases are there for the 20 amino acids? That is, how many different three-letter word made up of the letters A, T, G, and C are there? (Note that there are more three-letter words than there are amino acids indicating that the coding for an amino acid may not be unique.)

   (b) How many expressions are there for the amino acids that do not contain any repetition in the bases?

   (c) If the bases are randomly chosen and ordered into three-letter sequences, what is the probability that a randomly chosen sequence would have no repetitions in the amino acids?

   (d) The sequences TCT, TCC, TCA, TCG, AGT, and AGC all code for the amino acid serine. What is the probability that a random arrangement of three bases will result in the amino acid serine?

   (e) How many different amino acid sequences of 50 amino acids can be formed from the 20 standard amino acids?

8. Since each protein is made up of a chain of amino acids and each amino acid is made up three letter words using A, T, G, and C, we can think of a protein as a long word made up of the letters A, T, G, and C. We then can talk about 'DNA sequences' which is a succession of the letters A, T, G, and C where the order in which the letters appear is relevant and each letter can appear multiple times.

   (a) How many different DNA sequences of 20 letters can be made up from the 4 bases?

   (b) How many different DNA sequences of 20 letters start with

      (i) adenine?
      (ii) the sequence ATGC?

   (c) Consider the short sequence TTAAAGCCCA. Determine:

      (i) how many different 10-letter DNA sequences can be made using these letters?
      (ii) how many different 10-letter DNA sequences can be made with these letters such that the 2 thymine nucleotides are together?
      (iii) how many different 10-letter DNA sequences can be made with these letters that the 2 thymine nucleotides are apart?

9. A simple sequence repeat is a DNA strand with a repeating sequence of 2 of the 4 nucleotides. For example, CGCG....CG is a simple sequence repeat.

   (a) Determine the probability of obtaining a simple sequence repeat with the adenine and thymine nucleotides in a random DNA sequence of 8 letters.

   (b) Determine the probability of obtaining a simple sequence repeat with any 2 nucleotides in a random DNA sequence of 8 letters. (Do this problem without listing all possible simple sequence repeats.)

Chapter 2 — An Introduction to Probability

10. To estimate the size of a population of a particular animal, researchers often use the mark-and-recapture method (see Seber (2002)). In this method, researchers will capture R animals and tag them. The animals are released and allowed to mix back into the population. Another n animals will be captured, and the number of tagged animals in this sample of size n are noted. Let N denote the population size and r the number of tagged animals in the sample of size n. The estimate for the population size N is based on the values of n, R, and r. Find

    (a) The number of samples of n captured animals.
    (b) The number of ways of capturing n animals in which r are tagged.
    (c) The probability of capturing n animals in which r are tagged.

11. Based on her course work, Rachel estimates that the probability she will get an A in her statistics course (event St) is 0.20; an A in her biology course (event B) is 0.70; and an A in at least one of these two courses is 0.85.

    Determine the following probabilities, showing your calculations and including a statement to explain what the probability represents for each. (A Venn diagram may be useful.)

    (a) $P(St \cap B)$
    (b) $P(\overline{St} \cap B)$
    (c) $P(\overline{St} \cap \overline{B})$
    (d) $P(\overline{St} \cup \overline{B})$

12. Use a Venn Diagram to show that $P(\overline{A} \cap \overline{B}) = 1 - P(A \cup B)$. Note that the Venn diagram would not suffice as a proof of this fact. You should use earlier results to prove this statement as well.

13. An experiment was conducted to determine whether a deficiency of carbon dioxide in soil affects the phenotype of peas, which can have either smooth or wrinkled skin and either yellow or green colour. See, for example, Bowler (1989) for a discussion of Mendel's experiments on peas. Sixty peas grown in such soil were selected at random from a large plot and categorized as follows:

	Brown	Green
Smooth	24	9
Wrinkled	12	15

    (a) Let Sm, W, B, G represent the events that a pea randomly selected from these 60 is smooth, wrinkled, yellow, or green, respectively. Determine the following probabilities. Provide a concluding statement in each case, interpreting each probability in terms of the phenotypes. Use appropriate notation.

        (i) $P(Sm \cap B)$
        (ii) $P(Sm)$
        (iii) $P(W \cap Sm)$
        (iv) $P(Sm \cup G)$
        (v) $P(W \cap G)$
        (vi) $P(B \cap G)$
        (vii) $P(Sm \cap \overline{G})$
        (viii) $P(B|Sm)$
        (ix) $P(W|G)$
        (x) $P(Sm|\overline{G})$

(b) Based on this information, are the events Smooth and Brown independent or mutually exclusive, or neither? Justify your answer mathematically.

(c) Two seeds are randomly selected from these 60. Determine the probability that both peas are Smooth and Brown, given that the seeds were selected:

   (i) with replacement;

   (ii) without replacement.

14. It is known that a patient with a disease will respond to treatment with probability equal to 0.7. Suppose that three patients with the disease are treated and respond independently. Find the probability that:

    (a) at least one will respond;

    (b) none will respond;

    (c) at most two out of the three will respond.

15. Show that for three events A, B and C, $P(A \cap B \cap C) = P(A)P(B \mid A)P(C \mid A \cap B)$. This result is obviously true if the three events are independent. Why?

16. Short outside front claws in rabbits is a recessive genetic trait while long claws is a dominant trait. Let S denote the dominant long claw allele and s the recessive short claw allele. Suppose that a male and a female heterozygous long-clawed rabbit are mated.

    (a) Determine all possible genotypes resulting from this mating.

    (b) Determine the probability that:

       (i) a resulting offspring has long front claws;

       (ii) a resulting offspring has short front claws;

       (iii) a resulting offspring is of genotype Ss;

    (c) Suppose that a number of matings between these two rabbits occur. What proportion of their offspring would we expect to have short front claws?

    (d) If the pair of rabbits produces 35 offspring in the rabbits' lifetime, how many short-clawed offspring might we expect to see? Would you expect to see this many short-clawed offspring from every mating between two heterozygous straight-haired rabbits producing 35 offspring? Explain.

17. In a study to determine the side effects of a drug to treat migraines, 150 individuals who experience migraines were selected. Seventy-five of the participants were given a 10-mg dose of the drug, and the remaining are given a placebo. Some participants experienced nausea during the trial. The results are given below.

	Drug	Placebo
Nausea	27	13
No nausea	48	62

    (a) What is the probability of randomly selecting a subject that reported having nausea?

    (b) What is the probability of randomly selecting a subject that reported having nausea and did not take the drug?

    (c) If two different subjects are selected, what is the probability that they were both given the drug?

    (d) If one subject is selected, find the probability that he or she was given a placebo given that this individual experienced nausea.

CHAPTER 3

DISTRIBUTIONS OF RANDOM VARIABLES

Synopsis

A random variable is a function on the outcomes and events arising from a random experiment. The distribution of a random variable is the assignment of probabilities to the values and sets of values of this function. General properties of **discrete** and **continuous** random variables and their distributions are developed. Distributions of two or more variables (**bivariate** or **multivariate** distributions, respectively), along with **marginal** and **conditional** distributions, are introduced. The **expectation** of a random variable or of a function of a random variable is defined. **Moments** and the **moment generating function** are introduced. Examples illustrate the results.

3.1 INTRODUCTION

Why do we study distributions? A quote from Damon Runyon in "More than Somewhat" (Runyon and Bentley, 2008) may help:

> The race is not always to the swift nor the battle to the strong, but that's the way to bet.

This modification of *Ecclesiastes 9:11* incorporates some of the fundamental ideas of probability, which was first developed in response to gamblers looking for ways to improve their chances of winning at various games. Since then, probability has become fundamental to many important problems in all of the sciences, as well as business and industry. Distributions organize important information about the possible outcomes of random experiments and allow for the systematic exploitation of this information.

Recall from Chapter 2 that the probability of some specific outcome (winning a battle, winning a race) is the relative rate we see this outcome occur out of all outcomes. Thus, if we have 200 battles and 25 of them ended in draws, then we would assign the probability a battle ends in a draw to be 0.125. If, in one hundred 100-meter races, the person deemed the fastest before each race won 92 of them, we would assign the probability 0.92 to the event that the fastest person will win the race. If we take an appropriate sample (a simple random sample, say) of 1,500 from a population, and find 75 have a particular defective gene, we would assign the probability that a randomly selected person has this gene to be 1/20. It should be noted in each of these cases that the given probabilities should be viewed as approximate, but this is one way to assign probabilities in the absence of either theoretical or other information. In the case of the defective gene problem, we would need to observe the entire population to state exactly what the probability of someone having this gene is. Another issue to consider is that there has to be an agreed-upon definition of fastest runner, or a draw in a battle (do we consider Pyrrhic victories wins, losses or draws?), or what a defective gene is. We have a formal definition of probability in the previous chapter, but these considerations make it clear we have to have full understanding of the population in question, the experimental procedures to be used, and the questions to be answered.

Given the different possible outcomes to some activity or experiment, which one is most likely to happen (which should occur most frequently for this kind of activity)? Further, how confident are we in this answer? We would have to know, in the case of a race, some past history that indicates who in the current race is the swiftest through some set of criteria. We can also use the historical records of such races to determine the relative frequency of the event that races have been won by the individual deemed to be the swiftest prior to a race, versus the event that such identified runners in past races did not in fact win. If we are to quantify the probabilities of outcomes, information is needed to generate at least approximate probabilities. If no information is available, often a uniform distribution is used, or someone may use "gut" feelings. Regardless, there are always conditions that we cannot control, so the race must be run to determine the winner.

The first question in the previous paragraph requires finding a measure of the relative frequencies of the outcomes, or, in our terms, the distribution of these relative frequencies over the (exhaustive) set of outcomes. We take as given that the probability one of our outcomes must happen to be one. Also, all probabilities should be between zero and one — in any fixed number of runs of the experiment, we would take the probability of outcome A to be (approximately) the ratio of the number of times A actually happens to the total number of runs. Thus probabilities are always non-negative, and an outcome cannot happen more frequently that the total number of runs. Probabilities are often expressed in percentages, so that if we say a given outcome has 10% probability of occurring, that is the same as saying the probability is 0.1. This can be interpreted as: in long runs of the particular activity/experiment, 10% (or nearly so) of all of the outcomes will be this particular one. Such activities or experiments can include wars, races, and searching for those in a population with a given characteristic. An outcome may be winning a war, winning a race, or actually finding someone in the population with the required characteristic within a fixed number of selections.

The second question: how confident are we in our answer to the first question? is often even more important than the first. Suppose it is known that 25% of patients with migraines who are administered placebos report relief in half an hour or less from the time the placebo is administered. We want to determine if, within half an hour of administering a new drug to migraine sufferers, more than 35% will report relief. A random sample of 100 people with a history of migraines is selected, and when they get a migraine, the drug is administered. In this experiment, 36 reported relief in the given time. The result is 36 percent in this sample gained relief, and so we can report that the drug attained the required condition for this sample. However, how confident are we in extending the fact that 36% of this sample attained relief to the entire population of migraine sufferers? Even with this relatively large sample from the population of migraine suffers, another researcher could follow exactly the same procedures, and for their sample of 100, the number reporting relief as specified might be 28, or even 20. It is then important to know how likely these different outcomes might be. It is from distributions that we can quantify our results.

In general, if a doctor administers a drug to control some disease, the drug may improve the condition, have no effect or in fact be detrimental. It is possible for a drug to relieve one symptom, but create another, so it can be viewed as both beneficial and harmful simultaneously. If the new symptom is relatively mild compared to the first, it may be that the drug is deemed to have sufficient value to use. Humans are essentially chemical soups with greater or lesser concentrations of ingredients. A drug may interact with some of these ingredients in ways not anticipated. Organs do not work precisely the same across all people, meaning organs may not react well to the drug, or the drug may not be processed properly. Thus, the effect of the drug cannot be determined with certainty until after it is administered. Can we determine in advance how someone might react to the drug, if we take various measurements and perform perhaps some genetic tests? One consequence of such investigations may be that the population to whom the drug can be administered changes to a subpopulation, rather than all that have the disease or condition. Probability helps us improve our level of certainty that someone will be helped by the drug, as well our certainty that it will not harm an individual.

The probability distribution of a random variable is also used to design experiments to get good information for the least cost. (In this course, cost will refer primarily to the number of subjects of an experiment used. However, other factors including other resources needed to carry out the experiment, often have to be taken into account but that is left for texts on, for example, experimental design and survey analysis.) If an experiment involves living subjects, then ethical considerations must be incorporated in our thinking and our procedures. As we shall see, from the assumed probability distribution, we can determine the number of subjects needed to get the level of precision necessary to answer a question with confidence.

Let S be any sample space with probability function P defined on it. Recall that this means that P is a function of the events in S. If X is any real valued function on S, i.e., the elements of S, then X is called a random or stochastic variable or variate. We will concern ourselves with two types of random variables: discrete and

Chapter 3 — Distributions of Random Variables

continuous. A discrete random variable is such that it takes on a finite or countably infinite number of values, while a continuous random variable is one with values filling in one (or more) intervals of nonzero length.

3.2 DISTRIBUTIONS

3.2.1 Discrete Distributions

Assume that the number, say $N \geq 1$, of possible values of the random variable X for which there is nonzero probability is finite. (There are also important cases where the number of distinct outcomes is countable, and examples will be discussed.) Let the possible values of X be denoted x_i, $1 \leq i \leq N$. It will be assumed that through empirical study or theoretical consideration, $P(X = x_i) = p_i > 0$, and $\sum_{i=1}^{N} p_i = 1$. For any other value x, define $P(X = x) = 0$. It will be convenient to classify distributions of random variable by the form of the function $P(X = x)$ so that common properties can be derived. By examining the underlying reasons why a random variable has a specific distribution, we can then recognize this distribution type as it arises in applications. Thus, by studying the theoretical properties of general classes of distributions, we do not have to repeat the calculations in different applications.

Definition 1: Let X be a discrete random variable. The **probability distribution function**, or simply **pdf**, of X is the function f defined by: $f(x_i) = P(X = x_i)$ for all x_i with nonzero probability, and $f(x) = 0$ otherwise.

The pdf is defined for all real numbers. In general, a function f may be a probability distribution function, or pdf, for a discrete random variable if $f(x) \geq 0$, $\forall x$, and $\sum f(x) = 1$, where the sum is over all x for which $f(x) > 0$, or equivalently, over all x for which X has nonzero probability.

If there are two (or more variables) under consideration, say X and Y, we write $f_X(x_i) = p_i$, $1 \leq i \leq N$, and $f_Y(y_j) = q_j$, $1 \leq j \leq M$ so that the subscripts X and Y differentiate the pdfs.

One way to write the probability distribution of X is in table form:

X	x_1	x_2	...	x_N
$P(X = x)$	p_1	p_2	...	p_N

The fact that $f(x) = 0$ for all other values of x should be stated, but often is simply tacitly understood. If there is a simple formula, we can just write out this formula, with the set of values for which there is nonzero probability, and again call this the probability distribution function, or simply pdf. Thus, the random variable Y such that $f(j) = 1/M$ for the natural numbers $1 \leq j \leq M$ (and 0 otherwise) has a uniform distribution. Note that $\sum_{j=1}^{M} \frac{1}{M} = 1$.

Related to the pdf is the **cdf** or **cumulative distribution function**:

Definition 2: The **cumulative distribution function**, or **cdf**, F for the random variable X with pdf f is defined by $F(x) = P(X \leq x)$ for all $x \in \mathbb{R}$.

In words, the cdf is the accumulated probability for all values of X, up to and including x. In some texts, notation such as $F(\infty)$ and $F(-\infty)$ is used. This is not proper since infinity is not a number and cannot be put into formulae. Limits of F of the appropriate type should be used: $\lim_{x \to \infty} F(x)$ and $\lim_{x \to -\infty} F(x)$, respectively.

The cumulative distribution function can be used to find probabilities of events. In particular, suppose we know F, the cdf of discrete random variable X with pdf f. If we want to compute $P(a < X \leq b)$, this will be $F(b) - F(a)$. Caution must be taken in the discrete case, because the probability of a single value may not be zero. For example, if X has a uniform distribution on the first $k = 10$ natural numbers, so that $f(x) = 1/10$ for $1 \leq x \leq 10$ ($x \in \mathbb{N}$), then $P(2 < x < 5) = F(4) - F(2)$. We would also have $P(2 \leq x < 5) = F(4) - F(1)$.

Example 1: Let the random variable Y have distribution

Y	1	2	3
$P(Y = y)$	1/3	1/3	1/3

The cdf G is given by:

$$G(y) = \begin{cases} 0 & y < 1 \\ 1/3 & 1 \leq y < 2 \\ 2/3 & 2 \leq y < 3 \\ 1 & y \geq 3 \end{cases}$$

Again, G is defined for all real numbers. Also note that the cdf is a function defined in pieces and has jump discontinuities at the numbers 1, 2, 3. This cdf is "flat" except where the jumps occur. Also note that the cdf stays at 0 until we reach $Y = 1$ and stays equal to 1 from $Y = 3$ onwards. Although, for example, $P(Y = 2.5) = 0$, we have $G(2.5) = 2/3$.

One important reason to look at cdfs is that we may have an idea of what kind of distribution a random variable has, and hence, an idea of what the cdf should look like. In long-running experiments, the researcher may keep track of the cdf of the observed values, updated as more data comes in. Again, this will be summarized in a cdf built empirically from the data. If the sample cdf has serious departures from the assumed distribution's cdf, the researcher is alerted to changes in conditions, for example. This is used in quality control studies to determine if a process is running properly or there are problems occurring that need to be corrected.

As noted above, graphing the pdf and/or cdf can give a clearer understanding of the information contained in the distribution, or if sample information is not conforming to expectations. Graphical representations of distributions are excellent ways to summarize information in a way easily conveyed to others. For example the pdf of a discrete random variable may be represented by a **bar chart** (described in Chapter 7, section 7.5.1). A **histogram** is similar, except it is used to represent the information in a sample from a continuous random variable. Many other charts and visual representations are used, and are part of **descriptive statistics**, in Chapter 7.

3.2.1.1 Discrete Distributions Based on Past Data

Suppose we have records of the outcomes of previous applications of a procedure applied to experimental units, or simply observations of the end result of some activity. Examples include: the outcome of battles; the outcome of races (of a particular type); the determination of whether someone has a particular genetic defect; the result of administering a drug to relieve a particular symptom; or the sex of a deer captured in a given habitat for tagging. As noted above, we can use this information to help find at least approximate values of the probability of future outcomes. We can also think of this as a (finite) population in its own right, and then the assignment of probabilities to outcomes is exact.

Example 2: Over the years, researchers have captured, tagged, and released 175 different adult deer in a large free range. Of these animals, 93 were male and 82 were female, and many other physical measurements were recorded for each. If 5 of these records are randomly selected, what is the probability the sample contains 3 males and 2 females? Answer: approximately 0.3338.

This may be the first step in preparing for a new experiment on the deer population, or simply an audit of the activities of the researchers.

Example 3: In a large shipment of electronic trackers, 2% are defective. The purchaser will accept the shipment only if in a random sample of 5, none are defective. What is the probability the purchaser accepts the shipment? Answer: 0.91 (approximately). This may be too high, and so the purchaser may use this to determine just how many of the devices should be tested. On the other hand, it may also be the case that the only way to test the device is to break it apart, meaning the more sampled, the costlier the process.

Example 4: In a large urban hospital, Emergency Room (ER) records of all patients from the past several years are available, including information on type of emergency (violence, poisoning, accident, other), and characteristics such as age, blood pressure, sex, medical history, etc. A new administrator is evaluating the efficient use of resources, and as a first step, randomly selects 15 ER records. If 12% of all ER records include violence as the reason for coming to the ER, what is the probability the administrator's selection will contain at least 3 records including violence? Answer: approximately 0.265.

In this example, if X is the number of records in the 15 selected that include violence, then we want $P(X \geq 3)$. However, it is much simpler to first note that $P(X \geq 3) = 1 - P(X \leq 2)$. These probabilities can be worked out using counting methods in this chapter. It is important to note that certain underlying characteristics can be systematized so one general method may apply to problems that arise in very different contexts. Chapters 4 and 5 introduce a number of special distributions commonly seen in applications from many different disciplines.

Note that in these last two examples, the number of records is assumed large, meaning in effect we can assume they are infinite. In problems where the number of records is relatively small, small population methods have to be used. Such finite population problems occur frequently in surveys, and the interested reader can find more details in, e.g., Yates (1981).

3.2.2 Continuous Distributions

When we measure something, the possible values of the variable (i.e., the measurement) fill at least one interval of nonzero length. These values are said to constitute or fill a **continuum**, usually all real numbers in an interval such as (a, b), $b > a$. In the case of continuous variables, the probability of specific outcomes is not of interest. In fact, $P(X = x) = 0$ is the only reasonable probability assignment for each and every outcome $x \in \mathbb{R}$. The proof of this is beyond the scope of this text as it relates to the general concept of measure theory. For the purposes of this work, we will assume we are interested in the probability that the continuous random variable X falls in some interval or union of intervals and call these our events of interest.

Notation: Any finite set or any set that can be put into one-to-one correspondence with the natural numbers is said to be **countable** and all other sets are called **uncountable**.

The natural numbers, integers, and rational numbers are all countable sets. Any interval of nonzero length is uncountable, and in fact, the interval (0,1) has as many elements as $\mathbb{R}$. The interval (0,1) has more elements than $\mathbb{Q}$. It is because of the difference in "size" between countable and uncountable sets that we study discrete and continuous variables separately.

Caution: Problems often require verifying some property on a given set. It is essential that we know the nature of the set: If we are considering [0,1] and integers in [0,1], then this is just 0 and 1. If we are considering the rationals in this set, then we have to look at all numbers of the form n/m where $m, n \in \mathbb{Q}$ with $0 \leq n \leq m$ (and $m > 0$) (m and n could be negative with $0 \neq m \leq n \leq 0$ but that does not include any new rationals). If we are considering the real numbers in this interval, then we are considering all decimal expansions with appropriate restrictions. The point is that it is essential to know what numbers are of interest in a given problem.

One approach to assigning probabilities in the continuous case is to build an **empirical** distribution by running the experiment many times and recording how frequently values fall into selected intervals. This information is often summarized in the form of a histogram (see Chapter 7). From this, a continuous and smooth function that approximates the histogram is selected from some family of appropriate functions. The most common way to generate the probability of events associated with a continuous random variable is to develop recognizable principles that in turn tell us the form of a function f, called the **probability density function** or **pdf**, that generates probabilities through integration over intervals: $\int_a^b f(x)dx = P(a < X < b)$. A formal definition of the pdf is given in 3.2.2.1. A simple example is:

Example 5: Let the continuous random variable U be such that the probability of any set outside [0, 1] is zero, and the probability of any interval contained in [0,1] of length w is the same as the probability of any other interval in [0, 1] of length w. It can then be shown that the pdf g is:

$$g(u) = \begin{cases} K & 0 \leq u \leq 1 \\ 0 & \text{otherwise} \end{cases}$$

where K is a constant.

We call such variables **uniform** random variables, or say U is uniformly distributed on [0,1]. This readily extends to variables that are uniformly distributed on, say, $[a, b]$, where $b > a$. To find the constant, we recognize that $P(-\infty < U < \infty)$ must be 1, and that by the definition of a pdf, $\int_{-\infty}^{\infty} g(u)du = P(-\infty < U < \infty) = 1$. Such integrals are called improper, and a brief discussion of how to evaluate them is given in Appendix B. We are given that g is 0 outside [0,1], and so our integral reduces to $\int_0^1 g(u)du = \int_0^1 Kdu$. Note that we could not replace $g(u)$ by K until *after* we reduced the limits of integration to 0 and 1, since the definition of g changes from interval to interval. Now we have $\int_0^1 Kdu = K(1-0) = K = 1$, which then completely specifies our density function.

More about continuous random variables and the calculations associated with them will be given in 3.2.2.1 and Chapter 5. In preparation for Chapter 5, the reader should review single variable calculus, including limits, derivatives, integrals, and improper integrals. Examples in Chapter 5 will include integrals of the form:

$$\int_0^1 x^2(1-x)^3 dx, \quad \int_0^{\infty} e^{-x} dx, \quad \int_{-\infty}^{\infty} \frac{1}{1+u^2} du, \quad \text{and} \quad \int_0^{\infty} xe^{-x} dx.$$

3.2.2.1 Continuous Distributions: The Probability Density Function

We assume all continuous random variables have values that fill out $\mathbb{R}$, although there may be intervals for which the probability X is in these intervals is 0.

Definition 3: For the continuous random variable X, assume there is a function f defined on $\mathbb{R}$ with the properties that:

1. $f(x) \geq 0$ for all $x \in \mathbb{R}$;
2. $\int_{-\infty}^{\infty} f(x) = 1$; and
3. $P(X \in I) = \int_I f(x)dx$, where I is any interval.

We call f a **probability density function**, abbreviated **pdf**, for the continuous random variable X.

In the above definition, I is a single interval, but property 3 above may be extended to unions of intervals, or even more complicated sets. For our purposes, however, we can assume I is an interval. Because the integral over a single point is always 0, $P(X \in (a, b)) \equiv P((a, b)) = P([a, b)) = P((a, b]) = P([a, b])$. Note the simplification in notation used here and throughout when dealing with a continuous random variable, namely $P(X \in (a, b)) \equiv P((a, b))$ and similar for other types of intervals. We can have different pdfs for the same random variable, by, for example, changing the definition of f at a few points. This does not change the value of integrals over intervals containing (or not containing) these points. However, we do try to choose the simplest form for f, usually so that f is continuous everywhere, or, at worst, has jump discontinuities at a few points. Any continuous function f with properties 1 and 2 above can be a pdf. We often classify distributions by the form of the pdf, with functions chosen to reflect certain properties of the distribution. For the most part, we will assume the function has been chosen and then use this to generate probabilities for the corresponding random variable.

Although for continuous random variables the density's value cannot be interpreted as a probability, there is a related expression that can be interpreted this way, and it is in fact a better way to think about densities.

Chapter 3 — Distributions of Random Variables

The idea is that if we form $f(x)dx$, that is, multiply the density at some number by the differential dx, this can be interpreted as the approximate probability that the random variable falls in an interval centred at x, of width dx. This is from the way integrals are defined: The probability X is in $(x - dx/2, x + dx/2)$ is given by:

$$P(X \in (x - dx/2, x + dx/2)) = \int_{x-dx/2}^{x+dx/2} f(t)dt.$$

This is approximately the length of the interval of integration times any value of the function in the interval (we assume f is at least continuous, and dx is "small"), so we can choose x as that value. Thus,

$$P(X \in (x - dx/2, x + dx/2)) = \int_{x-dx/2}^{x+dx/2} f(t)dt \approx f(x)dx.$$

The smaller dx is, the better this approximation. We can then think of $f(x)dx$ as the (approximate) probability X falls in a region (of length dx) containing x. Also note that if we define $y = F(x)$ then the differential dy is given by $dy = f(x)dx$. This relationship is used to help change from one variable to another, a process used in several important problems in later chapters.

Throughout, pdfs are assumed to be defined on the entire real line, even if the values of the random variable associated with nonzero probability do not fill out $\mathbb{R}$. For example, it may be that we are measuring weights of adult humans. If W is this measurement, negative values are not possible, nor are very low weights in this case. If we are considering healthy adults, there is likely an upper bound to such weights as well, and so we expect $W \in (a, b)$ for some numbers $0 \leq a$ and $a < b$. Outside of this range, we take the pdf for W to be zero.

Definition 4: Let X be a continuous random variable with pdf f. The **cumulative distribution function**, **cdf**, is the function F defined by $F(x) = P(X \leq x)$ for all $x \in \mathbb{R}$.

The derivative of the cdf F is f, as long as f is continuous. There are random variables that have cdfs, but not continuous pdfs. These occur in finance for example, and will not be considered in this text.

Just as in the discrete case, the cdf can be used to compute probabilities. It is in fact simpler in the continuous case because the probability of a single value is always 0. Thus, if X is a continuous random variable with cdf F, then for all $a, b \in \mathbb{R}$, with $a < b$: $P(a < X < b) = P(a \leq X < b) = P(a < X \leq b) = P(a \leq X \leq b) = F(b) - F(a)$.

A review of basic integration techniques is given in Appendix B. This includes: integrating polynomials, exponentials, and other elementary functions; improper integration; techniques of integration such as integration by parts, substitution, and others. A thorough review of this material will help the reader with much of the material in later chapters.

Example 6: Let X have pdf given by $f(x) = 2x$ on $[0,1]$ and 0 otherwise. The probability $X \in (1/4, 3/4)$ is given by:

$$P(X \in (1/4, 3/4)) = \int_{1/4}^{3/4} 2x \, dx = x^2 \Big|_{1/4}^{3/4} = \frac{9}{16} - \frac{1}{16} = \frac{8}{16} = \frac{1}{2}.$$

Also,

$$P(X > 1/3) = 1 - P(X \leq 1/3) = 1 - \int_{0}^{1/3} 2x \, dx = 1 - x^2 \Big|_{0}^{1/3} = 1 - \left(\frac{1}{9} - 0\right) = \frac{8}{9}.$$

X has cdf F given by:

$$F(x) = \begin{cases} 0 & x < 0 \\ x^2 & 0 \leq x \leq 1 \\ 1 & x > 1 \end{cases}.$$

Note that we do not express our answers to these type of questions in decimal form, unless required to do so. Also note that the cdf in this case has three pieces, and a complete answer to calculating a cdf requires a function defined for all real numbers.

Example 7: Let Y have pdf given by $g(y) = 1/4$ on $[0, 4]$. This is an example of a uniformly distributed random variable (or simply uniform variable), and it is discussed in more generality and detail in Chapter 5. Note that:

$$\int_0^1 g(y)dy = \int_2^3 g(y)dy = \int_{2/3}^{5/3} g(y)dy = \int_{2.5}^{3.5} g(y)dy,$$

and in general we would find that for intervals of the same length contained in $[0,4]$, the probability is constant. In this case, the cdf is:

$$G(y) = \begin{cases} 0 & y < 0 \\ x/4 & 0 \leq y \leq 4 \\ 1 & y > 4 \end{cases}.$$

Example 8: Let U have pdf given by $h(u) = ke^{-4u}$ on $[0, \infty)$ and 0 otherwise. One standard question is, what value of k makes this a pdf? To determine this value, remember the properties of all pdfs: The function must be non-negative, and its integral over the entire real line must be 1. Thus:

$$1 = \int_{-\infty}^{\infty} h(u)du = \int_0^{\infty} ke^{-4u}du = k\int_0^{\infty} e^{-4u}du = k\left[\lim_{A \to \infty}\int_0^A e^{-4u}du\right].$$

The introduction of the limit is due to the fact that this is an improper integral, in that it is an integral over an infinite region. The integration you studied in calculus is called the Riemann integral, and it is assumed that the interval over which we are performing (Riemann) integration is finite in length and that the function has no infinite limit values in the interval. When these conditions are violated, we call the integral improper and introduce limits as needed. To complete the calculation:

$$1 = k\left[\lim_{A \to \infty} \frac{e^{-4u}}{-4}\Big|_0^A\right] = k\left[\lim_{A \to \infty}\left(\frac{e^{-4A}}{-4} + \frac{1}{4}\right)\right] = \frac{k}{4}.$$

Finally, then, $k = 4$. Now that we know k, we can calculate probabilities and the cdf:

$$P(U \geq 3) = 1 - P(U < 3) = 1 - 4\int_0^3 e^{-4u}du = 1 - 4\left(\frac{e^{-12}}{-4} + \frac{1}{4}\right) = e^{-12},$$

a very small number. The cdf will take the form:

$$H(u) = \begin{cases} 0 & u < 0 \\ 1 - e^{-4u} & u \geq 0 \end{cases}.$$

This is an example of an exponential distribution, which is discussed in more detail in Chapter 5.

Sometimes we will write out the formula for a density on the interval or intervals where it takes on nonzero values, and then assume it is 0 elsewhere. Similarly, in the discrete case, we may specify the probabilities through a probability distribution function defined at the random variable's values with nonzero probability. It is good practice, however, to continue to write "and 0 otherwise" to reinforce the full definition.

For the last example, it is wrong to write $\int_{-\infty}^{\infty} ke^{-4u} du$ when finding k, because the density is 0 on $(-\infty, 0)$. We write $\int_{-\infty}^{\infty} h(u) du$ first, eliminate the intervals where the density is 0, and replace this with $\int_{0}^{\infty} ke^{-4u} du$ for this problem.

3.3 BIVARIATE AND MULTIVARIATE DISTRIBUTIONS

In many circumstances, we measure or record the values of two or more variables. For example, we may record: the blood pressure, age, and the value of a general health index of an individual in for a checkup; the shape of leaf, height of plant, color of flowers, of a species of plant; the length of various bones, thickness of bones, etc., that a particular animal has. In these cases, we have two (or more) random variables for the same experiment. Note that not all of the variables have to be of the same type: some may be discrete, and some may be continuous in the same experiment. We call problems with one variable univariate; two variables bivariate; and in general, two or more variables multivariate. In this section, we will concentrate on two variables; but the ideas extend to the multivariate situation, and this generalization plays an important role in developing experimental procedures.

Let our random experiment be bivariate, with random variables X and Y to be recorded. Although one variable could be continuous and the other discrete (and often are), we will assume both are either continuous or both are discrete. For the discrete case, we have coded the X and Y values (if necessary) so the X values are $x_1, \ldots, x_m$, with $x_1 < x_2 < \cdots < x_m$ and the Y values are $y_1, \ldots, y_n$ with $y_1 < y_2 < \cdots < y_n$. In this case, both X and Y are finite, but the ideas extend to countably infinite cases as well. We then want to determine $P(X = x_i, Y = y_j) = f(x_i, y_j) = p_{ij}$ = probability that $X = x_i$ **and** $Y = y_j$, over all combinations of the values of the variables. The notation $P(X = x_i, Y = y_j)$ means the probability of the intersection of the two events $X = x_i$ and $Y = y_j$. The listing of these probabilities is called the **joint probability distribution** of X and Y. The function f (of two variables) defined by $f(x_i, y_j) = P(X = x_i, Y = y_j)$ for all $1 \leq i \leq m$ and $1 \leq j \leq n$, and 0 otherwise, is called the **joint probability distribution function**, or **jpdf**. Often, when there are only a few different values of X and Y, the probabilities are written in table form. Note that, in applications, we may have to compute double sums such as $\sum_{i=1}^{m} \sum_{j=1}^{n} f(x_i, y_j) = 1$.

If both X and Y are continuous, it is assumed there is a function of two variables, say f, such that $f(x, y) \geq 0$ for all $(x, y) \in \mathbb{R}^2$; $\int_{-\infty}^{\infty} \int_{-\infty}^{\infty} f(x, y) dx \, dy = 1$; and if A is any open region in $\mathbb{R}^2$, then $P((X, Y) \in A) = \iint_A f(x, y) dx dy$. In general, it is sufficient to think of A as an open disc, rectangle, triangle, etc. In two dimensions, if we integrate over anything that is one dimensional, the integral will be zero, similar to the fact that, in one dimension, integrals over single points are zero. We will not be overly concerned about the theoretical aspects of integrating in two or more dimensions, beyond extending methods for integrating in one dimension to special two (and higher) dimensional integrals.

Related definitions include:

Definition 5: (Discrete Variables): Let X and Y be (finite) discrete random variables with joint probability distribution function f. The **joint cumulative distribution function**, abbreviated **jcdf**, is the function F defined by:

$$F(x, y) = P(X \leq x \text{ and } Y \leq y) = \sum_{t \leq x} \sum_{u \leq y} P(X = t, Y = u).$$

The values t and u are the different values of X and Y less or equal x and y, respectively, for which there is nonzero probability in the joint distribution.

The **marginal distribution** of X is the distribution resulting from summing out the influence of Y, namely the set of values:

$$P(X = x_i) = \sum_{j=1,2,\ldots,n} p_{ij} = \sum_{j=1,2,\ldots,n} P(X = x_i, Y = y_j)$$

computed for all $1 \leq i \leq m$. A similar definition holds for the marginal distribution of Y, with the sums now over the different x_i values, for all of the different values $1 \leq j \leq n$.

The **conditional distribution** of X, given $Y = y_j$, is the distribution given by $P(X = x_i, Y = y_j)/P(Y = y_j) \equiv f_{X|Y=y_j}(x_i)$, $1 \leq i \leq m$. We call $f_{X|Y=y_j}$ the conditional distribution function of X, given $Y = y_j$. For each different value of j, $1 \leq j \leq n$, there is a conditional distribution of X given $Y = y_j$. In a similar fashion, we define the conditional distribution of Y, given $X = x_i$, for each $1 \leq i \leq m$.

As an exercise, the reader should extend these definitions to the cases where one or both of X and Y is countable.

Definition 6: (Continuous Variables): Let X and Y be continuous random variables with joint probability distribution function f. The **joint cumulative distribution function**, abbreviated **jcdf**, is the function F defined by:

$$F(x, y) = P(X \leq x \text{ and } Y \leq y) = \int_{-\infty}^{y} \int_{-\infty}^{x} f(u, v)\, du\, dv.$$

This is the integral of the density function over the region in $\mathbb{R}^2$, or the $(u.v)$ plane, bounded by the lines $u = x$ and $v = y$. Note the use of different letters for the variables of integration when the limits involve x and y, and the integral as written means we do the u integration followed by the v integration.

The **marginal distribution** of X is the distribution resulting from integrating out the influence of Y. This yields a a univariate distribution with variable X, denoted X:

$$f_X(x) = \int_{-\infty}^{\infty} f(x, y)\, dy.$$

A similar definition holds for the marginal distribution of Y, with the integral now over x from $-\infty$ to ∞.

The **conditional density function** of the distribution of X, given $Y = y$, is given by:

$$f_{X|Y=y}(x) = f(x, y)/f_Y(y).$$

It is assumed that $f_Y(y) > 0$. The reader should verify that this in fact is a proper density. In a similar fashion, we define the conditional density function $f_{Y|X=x}$ of the distribution of Y, given $X = x$, assuming $f_X(x) > 0$.

As in the case of univariate continuous distributions, the joint pdf cannot be interpreted as probability. However, as we did in the univariate case, using the fundamental definition of the Riemann integral, we can interpret $f(x, y)dxdy$ as the approximate probability (X, Y) falls in a rectangle centred at (x, y), where the rectangle has side lengths dx and dy, respectively. We can also think of $f(x, y)dxdy$ as a differential in two variables, and then this form tells us how to change to a new set of variables. This will be explored later.

Example 9: Let the discrete random variables X and Y have a joint distribution, as given in the following table:

		\multicolumn{3}{c}{X}		
		0	1	2	
Y	0	0.09	0.12	0.05	0.26
	1	0.25	0.25	0.03	0.53
	2	0.13	0.05	0	0.18
	3	0	0.03	0	0.03
		0.47	0.45	0.08	1.00

The marginal distributions are written in the "margins" of the table, so that the marginal distribution of X is:

X	0	1	2
$P(X=x)$	0.47	0.45	0.08

Chapter 3 — Distributions of Random Variables

The conditional distribution of Y, given $X = 2$, is:

Y	0	1	2	3
$P(Y \mid X=2)$	$.05/.08 = 5/8$	$.03/.08 = 3/8$	$0/.08 = 0$	$0/.08 = 0$

Normally, of course, we do not write the calculations in the array, just the simplified numbers.

Additional exercises for this data set the reader can complete include verifying the marginal distributions of X and Y, and determining, say, the conditional distribution of X given $Y = 1$.

Definition 7: Let X and Y have a discrete bivariate distribution. We say the variables are independent iff $P(X = x, Y = y) = P_X(X = x)P_Y(Y = y)$ over all possible values of X and Y. If there is at least one exception to this requirement, we say X and Y are **dependent**, or **dependent random variables**.

Notation: Let X and Y be random variables. We typically replace $P(X = x, Y = y)$ by $f(x, y)$, where f is the bivariate jpdf. We denote the marginal distribution function of X by f_X and the marginal distribution function of Y by f_Y. Sometimes, to increase clarity, we may use different letters for these marginals, say $g(x)$ in place of $f_X(x)$, and $h(y)$ in place of $f_Y(y)$. For example, with these definitions, X and Y are independent iff $f(x, y) = g(x)h(y) \; \forall (x, y) \in \mathbb{R}^2$. The conditional distribution function of X, given $Y = y$, is denoted $f_{X|Y=y}$ and then a value is $f_{X|Y=y}(x)$, or we use the notation $f(\bullet|Y=y)$ to denote the function with a value denoted $f(x|Y=y)$. Similar definitions hold for the conditional distribution function of Y given $X = x$, and for extensions to multivariate cases.

In problems involving many variables (and there are problems with thousands of variables), we use subscript notation. For example, suppose we have three random variables, X, Y, and Z. Instead, we could write X_1, X_2, and X_3, respectively. Essentially, the subscript indicates the *order* of the variables, and we assume the values of the three variables are recorded simultaneously in ordered triples.

We use the abbreviation jpdf in both the continuous and the discrete cases, but they have very different meanings. For discrete variables, the jpdf is actually a listing or formula for the probability of specific combinations of values of the variable. The jpdf for continuous variables is not the probability of specific combinations of the variables, but a function used to generate probabilities through integration.

A very common error is to mix up discrete and continuous variables: summation is used where integration is necessary, or integration where summation is necessary. It is essential that the type of variable be identified (continuous variables involve measurements, discrete variables typically relate to counts of things) so that the correct methods are used in calculations.

Example 10: Consider the function $f(x, y) = kxy$ for $0 \leq x, y \leq 1$, and 0 otherwise. Suppose we want the value k that makes this a jpdf for random variables X and Y, and then we want to know if X and Y are independent. First:

$$\int_0^1 \int_0^1 xy \, dy dx = \int_0^1 \left[\int_0^1 y \, dy \right] x \, dx = \int_0^1 \frac{1}{2} x \, dx = \frac{1}{4}$$

which means $k = 4$. To find the marginal distributions, we integrate out the other variable, that is, if we want the marginal distribution of X we then integrate over the Y values. If $g(x)$ is the marginal density of X and $h(y)$ is the marginal density of Y, then:

$$g(x) = 4 \int_0^1 xy \, dy = 2x, \; 0 \leq x \leq 1 \quad \text{and} \quad h(y) = 4 \int_0^1 xy \, dx = 2y, \; 0 \leq y \leq 1$$

with these functions equal zero everywhere else. Clearly, since the product of the marginals is exactly the same as the jpdf, then X and Y are independent.

Example 11: Suppose X and Y are continuous random variables, with jpdf $f(x, y) = 3x(y + x)/5$ on $0 \leq x \leq 1$ and $0 \leq y \leq 2$. Are X and Y independent? Let $g(x)$ and $h(y)$ be the respective marginals. Then:

$$g(x) = \frac{3}{5}\int_0^2 x(y+x)dy = \frac{3}{5}\left[\frac{1}{2}xy^2 + x^2 y\Big|_0^2\right] = \frac{6}{5}\left(x+x^2\right), \text{ while } h(y) = \frac{3}{5}\left(\frac{1}{4}y^2 + \frac{1}{3}y\right).$$

As usual, these functions are 0 everywhere else.

If we take the product of g and h, we will not get f. One way to determine this is that two polynomials are equal if and only if the corresponding terms are the same, which will not be the case here.

In most instances, the domains are also linked, such as $0 \leq x \leq y \leq 1$. This will mean that our integrals will not be as simple as those in the above examples. These examples illustrate the main ideas of this section and more complicated examples are left for advanced courses.

The definitions introduced so far readily extend to many variables, but actual calculations of the above type will be restricted to two variables (i.e., marginals and conditionals) with simple domains. It is very useful, however, to be familiar with the concepts as they would apply to more general cases, if not the methods of calculation, since a number of important results are based on multivariate frameworks.

3.4 EXPECTATION

There are a number of standard calculations, other than calculating probabilities of events, associated with distributions.

Definition 8: We define the **expected value** of X to be $E(X) = \sum_{i=1}^n x_i p_i = \mu_X$ when X is a discrete random variable with values $x_1, \ldots, x_n$, and $E(X) = \int_{-\infty}^{\infty} xf(x)dx = \mu_X$ when X is continuous. If there is only one random variable X, we often write μ in place of μ_X. If the distribution of X is the marginal distribution of X from some bivariate or multivariate distribution, then we call μ_X the marginal expectation of X.

Let $g(X)$ be a function of X. We define the **expected value of** $g(X)$ to be $E(g(X)) = \sum_{i=1}^n g(x_i)p_i$ when X is discrete, and $E(g(X)) = \int_{-\infty}^{\infty} g(x)f(x)dx$ when X is continuous.

If $g(X) = e^{tX}$, $E(e^{tX})$ is called the **moment generating function** of X or of the distribution, whenever it exists for t on an open interval containing 0. It is denoted $M_X(t)$, or simply $M(t)$.

In the special case $g(X) = (X - \mu)^2$, define the **variance** of X, or the variance of the distribution, as:

$$var(X) = V(X) = E((X-\mu)^2) = \sigma_X^2 = \sigma^2 = \sum_{i=1}^n (x_i - \mu)^2 p_i$$

$$= \sum_{i=1}^n x_i^2 p_i - \left(\sum_{i=1}^n x_i p_i\right)^2 = E(X^2) - (E(X))^2$$

when X is discrete, and when X is continuous,

$$var(X) = V(X) = E((X-\mu)^2) = \sigma_X^2 = \sigma^2 = \int_{-\infty}^{\infty}(x-\mu)^2 f(x)\,dx$$

$$= \int_{-\infty}^{\infty} x^2 f(x)\,dx - \left(\int_{-\infty}^{\infty} xf(x)\,dx\right)^2 = E(X^2) - (E(X))^2.$$

If the distribution of X is the marginal distribution from some bivariate or multivariate distribution, we call σ_X^2 the marginal variance of X.

The square root of the variance is called the **standard deviation** of X.

Let $n \in \mathbb{N}$. $E(X^n)$ is called the nth **uncorrected** or **non-central moment** of X and $E((X - \mu)^n)$ the nth **central** or **mean corrected moment** of X. $E(X^0) = 1$ by definition.

In the discrete case in the above definitions, n is the number of distinct X values for which $P(X = x) \neq 0$ and can be finite or countably infinite. In the countably infinite case, $\sum_{i=1}^{n}$ is replaced by $\sum_{i=1}^{\infty} = \lim_{M \to \infty} \sum_{i=1}^{M}$. The value M is a positive integer, and the limit (if it exists) is the limit of the partial sums of whatever expression follows the summation sign.

Definition 9: $E()$ (or $E(\bullet)$) is called the **expectation operator**.

The moment generating function plays an important role in probability, as it characterizes the distribution. In fact:

Theorem 1: Uniqueness of the mgf. Let the random variable X have moment generating function M_X such that $M_X(t)$ is defined for all $t \in (-h, h)$, where $h > 0$. Then M_X is unique to X and X is unique to M_X.

Basically, if we have a moment generating function M under the conditions in the theorem, and we also know that a particular random variable X has that moment generating function, then X must have been the random variable that produced M. The proof is beyond the scope of this text.

Theorem 2: Under the conditions of the uniqueness theorem, the moment generating function M_X will have derivatives of all orders. Further,

$$\frac{d^k M_X(t)}{dt^k}\Big|_{t=0} = E(X^k)$$

that is, the kth derivative of the moment generating function with respect to t, evaluated at 0, is the kth non-central moment of X.

Definition 10: Let X and Y have joint pdf f_{XY}. We define the joint moment generating function of X and Y to be:

$$M_{XY}(t, s) = E\left(e^{tX + sY}\right)$$

whenever this is defined for $(t, s) \in R$, where R is a rectangle centred at $(0,0)$, with $t \in (-h, h)$ and $s \in (-k, k)$ for some numbers $h, k > 0$.

Theorem 3: For the random variables X and Y in the last theorem, define $Z = X + Y$. Assume also that X and Y are independent. Then:

$$M_Z(t) = M_X(t) M_Y(t).$$

In general, let $X_1, \ldots, X_n$ be n independent random variables, and define $Z = \sum_{i=1}^{n} X_i$. Then the moment generating function of Z is:

$$M_Z(t) = \prod_{i=1}^{n} M_{X_i}(t).$$

In particular, if all of the X_is have the same distribution, and hence the same moment generating function, M_X say, then:

$$M_Z(t) = M_X^n(t).$$

The proofs of the last two theorems are left as exercises.

Example 12: Let the random variable X have mgf M with $M(t) = \exp(2t^2)$. Determine the first two non-central moments and the variance of X.

Solution: By calculus,

$$M'(t)|_{t=0} = 4te^{2t^2}|_{t=0} = 0, \text{ and } M''(t)|_{t=0} = 4e^{2t^2} + e^{2t^2}(4t)^2|_{t=0} = 4.$$

Hence, the mean is 0 and the variance is 4.

Example 13: A random variable Y with mgf of the form $M(t) = \exp(a^2 t^2/2)$ has a normal distribution with mean 0 and variance a^2. Consider the random variable X from the previous example. Let $X_1, \ldots, X_n$ be independent random variables, each with same distribution as X. Determine the distribution of $\overline{X} = \frac{1}{n}(X_1 + \cdots + X_n)$.

Solution: We can write the moment generating function of $\overline{X}$ as:

$$M_{\overline{X}}(t) = E\left(\exp\left(t\left(\frac{1}{n}\sum_{i=1}^{n} X_i\right)\right)\right) = E\left(\exp\left(\frac{t}{n}\sum_{i=1}^{n} X_i\right)\right)$$

and this then is the moment generating function of the sum of the independent random variables evaluated at t/n. But by a previous result,

$$M_{\overline{X}}(t) = \prod_{i=1}^{n} \exp(2t^2/n^2) = \exp(4nt^2/(2n^2)) = \exp(4t^2(2n)).$$

By comparison, then, this is the moment generating function of a normal random variable with mean 0 and variance $4/n$.

This last example illustrates the power of the moment generating function: we can often determine the distribution of a function of random variables by the form of the resulting mgf, by comparing it to known mgfs.

Example 14: Let the moment generating function of X be $M(t) = e^t/6 + e^{2t}/3 + 3^{3t}/2$. Determine X.

Solution: In general, sums of exponential expressions in a moment generating function indicate the variable was discrete:

$$M(t) = \sum_{i=a}^{b} e^{it} p(i).$$

By comparison, $i = 1, 2, 3$. The only way we can have $e^t/6 + e^{2t}/3 + 3^{3t}/2 = e^t p(1) + e^{2t} p(2) + 3^{3t} p(3)$ is to take $p(1) = 1/6$, $p(2) = 1/3$ and $p(3) = 1/2$. By the uniqueness theorem, the random variable X must have distribution:

$X = i$	1	2	3
$p(i)$	1/6	1/3	1/2

Certain moments are important in determining the nature of a given distribution: The first moment is just the mean, a measure of the centre of the distribution, when it exists. The second central moment is the variance and indicates how the probability is spread around the mean. Higher moments are used to measure other characteristics, such as skewness and kurtosis, and are discussed in section 3.5.

Chapter 3 — Distributions of Random Variables

Example 15: Suppose the random variable X had the following probability distribution:

X	1	2	3	4	5
$P(X=x)$	1/5	2/5	1/10	1/10	1/5

Then $E(X) = \mu = 1(1/5) + 2(2/5) + 3(1/10) + 4(1/10) + 5(1/5) = 27/10$ and
$Var(X) = \sigma^2 = E((X-\mu)^2) = E(X^2) - \mu^2 = 93/10 - 27^2/100 = 201/100$.

3.4.1 Properties of Expectation

Theorem 4: Let X be a random variable, and let a, b, and c be constants. Some easily established rules for expectation, regardless of the type of distribution, include:

(i) $E(X + b) = E(X) + b$.
(ii) $E(c) = c$.
(iii) $E(aX) = aE(X)$.
(iv) $Var(X + b) = Var(X)$.
(v) $Var(aX) = a^2 Var(X)$.

Definition 11: Let the random variable X have mean μ and variance σ^2. Define the random variable Z by $Z = (X - \mu)/\sigma$. Z is called the **standardization** of X, and we say we are **standardizing** X.

If the random variable Z is the standardization of another random variable, then $E(Z) = 0$ and $Var(Z) = 1$, and further Z has no units. This calculation is used to eliminate dependence on location and scale, so that different distributions can be compared. Other functions of variables are also of interest.

Example 16: Let the random variable Y have distribution given by:

Y	0	2	4	8
$P(Y=y)$	0.1	0.1	0.5	0.3

Determine each of the following for Y: (a) the mean; (b) the variance; (c) $var(2Y - 1)$; (d) $E(4Y - 3)$, and (e) the standardization of $Y = 4$.

Solution:

(a) $E(Y) = 0.1 \times 0 + 0.1 \times 2 + 0.5 \times 4 + 0.3 \times 8 = 4.6$. Note that an expected value does not have to be one of the actual values of the random variable.

(b) $E(Y^2) = 0.1 \times 0^2 + 0.1 \times 2^2 + 0.5 \times 4^2 + 0.3 \times 8^2 = 27.6$, and hence, $var(Y) = 27.6 - 4.6^2 = 27.6 - 21.16 = 6.44$. The standard deviation would be $\sqrt{6.44} = 2.54$, approximately.

(c) $var(2Y - 1) = var(2Y) = 4var(Y) = 25.76$.

(d) $E(4Y - 3) = 4E(Y) - 3 = 4(4.6) - 3 = 18.4 - 3 = 15.4$.

(e) If we let Z be the standardization of Y, then when $Y = 4$ the standardized value is $z = (4 - 4.6)/2.54 = -0.6/2.54 = -0.236$.

What would the conclusion be if for a random variable X, calculation yielded $E(X^2) - E^2(X) = -4$? It has to that the calculation has gone horribly wrong. If the pdf is properly defined, variances can never be negative. What often happens is the order of the terms is reversed due to faulty memorization.

Example 17: Let the random variable U have the uniform distribution on $[0,1]$. Determine each of the following for U: (a) the mean; (b) the variance; (c) $var(2U-1)$; (d) $E(4U-3)$ and (e) the standardization of $Y = 3/4$.

Solution: Let g be the density, so that $g(u) = 1$ when $0 \leq u \leq 1$ and is 0 otherwise.

(a) $E(U) = \int_{-\infty}^{\infty} ug(u)du = \int_0^1 u \times 1 du = \frac{u^2}{2}\big|_0^1 = 1/2$.

(b) $E(U^2) = \int_{-\infty}^{\infty} u^2 g(u) du = \int_0^1 u^2 \times 1 du = \frac{u^3}{3}\big|_0^1 = 1/3$. Thus the variance is $1/3 - (1/2)^2 = 1/12$. The standard deviation would be $\sqrt{1/12} = 1/\sqrt{12}$.

(c) $var(2U - 1) = 4 \, var(U) = 4/12 = 1/3$.

(d) $E(4U - 3) = 4/2 - 3 = -1$.

(e) The required standardization is $(3/4 - 1/2)/\sqrt{1/12} = \sqrt{3}/2$.

Example 18: Teams of 5 researchers go into remote areas of a forested area, and some or all may contract poison oak. It costs \$100 to treat each person who gets poison oak, and the organizer needs to determine how much money to have available to treat those afflicted. From past trips, the random variable V, the number in a team who get poison oak, has distribution:

$$P(V = i) = \binom{5}{i}(0.4)^i(0.6)^{5-i}$$

$0 \leq i \leq 5$, and is 0 otherwise. Find the expected number of researchers who will get poison oak (in a team of 5) and the variance of V.

Solution: From this information, we need to evaluate:

$$\sum_{i=0}^{5} i \binom{5}{i}(0.4)^i(0.6)^{5-i} \text{ and } \sum_{i=0}^{5} i^2 \binom{5}{i}(0.4)^i(0.6)^{5-i}.$$

The first is exactly 2, and the second value is 5.2. Thus, $E(V) = 2$ and $var(V) = 5.2 - 4 = 1.2$. The probability that all 5 will contract poison oak is about 0.01. This information may suggest that being prepared to treat up to 4 researchers would be prudent.

We can extend the definition of expectation to bivariate and multivariate cases:

Definition 12: Let the random variables X and Y have a bivariate distribution with jpdf f. Let g be a function of X and Y. Then the expectation of g is denoted $E(g(X, Y))$, with:

$$E(g(X,Y)) = \sum_{j=1}^{m}\sum_{i=1}^{n} g(x_i, y_j) f(x_i, y_j)$$

in the discrete case, with the usual definition of the values x_i and y_j, and by:

$$E(g(X,Y)) = \int_{\mathbb{R}}\int_{\mathbb{R}} g(x,y) f(x,y) dx \, dy$$

in the continuous case. These definitions extend in a natural way to several random variables.

Note the use of $\int_{\mathbb{R}}$ in place of $\int_{-\infty}^{\infty}$. We can also replace $\int_{-\infty}^{\infty}\int_{-\infty}^{\infty}$ by either $\int_{\mathbb{R}}\int_{\mathbb{R}}$ or by $(\int\int)_{\mathbb{R}^2}$.

Chapter 3 — Distributions of Random Variables

Theorem 5: Let X and Y be random variables with a joint distribution. Then for all constants a, b, c: $E(aX + bY + c) = aE(X) + bE(Y) + c$. In particular, $E(X + Y) = E(X) + E(Y)$. Proof: Exercise.

Another way to say this last result is that $E()$ is a **linear operator**. In fact, the expected value of any sum is just the sum of expected values:

$$E\left(\sum_{i=1}^{k} a_i X_i\right) = \sum_{i=1}^{k} a_i E(X_i).$$

Caution: The expected value of a **product** or **quotient** of random variables is not the product or quotient, respectively, of expected values generally. If X and Y are independent, however, $E(XY) = E(X)E(Y)$. Even in the case of independence, the expected value of a quotient of random variables is not the quotient of the expected values, except in a very special case. As an exercise, determine the conditions under which the expected value of a quotient will be the quotient of expected values.

Another standard calculation is that of the **covariance** between two variables.

Definition 13: Let X and Y have a joint distribution. Then the covariance between X and Y is denoted $cov(X, Y)$, and is given by $cov(X, Y) = E((X - \mu_X)(Y - \mu_Y)) = E(XY) - E(X)E(Y)$. Here, μ_X and μ_Y are the marginal expectations of X and Y, respectively.

Note that when X and Y are independent, $cov(X, Y) = 0$ because $E(XY) = E(X)E(Y)$. However, $cov(X, Y) = 0$ does not mean X and Y are independent.

The definition of covariance does not require X and Y to be distinct variables. Thus, we have $cov(X, X) = E(X^2) - E^2(X) = var(X)$, so covariance can be thought of as a generalization of variance.

Definition 14: The value $cov(X, Y)/(\sigma_X \sigma_Y) \equiv \rho$ is called the (linear) **correlation** between (of) X and Y. This can be shown to be a number between -1 and 1 always, it is free of units of X and Y and allows comparisons of different distributions. This value plays an important role in regression: The closer ρ is to $+1$ or -1, the stronger the linear relationship is between the variables. The closer ρ is to 0, generally, the "nearer" X and Y are to being independent.

The value ρ is often called the Pearson correlation coefficient. It is used in determining the strength of linear relationships between variables, but does not provide good information should the relationship be non-linear. Linear regression is studied in Chapter 11, and non-linear regression is beyond the scope of this text.

Example 19: Let the random variables X and Y have a joint distribution, with $E(X) = 5$, $E(Y) = 7$, $E(XY) = 32$, $E(X^2) = 30$ and $E(Y^2) = 57$. Determine (a) $cov(X, Y)$; and (b) $corr(X, Y)$.

Solution: From the given information, $\sigma_X^2 = 5$, $\sigma_Y^2 = 8$. Thus:

(a) $cov(X, Y) = 32 - 35 = -3$.
(b) $\rho = corr(X, Y) = -3/\sqrt{40} \approx -0.474$.

In this example, the covariance is -3, so we know that as X increases, Y will, on average, decrease. However, the covariance does not tell us how strong this relationship is. The correlation tells us that this relationship is not that strong: A measure of the strength of the linear relationship is ρ^2, called the **coefficient of determination**. (There are other definitions of the coefficient of determination, but this is the one we will use.) An interpretation is that a linear relationship explains about $\rho^2 \times 100\%$ of the variability seen in Y, explained by the linear relationship with X (discussed in detail in the chapter on linear regression). In this case, the coefficient of determination is 0.225, so the linear relationship of Y to X explains only 22.5% of Y's variability. A rule of thumb is that this relationship should explain at least 50% of Y's variability to consider the linear relationship at least acceptable, so we would not consider Y to be strongly linearly related to X. As noted earlier, there may be non-linear relationships that would be more useful.

Example 20: Let the discrete random variables X and Y have a joint distribution, as given in the following table (reproduced from earlier):

		\multicolumn{3}{c}{X}		
		0	1	2	
Y	0	0.09	0.12	0.05	0.26
	1	0.25	0.25	0.03	0.53
	2	0.13	0.05	0	0.18
	3	0	0.03	0	0.03
		0.47	0.45	0.08	1.00

Some summary values include (assuming this is a population): $E(X) = 0.61$; $E(Y) = 0.98$; $E(X^2) = 0.77$; $E(Y^2) = 1.52$ and $E(XY) = 0.50$. From these numbers: $\sigma_X^2 = 0.3979$; $\sigma_Y^2 = 0.5596$; and $cov(X, Y) = -0.0978$. Based on these values, the correlation coefficient is $\rho = -0.207$.

Note: A common question is: how many decimals should be kept in calculations? A full error analysis would dictate that, but in this text, unless otherwise specified, when data is given in decimal form, we keep several extra decimals in our calculations and express the end results to no more than two more decimals than given in the data.

We often need to have some idea of the chance or the probability we would see values of a random variable in some interval, whether continuous or discrete. One fundamental relationship is given in Chebyshev's Theorem, which also gives a link between the mean and variance of a distribution.

Theorem 6: Chebyshev's Theorem: This is also called **Chebyshev's Inequality**. If the random variable X has finite mean μ and variance σ^2, then for any $k > 0$, $E(|X - \mu| < k\sigma) \geq 1 - 1/k^2$; that is, the probability of being within k standard deviations of the mean is at least $1 - 1/k^2$.

Example 21: Consider the continuous random variable X with pdf:

$$f(x) = 2\sqrt{2}/\left[\pi(1 + 2x^2)^2\right], \quad x \in \mathbb{R}.$$

This distribution has mean 0 and variance 1/2. Thus, by Chebyshev's Inequality: at least 0% of the distribution is within $1/\sqrt{2}$ of 0, i.e., in the interval $(-1/\sqrt{2}, 1/\sqrt{2}) = (-0.707, 0.707)$; at least 75% in $(-\sqrt{2}, \sqrt{2}) = (-1.414, 1.414)$; and at least $(8/9) \times 100\% \approx 89\%$ in $(-3/\sqrt{2}, 3/\sqrt{2}) = (-2.121, 2.121)$. What fraction of the distribution is actually in these intervals? 0.818, 0.959, and 0.986, respectively.

Example 22: Let the random variable Y have distribution given by:

Y	-2	0	2
$P(Y = y)$	0.0125	0.975	0.0125

The mean is 0, and the standard deviation is 0.316. Using Chebyshev's Inequality, we are guaranteed at least 97.2% of the distribution within $k = 6$ standard deviations of the mean, and in fact, we have 97.5% in that interval. Note that there is 97.5% of the distribution within 1 standard deviation, and this does not change until $k \approx 6.33$.

It should be pointed out that Chebyshev's Inequality is true for discrete as well as continuous variables, and it holds true for samples as well as populations.

If we have a distribution that has the classic bell-shape (see Figure 3-1), then we can say more:

Empirical Rule. If X has a bell-shaped distribution, then the probability of observing X within one standard deviation of its mean is approximately 67%; within 2 standard deviations, it is approximately 95%; and within 3 standard deviations, it is approximately 99.7% of the observations on X.

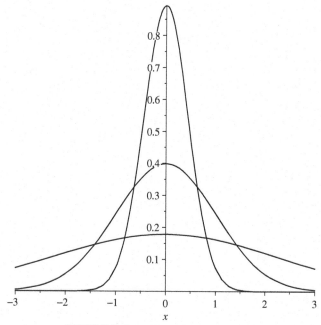

FIGURE 3-1. Bell-Shaped Distributions.

In the last example, it took more than six standard deviations from the mean to exceed 97.5% of the distribution.

This rule is based on the fact that bell-shaped distributions are essentially **normal distributions** (discussed in chapter 5), and the normal has a well-defined and computable probability for each of the values k in Chebyshev's inequality. We will examine the normal distribution in detail, since it underlies many of the statistical and probabilistic calculations in quantitative analysis.

3.5 OTHER CHARACTERISTICS OF DISTRIBUTIONS

In the preceding sections, we computed moments (including the mean and the variance), the probability density/distribution function (pdf), cumulative distribution function (cdf) and the moment generating function (mgf). Such values and functions help us characterize distributions. For continuous or discrete random variables, the pdf, cdf, and mgf contain all of the information about a distribution, while specific moments (the mean and variance, for example) provide measures of specific aspects of the distribution. For example, we know from Chebyshev's Inequality that the smaller the variance, the more tightly the probability is distributed around the mean. Other characteristics can be used as well. In this section, we will introduce some of the more common values and descriptions used in describing distributions, and provide an expanded discussion in Chapter 7.

Definition 15: Let the random variable X have pdf f. If there exists a unique value, μ_o say, that maximizes f, we call it the **mode**. If X is a continuous random variable and its density has two (or more) local maxima, then we say the distribution is **bimodal** (or **multimodal**). In discrete distributions covered in this text, there may be two values that maximize the pdf, and each is called a mode. In the case of uniform distributions, we do not report a mode.

Definition 16: Let X be a random variable. The **median** of the distribution of X is a value m such that $P(X \leq m) \geq 1/2$ and $P(X \geq m) \geq 1/2$. By convention for discrete variables, if u and v are such that $u < v$, both $P(X = u)$ and $P(X = v)$ are positive, and $P(X = x) = 0 \, \forall \, u < x < v$, each satisfying the definition of being a median, then the median is taken to be $(u + v)/2$.

Definition 17: Let X be a random variable, continuous or discrete. The kth **percentile**, $1 \leq k \leq 99$ and integer, of a distribution is a value X_k such that $P(X \leq X_k) \geq k/100$ and $P(X \geq X_k) \geq (100 - k)/100$. The 25th percentile is called the **first quartile**; the 50th percentile is called the **second quartile** (also the median); the 75th percentile is called the **third quartile**. The **first decile** is the 10th percentile, with similar definitions for the second up to the ninth deciles. Collectively, percentiles, deciles, and quartiles are called **quantiles**.

As with the median, if a percentile, quartile or decile can be any number between two values, we will take the average of these two values as the appropriate quantile. In the case of continuous random variables, quantiles are uniquely determined, so this approximation is necessary only for discrete variables.

Consider finding the median of the continuous random variable X with density f such that $f(x) > 0$ on interval I and $f(x) = 0$ elsewhere, then the median will be the number $m \in I$ such that $P(X \leq m) = 1/2$. This follows from the fact that the density is assumed continuous and positive on an interval. If we define F to be the cdf, then F is monotonically increasing on I from the left endpoint of I to the right endpoint. By the Intermediate Value Theorem, there must be a value $m \in I$ for which $F(m) = 1/2$. For any other value of $x \in I$, $F(x) > 1/2$ or $F(x) < 1/2$; so, in contrast to discrete variables, there cannot be more than one value that satisfies the definition of being a median in this case. This argument generalizes to all quantiles.

Example 23: Let the discrete random variable have distribution:

$$f(x) = \begin{cases} 1/6 & x = 0, \ 1, \ 2, \ \text{or} \ 4 \\ 1/3 & x = 3 \end{cases}.$$

The mode is 3 in this case. The median is 2.5, which is the same as the 50th percentile or 5th decile. The 10th percentile is 0, while the third quartile is 3.

Let the continuous random variable X have pdf given by $f(x) = xe^{-x}$, for $x \geq 0$ and $f(x) = 0$ otherwise. From calculus, $f'(x) = e^{-x}(1 - x)$ on $x > 0$. The only critical number is $x = 1$, and since f tends to 0 as $x \to 0^+$ and $x \to \infty$, $f(1)$ must be the absolute maximum. There are no other critical numbers, and hence, no other relative extrema. The mode is $e^1 = e$, occurring at $x = 1$. The median m satisfies $\int_{-\infty}^{m} f(x) dx \equiv \int_{0}^{m} xe^{-x} dx = 1/2$, or $1 - e^{-m}(m + 1) = 1/2$. This example illustrates that, in many calculations, an exact value is not possible. In this case, we need to solve $1 = 2e^{-m}(m + 1)$, and to 5 decimals of accuracy, this is 1.67835.

The quantiles are usually computed when there are a large number of possible outcomes, such as in the case of continuous variables, or discrete variables with many mutually exclusive (and exhaustive) categories that are numerically coded. One use of such values is to allow comparisons between different populations where the same experimental procedure is applied. Quantiles remove absolute scale values so that comparisons are based on positions within the different populations.

Definition 18: Let the random variable X have pdf f. We say the distribution is **symmetric** about the mean, or simply symmetric, if $f(\mu - x) = f(\mu + x)$ for all x.

Definition 19: The part of the distribution to the left of the mean is called the **left tail** of the distribution, and the part to the right of the mean is called the **right tail**. If a distribution does not have a mean, then use the median as the centre in defining the tails.

Definition 20: Consider a unimodal distribution. If the left tail of such a distribution extends further to the left of the mean than the right tail extends to the right of the mean, we call the distribution **skewed to the left**. If the reverse is true, we say it is **skewed to the right**.

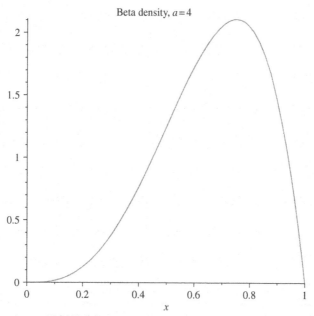

FIGURE 3-2. Skewed to the Left Distribution.

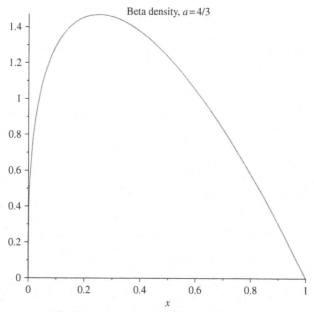

FIGURE 3-3. Skewed to the Right Distribution.

Example 24: Consider a continuous random variable X with probability density function $f(x) = a(a + 1)x^{a-1}(1 - x)$, $0 < x < 1$ and $f(x) = 0$ otherwise. Here, $a > 0$ is a constant, or parameter. This is a specific case of the beta family of distributions, and is used in modeling proportions. If $a = 4$, the mean is 2/3, the median is 0.6862 (approximately) and the mode is 3/4. Also, the first quartile is 0.546, and the third quartile is 0.806, which are not symmetric about the mean.

If $a = 4/3$, the mean is 2/5, the median is 0.37802 (approximately), and the mode is 1/4. Diagrams are given in the next two figures. From the diagrams, it is clear that X's distribution is skewed to the left when $a = 4$, and it is skewed to the right if $a = 4/3$. As an exercise, determine the first and third quartiles when $a = 4/3$ and compare the values to the mean.

Definition 21: Measure of Skewness: Let the random variable X have finite third moments. Set $\gamma_1 = E((X - \mu)^3)/\sigma^3$. Then γ_1 is a measure of the skewness of the distribution of X. We also define $\beta_1 = \gamma_1^2$.

Definition 22: Measure of Kurtosis: Let the random variable X have finite fourth moments. Set $\gamma_2 = E((X - \mu)^4)/\sigma^4 - 3$. Then γ_2 is a measure of the kurtosis of the distribution of X. We also define $\beta_2 = \gamma_2 + 3$.

These are not the only measures of skewness and kurtosis, but we will use them in this book. They are often called the Pearson coefficients of skewness and kurtosis, respectively. The reason for dividing by the appropriate power of the variance is to make the measures unitless, so different distributions can be compared. These overall measures of shape are not affected by changes in scale. Kurtosis is a measure of how peaked the distribution is near the mean. Bell-shaped distributions, i.e, normal distributions, have a kurtosis of 3, and so if $\gamma_2 > 0$ the distribution has thicker tails, while if $\gamma_2 < 0$ it has thinner tails than the normal.

Example 25: Determine γ_1 and γ_2 for the continuous uniform random variable X with density $f(x) = 1, 0 \leq x \leq 1$ and $f(x) = 0$ otherwise.

Solution: We can readily show that $\int_0^1 x^n \, dx = 1/(n+1)$ so the mean is 1/2, $\sigma^2 = 1/12$, $E((X - \mu)^3) = 0$, and $E((x - \mu)^4) = 1/80$. Thus, $\gamma_1 = 0$, and this indicates the distribution is symmetric about the mean (and a graph of the density confirms this). Also, $\gamma_2 = -6/5$ indicating the tails of the uniform are thinner than those of the normal, and again, a graph of the density verifies this — the tails actually have 0 thickness beyond the interval [0,1].

Exercises

Throughout these exercises, it is good practice to identify the type of distribution under consideration. As mentioned earlier, a common error is to use integration when summation is appropriate, and vice versa.

1. Consider $f(x) = (k + 1)x$ $x \in \{2, 4, 6, 8, 10, 12, 14\}$, and $f(x) = 0$ otherwise.

    (a) Determine the (exact) value of k that makes f a probability distribution function for the discrete random variable X, where the values of X are 2, 4, 6, 8, 10, 12, 14.

    (b) For X in (a), find $E(X)(= \mu)$.

    (c) Find the variance of X in (a) by:

    (i) $var(X) = \sum_{i=1}^{n}(x_i - \mu)^2 f(x_i)$
    (ii) $var(X) = E(X^2) - (E(X))^2$.

    Which procedure for calculating variances is the simplest?

    (d) Determine the median and the first and third quartiles for this distribution.

2. A random variable X can take on the values 1, 2, 3, 4, and 5, according to the probability distribution function given by $f(x) = (kx)/(2 + x)$.

    (a) Determine the value of k that ensures that f is a probability distribution function, and write the distribution of X in table form.

    (b) Determine the expected value and variance of X.

Chapter 3 — Distributions of Random Variables

(c) Determine the probability that X is:

  (i) no more than 2;
  (ii) at least 2;
  (iii) more than 1 and at most 3;
  (iv) at most 2 or greater than 3;

(d) Plot the distribution of X.
(e) Find the cumulative distribution function of X.
(f) Plot the cumulative distribution function.

3. Given the cumulative distribution function:

$$F(x) = \begin{cases} 0 & x < -1 \\ 0.15 & -1 \leq x < 2 \\ 0.25 & 2 \leq x < 4 \\ 0.6 & 4 \leq x < 6 \\ 0.87 & 6 \leq x < 7 \\ 1 & x \geq 7 \end{cases}$$

for the discrete random variable X, determine:

(a) the probability distribution function of X, and then plot it;
(b) the expected value and variance of X;
(c) the probability that X is within two standard deviations of the mean, and compare this to the value guaranteed by Chebyshev's Inequality;
(d) if events A and B are: $A = \{X | X \leq 3\}$ and $B = \{X > 1\}$, then evaluate each of the following probabilities:

  (i) $P(A \cup B)$;
  (ii) $P(A \cap B)$;
  (iii) $P(\bar{B})$;
  (iv) $P(B \cap \bar{A})$;
  (v) $P(\overline{A \cap B})$.

4. Let the discrete random variable X be uniformly distributed over the set $\{-1, 0, 1, 2, 3, 4, 5, 6\}$. See also §2.4.

(a) State the probability distribution function of X as a formula and in table form.
(b) Determine $E(X)$ and $var(X)$.
(c) Determine $E(X^2)$ and $E(X(X-1))$.
(d) Determine the cumulative distribution function of X. Write it as a piecewise defined function.
(e) Determine the median, the second quartile, and the sixth decile of this distribution.

5. Fill in the following tables under the assumption X and Y are independent variables. From your tables, calculate $E(XY)$, $E(X)$, and $E(Y)$ directly, and show $cov(X, Y) = 0$.

(a)

	X		
Y	1	2	
0			1/3
1			2/3
	3/4	1/4	1

(b)

	X			
Y	1	2	3	
1				1/5
2				2/5
3				2/5
	1/6	1/6	2/3	1

6. Determine the marginal distributions of X and Y in each of the following tables. In each case, determine:

   (a) the mean, variance, mode and median of X and Y;
   (b) both $cov(X, Y)$ and $\rho = corr(X, Y)$; and
   (c) the conditional distribution of Y, given $X = 2$, as well as the conditional mean and variance of Y, given $X = 2$.

   (i)

		X	
		1	2
Y	0	0.2	0.1
	1	0.3	0.4
			1.00

   (ii)

		X		
		1	2	3
	1	0.10	0.15	0.15
Y	2	0.05	0.10	0.05
	3	0.15	0.15	0.10
				1.00

7. Assume each of the following functions is the pdf of a continuous random variable X. Assume the function is 0 for all other values of x not specified. [It may be helpful to review Appendix B before attempting these problems.] For each, find:

   (a) the constant k that makes the given function a probability density function;
   (b) the mean and variance;
   (c) the cdf; and
   (d) the indicated values. If an exact value is not possible, give your answer to three decimals of accuracy.

   (i) $f(x) = k/x, 2 \leq x \leq 3$. Find $P(2.5 < X \leq 3)$, the median, and the mode.
   (ii) $g(x) = k \sin x, x \in [0, \pi/2]$. Find $P(\pi/4 < X < \pi/3)$, the median, and the mode.
   (iii) $h(x) = kx(1-x)^2, 0 \leq X \leq 1$. Find $P(X < 3/4)$, the 10th percentile, and the 3rd decile.
   (iv) $w(x) = kx^2 e^{-x/2}, x \geq 0$. Find $P(X > 1)$, the mode, and the median.

Chapter 3 — Distributions of Random Variables

8. Suppose the random variable U is such that its mean is always 0, but its variance is σ^2/n, where $\sigma^2 > 0$ is constant and n is a natural number. Use Chebyshev's Inequality to show that for any given $t > 0$, $P(|U| \leq t) \to 1$ as $n \to \infty$.

9. Using a graphing calculator or other computer software, plot the density functions in 7, and comment on the skewness of the resulting distribution.

10. On the same axes, plot the probability density functions $f(x) = \sqrt{2}/[\pi(1 + x^4)]$ and $g(x) = e^{-x^2/2}/\sqrt{2\pi}$ on the interval $[-3,3]$. The functions are defined for all real numbers, and they are pdfs of random variables that have mean 0 and variance 1. Use a graphing calculator or appropriate computer software to do the plotting. From your plot, comment on the skewness of each and on the comparative kurtosis of the two distributions.

11. Let X and Y be random variables, with $E(X) = 10$, $E(Y) = 7$, $var(X) = 3$, and $var(Y) = 2$.

    (a) Determine $E(2X - 3Y + 3)$.
    (b) Determine $var(2X - 3Y + 3)$ if X and Y are independent.
    (c) Determine $var(2X - 3Y + 3)$, if X and Y are dependent, with $cov(X, Y) = -2$.
    (d) Under the conditions in (c), determine the correlation between X and Y.

12. Determine the total number of possible outcomes that result from recording a card drawn from a well-shuffled deck, and if the suit is red, a die is rolled, and if the suit is black, a coin is tossed. An individual outcome takes the form (card drawn, number on die) or (card drawn, either heads or tails). A tree diagram may be helpful.

13. It has been observed that the average number of a specific mutation seen in 10,000 bacteria is 2, and it can then be shown that if X is the number of this mutation observed in 10,000 bacteria then $P(X = x) = \exp(-2)2^x/x!$, where $x = 0, 1, \ldots$, that is, x is a non-negative integer. Determine the probability of observing at least 4 mutations in 10,000 bacteria. Give your answer to three decimals of accuracy.

14. In some diseases, the fraction X of damaged white cells per cc of blood drawn of all white cells varies from hour to hour, according to the continuous distribution with density $f(x) = 12x^2(1 - x), 0 < x < 1$. Some damaged cells will be removed by the body, while more damaged cells may be created by the effects of the disease. Numerous hourly measurements are taken.

    (a) Determine the mean or average fraction of damaged white cells seen in hourly samples.
    (b) Determine the mode of X, if it exists.
    (c) Determine the probability that the observed fraction X is at least .90.
    (d) Determine the probability that the observed fraction X is less than 0.10.

15. Let the continuous random variables X and Y have jpdf $f(x, y) = kxy$, where $0 \leq x \leq y \leq 1$, and is 0 elsewhere.

    (a) Find the value of k that makes this a jpdf;
    (b) Determine the marginal distributions of X and Y;
    (c) Determine the (marginal) means and variances of X and Y;
    (d) Determine the correlation between X and Y; and
    (e) Determine the conditional density of Y, given $X = 1/2$.

    To do the integrals involved, set them up as iterated integrals; that is, integrate with respect to y first, then x [or vice-versa]. The variable y is not free — it must be greater than or equal to the x value. For example, if R is a triangle defined by $0 \leq a \leq x \leq y \leq c \leq 1$, then:

    $$\int \int_R f(x, y) dx dy = \int_a^c \left(\int_x^c f(x, y) dy \right) dx$$

CHAPTER 4

SPECIAL DISCRETE RANDOM VARIABLES AND THEIR DISTRIBUTIONS

Synopsis

A number of special discrete random variables and their distributions are introduced, along with the underlying properties that distinguish them. Distributions considered include: Uniform; Bernoulli; Binomial; Negative Binomial; Geometric; Hypergeometric and Multinomial Hypergeometric; Poisson; and Multinomial. Some important relationships between distributions are discussed. Examples and problems illustrate the results.

4.1 INTRODUCTION

An important problem in statistics is to identify the distribution of a random variable X. The possible outcomes of the experiment, and hence of X, determine whether X is continuous or discrete. Further, if we know X's pdf up to some constants, called **parameters**, then we can readily compute characteristic values of the distribution, such as means and variances, from formulae generated for all such pdfs. Recall that we assume a random variable, discrete or continuous, has pdf defined on the entire real line, and that we can assume its values where it has nonzero probability are real numbers by coding values if necessary. For example:

Definition 1: Let the discrete random variable X have nonzero probability at each number $x_1 < x_2 < \cdots < x_n, n \in \mathbb{N}$, with $P(X = x_i) = 1/n$. Then we say X has a **uniform distribution**, or is **uniformly distributed**, on the values $x_1, x_2, \ldots, x_n$. As usual, even though not written explicitly all the time, $P(X = x) = 0$ for all other real numbers x.

Example 1: Let the random variable X be **uniformly distributed** on the set of n consecutive integers $a, a+1, a+2, \ldots, a+n-1 \equiv b$. The values a and n are the parameters of this distribution. Because X is uniform, then $P(X = i) = 1/n$, $a \leq i \leq b$, and $P(X = x) = 0$ for all other real numbers x. If we let u be the pdf, we can write the probability assignments as $u(x) = 1/n$ for the natural numbers $x = a, \ldots, a+n-1$, and $u(x) = 0$ otherwise. The mean is given by

$$\mu = \frac{1}{n}\sum_{i=a}^{b} i = \frac{1}{n}\sum_{j=1}^{n}(a+j-1) = \frac{n(a-1)}{n} + \frac{n(n+1)}{2n} = a - 1 + \frac{n+1}{2}.$$

To find the variance, we first show:

$$\sum_{i=a}^{b} i^2 = n(a-1)^2 + (a-1)n(n+1) + n(n+1)(2n+1)/6$$

from which:

$$\sigma^2 = var(X) = (n^2 - 1)/12.$$

The details are left as an exercise. If $a = 1$, then $\mu = (n+1)/2$ and $\sigma^2 = (n^2-1)/12$. If $a = 3$, then $\mu = 2 + (n+1)/2$ and $\sigma^2 = (n^2-1)/12$.

An interesting observation about the last example is that the variance does not depend on a, the starting point, only on n. Is this unusual? No — variance is a measure of the spread of the distribution around the mean, so if we shift all values by the same amount, this will not change the spread of the distribution relative to the new mean. Changing the starting point to some new integer k with $k, k+1, \ldots, k+n-1$ the new values for which the random variable has nonzero probability, and assuming the distribution is uniform, produces a mean of $k-1+(n+1)/2$ and a variance of $(n^2-1)/12$.

Another aspect of the last example was the **change of index of summation** used to change the original sums into more recognizable expressions. This is similar to the **change of variable** procedure used in calculus, with the index of summation playing a similar role to the differential. In our case, we introduced a new index, j, defined by $j = i + 1 - a$. Where does this come from? We have $a \leq i \leq a + n - 1$. If we add $1 - a$ to each side (and recalling adding constants to equations or inequalities does not change the solution set of the equation or inequality), we have $1 \leq i + 1 - a \leq n$ and $i + 1 - a$ is still an integer. We simply give it a new name: j. Now $1 \leq j \leq n$. Wherever i occurs in the original formula, we can replace it by $a + j - 1$, since $i = a + j - 1$. This change of index of summation plays a significant role in many sums involving consecutive sets of integers.

Our goal in this chapter is to classify random variables by the form of the pdf and by experimental conditions. The pdf (probability distribution function in this chapter) is a function that has certain **parameters** that are fixed for a given problem, but may be different from problem to problem. For example, if the pdf of Y is:

$$f(y) = \binom{n}{y} \pi^y (1-\pi)^{n-y}, \ y \in \mathbb{Z}, \text{ with } 0 \leq y \leq n$$

where $\pi \in [0, 1]$ and $n \in \mathbb{N}$ are constants, and $f(y) = 0$ otherwise, then n and π are parameters of this type of distribution (called the binomial distribution). ($\pi = 0$ or 1 are not particularly interesting cases, but are included for completeness.) In most problems, n is given, but it is frequently the case that π is not. The value π is, in most problems, the proportion of an infinite population (a very large population suffices) that has a given characteristic. A significant problem is to estimate π from sample information, and we will discuss this in detail in the chapter on parameter estimation.

For Y in the previous paragraph, it can be shown that $E(Y) = n\pi$ and $var(Y) = n\pi(1 - \pi)$. Thus, no matter the context in which this distribution arises or what the specific values of the parameters are, we know the mean and variance without having to derive them each time.

Results for the distributions discussed are, for the most part, derived. Others are left to the reader to complete the calculations. It is important to be able to carry out such derivations for at least three reasons:

1. to gain mastery of the material in this book, a level of comfort with mathematical calculations is essential;
2. not every result and not every distribution can be covered in this book; and
3. to connect mathematics learned in other courses with important statistical procedures.

Often, researchers need to read through the details of research papers and manuscripts, and having a good level of familiarity with algebra, calculus, and proof methods will aid in determining where results come from, and, in fact, if they are valid. The distributions introduced in this book are perhaps the most commonly occurring distributions in the life sciences, but there are many others used in specialized circumstances. It may very well be that for a particular research project, a distribution not covered in this book becomes critical to the analysis. Working through the details of the results in this text will make it easier to access new information. Practicing mathematical arguments helps develop strong writing in both prose and quantitative analyses, improving communication of results to other researchers.

We now consider some specific discrete distributions and their parameters.

4.2 SPECIAL DISCRETE DISTRIBUTIONS

Throughout this chapter, the random variables are assumed discrete, and the probability distribution functions will be assumed defined on $\mathbb{R}$. Typically, the values of a random variable for which the probability is nonzero are consecutive integers, often starting at 0 or 1. In some cases, however, describing the possible outcomes in

Chapter 4 — Special Discrete Random Variables and their Distributions

words may increase clarity. The distribution will be specified by a function, the pdf, rather than a list of values in a table.

4.2.1 The Uniform Distribution

Definition 2: Let the random variable X be such that it has a finite number of values at which its pdf is positive, and at these values, the probability is the same number. We say X has a **uniform distribution**. This is also stated as the variable is **uniformly distributed**, or simply X is **uniform**. (This repeats Definition 1 and is put in this section for completeness.)

Conditions for a uniform random variable: Experiments that give rise to a uniformly distributed random variable are characterized by:

1. there is a finite number, k, of possible outcomes;
2. the experiment is to select one experimental unit at random, and the random variable records the particular outcome of this selection; and
3. each outcome is as likely to occur as any of the other outcomes, and so the probability of any outcome is $1/k$.

Typically, the outcomes are coded by the consecutive natural numbers 1 to k, although other mnemonics can be used.

An equivalent way to state these requirements is that there are k mutually exclusive and exhaustive categories in the population. Each experimental unit belongs to one (and only one) of these categories, and the proportion of units falling into each category is the same. An experimental unit is to be selected, and the random variable records the category. Categories are usually designated 1 to k, so that the uniform random variable X has nonzero probability for each natural number i, with $1 \leq i \leq k$. Thus, $P(X = i) \equiv u(i) = 1/k$, $1 \leq i \leq k$ with $u(x) = 0$ otherwise. The function u is the pdf of the uniform.

Example 2: Let an experiment consist of recording the face up on a fair die after it is rolled on a flat surface. Thus the sample space is:

$$S = \{1, 2, 3, 4, 5, 6\}.$$

The random variable is the value on the face showing up. The requirement that the die is fair means that $P(X = i) = P(X = j)$ for all $1 \leq i, j \leq 6$, and so $P(X = i) = 1/6$, $1 \leq i \leq 6$. [For all other values, $P(X = x) = 0$.]

General Results for the Uniform: (See also the example of a uniform variable in 4.1.) Let X be uniformly distributed on the set of natural numbers i, $1 \leq i \leq k$. Then:

$$P(X = i) = \frac{1}{k}, \quad E(X) = \frac{k+1}{2}, \quad \text{and} \quad var(X) = \frac{1}{12}(k-1)(k+1).$$

In particular, if $k = 10$ then $E(X) = 5.5$ (which is exactly halfway between 1 and 10) and $var(X) = 8.25$. These results are based on the following identities:

$$\sum_{i=1}^{k} i = \frac{1}{2}k(k+1) \quad \text{and} \quad \sum_{i=1}^{k} i^2 = \frac{1}{6}(2k+1)(k+1)k.$$

When a sum depends on an arbitrary natural number, and is equal to an expression involving only one or a few terms, we call the resulting expression a **closed formula for the sum**. The formulae above for summing the first k natural numbers and their squares are examples of closed formulae, and there are general methods for finding such expressions for $\sum_{i=1}^{k} i^m$, where m is a fixed natural number.

The moment generating function for a uniform variable X taking nonzero probability on $1, 2, \ldots, k$ is

$$M_X(t) = \frac{e^t}{k} \frac{1 - e^{kt}}{1 - e^t}, \ t \neq 0 \text{ and } M_X(0) = 1.$$

In this case, the moment generating function is defined for all $t \in \mathbb{R}$.

If X is uniform on any finite set of numbers, similar results can be derived, but they may not be as compact. For example, suppose Y is uniform on 1,2,4,8,16. Then the pdf of Y, g, satisfies $g(y) = 1/5$ whenever $y = 1, 2, 4, 8,$ or 16, and is 0 otherwise. From this, $E(Y) = 31/5 = 6.2$, and $var(Y) = 351/5 - (31/5)^2 = 351/5 - 961/25 = 794/25 = 31.76$.

Example 3: Let X be uniformly distributed on the natural numbers $1, 2, \ldots, k$. Prove the formula for the moment generating function.

Solution: It is always the case that $M_X(0) = 1$ for any distribution. Assume that $t \neq 0$. Then:

$$M_X(t) = \sum_{i=1}^{k} e^{it} \frac{1}{k} = \frac{1}{k} e^t (1 + e^t + \cdots + e^{(k-1)t})$$

$$= \frac{e^t}{k} \frac{1 - e^{kt}}{1 - e^t}.$$

The last equality comes from the fact that if we write $e^t = a$, then we need to be able to sum $1 + a + a^2 + \cdots + a^{k-1}$. But this is a finite geometric series, and it has closed formula $(1 - a^k)/(1 - a)$. [Note that since $t \neq 0$, then $e^t \neq 1$, so the formula is valid.] Replacing a by e^t completes the proof.

One reason for carrying out a proof is to illustrate the use of mathematical formulae and methods that the reader should have encountered elsewhere. These techniques show up in many different contexts and problems, so having exposure to their use will help develop skills useful in any quantitative analysis. In evaluating certain quantities related to a uniform variable, closed forms of general finite sums were valuable in establishing important results.

Example 4: Consider the random variable Y that records the face up on a fair die, after it has been rolled. Y is uniformly distributed, with $k = 6$, and so $E(Y) = 7/2 = 3.5$ and $var(Y) = 35/16$.

Note: In the last example and elsewhere, values that are exact should be left in fractional form, unless there is an equivalent decimal with very few decimals. For example, if the result is 22/7, it should be left in that form.

4.2.2 The Bernoulli Distribution

Definition 3: Let X be a dichotomous random variable, that is, X takes on exactly two values, typically "success" and "failure" or 1 and 0. For the purpose of this definition, we assume the values of X are 1 and 0. X is said to have a **Bernoulli distribution**, or to be a **Bernoulli random variable**, if $P(X = 1) = \pi$ and $P(X = 0) = 1 - \pi$ for some constant $\pi \in [0, 1]$.

Conditions for a Bernoulli random variable: Bernoulli random variables are associated with **Bernoulli experiments**. A Bernoulli experiment is characterized by:

1. the population is divided into two mutually exclusive and exhaustive categories, labeled 0 and 1, say;
2. the proportion of the population falling in category 1 is π, and hence, the proportion falling in the other category is $1 - \pi$;

3. one of the experimental units is to be randomly selected from the population, and the outcome of the experiment is to determine the category it falls in; and

4. the Bernoulli random variable is the value (0 or 1) of the category the selected unit falls in.

One run of the experiment is called a **Bernoulli trial**.

Examples of dichotomous variables include recording: success vs. failure; male vs. female; a veteran hockey player scores on a given shot on goal, or doesn't; a vote for a candidate; heads or tails as the result of flipping a coin; if an individual selected from a large population has a given characteristic or does not; whether a 6 does or does not show on the face up on a die. In all cases, there are two mutually exclusive and exhaustive categories, and all members of the population must fit into exactly one of them. One category is coded 1, the other 0, and we record the observed number.

In some problems, it may be more useful to retain the names of the original categories, rather than label them 1 and 0. For example, if the experiment is the determination of the sex of an individual, retaining the names male and female for the categories retains essential information about the experiment. Similarly, a success or failure indicates something about the nature of the experiment. The use of 1 and 0 helps remove subjectivity in the results and makes it easier to write the analysis of Bernoulli and other experiments. Probability distribution functions are defined on the real numbers, so introducing the coding above makes the definition of the pdf straightforward. However, it is often the case that retaining the names of the original categories provides a clearer analysis in multistep problems.

There is no implication that whatever is coded 1 is somehow better than 0. For example, the experiment may be to select an individual from a large population and determine if that person has a congenital condition, perhaps in preparation for a study of this condition. For a researcher, it may be viewed as a success if the person chosen has the congenital condition, and this would be coded 1. Not having the condition would then be coded 0. One reason for the given coding is that if the researcher needs a certain number of individuals with the condition for a study to be carried out, and she needs 20 people with the congenital problem, then she simply needs to keep sampling until she gets 20 1s. This type of problem is discussed in detail under the **negative binomial distribution** later in this chapter.

General results for the Bernoulli random variable: For a Bernoulli random variable, we have:

$$E(X) = 1(\pi) + 0(1-\pi) = \pi,$$
$$var(X) = 1^2(\pi) + 0^2(1-\pi) - \pi^2 = \pi(1-\pi),$$
$$M_X(t) = 1 - \pi + e^t \pi.$$

Example 5: Baseball is a game of statistics, and managers often use statistics to decide game day pitching match-ups, who to call on in a pinch-hit situation, who to put in the outfield for defensive purposes, etc. Consider a veteran hitter on the bench, with a batting average of .300. This means, over a long period of time, the hitter on average gets a hit 3 times out of 10 times at bat (excluding walks). Another batter on the bench has hit 0.250 over his career. The manager is considering one of the two as a pinch hitter. Who does the manager choose? If this is all of the information available, for each batter we can view the probability of getting a hit as a Bernoulli random variable, and then the choice must be the batter with the better overall average.

However, baseball keeps statistics on everything. Suppose the pitcher is left-handed, and the batting average of the first hitter against left-handers is 0.100, while the second batter has an average of 0.275 against left-handers. Now the manager is more likely to use the second batter in this pinch-hit situation. This is an example of making use of Bayes' theorem to revise probabilities as more information becomes available.

One reason the Bernoulli experiment is important is that it provides a basis for the Binomial Distribution.

4.2.3 The Binomial Distribution

Definition 4: Let a Bernoulli trial be repeated n independent times on a given population, with probability π of observing 1 on each trial. Let X count the number of 1s observed in the n trials. Then we say X has a **binomial distribution**, and write $X \sim B(n, \pi)$ and denote its pdf as $b(x; n, \pi)$.

Conditions for a binomial random variable: A **binomial experiment** consists of n Bernoulli trials on the same population, such that:

1. the n trials are independent;
2. the conditions for each trial are the same from trial to trial;
3. the outcome is either a 1 or 0 on each trial;
4. the probability of observing a 1 stays constant, $= \pi$ say, from trial to trial.

If X counts the number of 1s observed in the n trials, then X is called a **binomial random variable**. The distribution of the probabilities for the possible values of X is called a binomial (probability) distribution. The values n and π are the parameters of the distribution.

The conditions for a binomial random variable apply to an infinite population with two mutually exclusive and exhaustive categories, one containing $\pi \times 100\%$ of the population, the other the rest. The conditions hold for a finite population when the experimental units are replaced in the population after each individual selection. When a population is large but finite, the conditions are met sufficiently well if $n \ll N$, where N is the population size. A rule of thumb is $20n \leq N$ is sufficient.

Binomial distribution pdf: Let X have a binomial distribution. Suppose we want to find $P(X = k)$, where $0 \leq k \leq n$. Then there are exactly k 1s and $n - k$ 0s in the n trials. Since the trials are independent, $P(\text{a particular order of } k \text{ 1s and } n - k \text{ 0s}) = \pi^k(1 - \pi)^{n-k}$. But there are $n!/k!(n - k)!$ ways of arranging k 1s and $n - k$ 0s, so that:

$$P(X = k) \equiv b(k; n, \pi) = \binom{n}{k} \pi^k (1 - \pi)^{n-k}.$$

In this derivation, we use the fact that each outcome is mutually exclusive of any other outcome. For example, if $n = 3$, the two outcomes $(1,1,0)$ and $(1,0,1)$ cannot happen simultaneously, and hence, are mutually exclusive. This fact allows us to compute the probability of a particular arrangement of 0s and 1s, and then multiply by the number of arrangements of the 0s and 1s.

Note that:

$$\sum_{k=0}^{n} \binom{n}{k} \pi^k (1 - \pi)^{n-k} = (\pi + 1 - \pi)^n = 1^n = 1$$

so these probabilities sum to 1 as required. This is a result of the Binomial Theorem, and it is from this that the name of the distribution is derived.

One way to get a better understanding of a discrete distribution is to plot it in a histogram. We take the horizontal axis (called the x-axis) to be the values of X, and the vertical axis (called the p-axis) to be the probability axis (with scale from 0 to 1). In the case of a binomial, we plot a rectangle with base length one and height equal to the probability $X = k$, over the interval $[k - 1/2, k + 1/2]$ on the x-axis. If $\pi = 1/2$, then the shape of this distribution is symmetric (see section 3.5). If $\pi > 1/2$, the distribution is skewed to the left, and if $\pi < 1/2$, it is skewed to the right.

General results for the binomial distribution: Because of the relation to the Bernoulli, we can easily determine the expected value, variance, and other moments of a binomial random variable. In all cases, the results are based on either trying to identify the binomial theorem in reverse, or we can use other distributions to help us. For example, the mean can be evaluated as follows:

$$E(X) = \sum_{i=0}^{n} i \binom{n}{i} \pi^i (1 - \pi)^{n-i} = \sum_{i=1}^{n} i \binom{n}{i} \pi^i (1 - \pi)^{n-i} = \sum_{i=1}^{n} i \frac{n!}{i!(n - i)!} \pi^i (1 - \pi)^{n-i}$$

$$= \sum_{i=1}^{n} \frac{n!}{(i - 1)!(n - i)!} \pi^i (1 - \pi)^{n-i} = \sum_{i=0}^{n-1} \frac{n(n - 1)!}{i!(n - i - 1)!} \pi^{i+1} (1 - \pi)^{n-i-1}$$

$$= n\pi \sum_{i=0}^{n-1} \binom{n - 1}{i} \pi^i (1 - \pi)^{n-i-1} = n\pi (\pi + 1 - \pi)^{n-1} = n\pi.$$

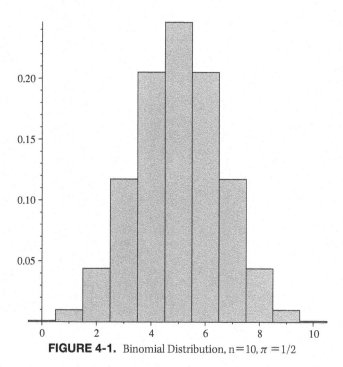

FIGURE 4-1. Binomial Distribution, n=10, $\pi = 1/2$

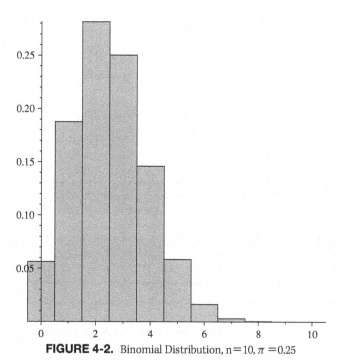

FIGURE 4-2. Binomial Distribution, n=10, $\pi = 0.25$

To calculate $E(X^2)$, we do the same except we need to break the sum into two pieces, to eliminate the i^2 factor. This is left as an exercise. The moment generating function $M(t) = E(\exp(tX))$ is given by:

$$M(t) = (pe^t + 1 - p)^n.$$

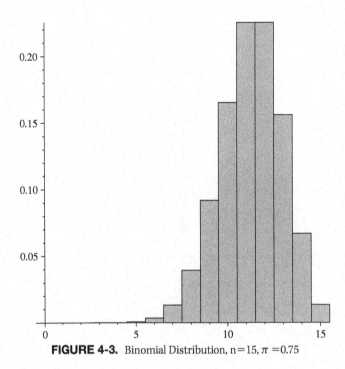
FIGURE 4-3. Binomial Distribution, n=15, π =0.75

Another approach is to use the observation that we can think of each trial in the n trials as associated with a Bernoulli random variable. Thus, we can write $X = X_1 + \cdots + X_n$, where each X_i is Bernoulli with parameter π, and they are independent. The fact that this can be done relies on the properties of the moment generating function. Given this relationship between the Bernoulli and the binomial, and that $E(X_i) = \pi$ and $var(X_i) = \pi(1 - \pi)$, we have $E(X) = n\pi$ and:

$$var(X) = var\left(\sum_{i=1}^{n} X_i\right) = \sum_{i=1}^{n} var(X_i) \text{ (from independence)} = n\pi(1 - \pi).$$

Example 6: The moment generating function of the random variable X defined by $X = X_1 + \cdots + X_n$, where the random variables X_i, $1 \leq i \leq n$ are independently and identically distributed Bernoulli random variables with common parameter π, is the same as the mgf of a binomial random variable with parameters n and π. The proof of this is left as an exercise. **Hint**: We know that if U and V are independent random variables, then the moment generating function of $U + V$ is the product of their moment-generating functions. This generalizes to the sum of n independent random variables, by mathematical induction.

Example 7: There is a large raccoon population in and around an urban area. It is known from past studies that 5% of adult raccoons in this population have rabies. In preparing for testing a new treatment, 30 adult raccoons are captured at random. What is the probability this sample contains 5 rabid raccoons? or no rabid raccoons?

Solution: We assume the captured raccoons form a random sample, and the population is at least 20 times the sample size, that is, 600 or more. A question for discussion is, how would a researcher actually carry out the experiment so that exactly 30 raccoons are captured to form a simple random sample? Let X count the number of raccoons with rabies. Then X is binomial with parameters $n = 30$ and $\pi = 0.05$. It is important to note in the findings the assumptions that the total number of captured raccoons is small compared to the population and that the sample is a simple random one.

Chapter 4 — Special Discrete Random Variables and their Distributions

With these assumptions, we have:

$$P(X = 5) = \binom{30}{5}(0.05)^5(0.95)^{25} \approx 0.0124, \quad P(X = 0) = \binom{30}{0}(0.05)^0(0.95)^{30} \approx 0.2146.$$

The average number we would expect to see is 1.5 (interpreted as: if we repeat this binomial experiment many times, and average the number of rabid raccoons seen in each of the samples of size 30, then this average would be 1.5). The variance in X is 1.425.

The last example illustrates that if we decide in advance how many Bernoulli trials to run, we may get few, if any, of those with the characteristic of interest. An alternate approach is **inverse sampling**, based on the **negative binomial distribution**:

4.2.4 The Negative Binomial Distribution

Definition 5: Consider a population that satisfies the conditions needed to run Bernoulli trials as often as necessary, with all resulting Bernoulli variables (recording either 1 or 0) independently and identically distributed with common parameter π. Let the random variable X count the number of Bernoulli trials needed to find k 1s. Then X is said to have a **negative binomial distribution**, with $X \geq k$. We write $X \sim B^*(k, \pi)$, and its pdf is denoted $b^*(x; k, \pi)$.

In contrast to a binomially distributed random variable, the total number of trials is not fixed, but the number k of required 1s is. Examples include: flipping a coin until the 2nd head appears; determining the probability that a habitual criminal will be caught for the second time on his/her 8th crime; in random testing, determining how many people must be tested for a specific gene before the 10th is found. The negative binomial random variable has (theoretically, at least) a countably infinite number of possible values of X with nonzero probability.

Conditions for a negative binomial random variable: A **negative binomial experiment** consists of an unknown number of Bernoulli trials on a given population, such that:

1. the trials are independent;
2. the conditions for each trial are the same from trial to trial;
3. the outcome is either a 1 or 0 on each trial;
4. the probability of observing a 1 stays constant, $= \pi$ say, from trial to trial; and
5. the experiment stops when the kth 1 occurs.

The pdf of a negative binomial variable: In general, if the kth 1 occurs on the xth trial, then there must be $k - 1$ 1s in the first $x - 1$ trials. The probability of this is $\binom{x-1}{k-1}\pi^{k-1}(1-\pi)^{x-k}$ since this the same as having $k - 1$ successes in $x - 1$ trials in a binomial distribution with $x - 1$ trials and probability of success or seeing a 1 is π. Therefore, the probability that the kth 1 occurs on the xth trial is:

$$b^*(x; k, \pi) = P(k - 1 \text{ 1s in the first } x - 1 \text{ trials and a 1 on the } x^{th} \text{ trial})$$

$$= P(k - 1 \text{ 1s in the first } x - 1 \text{ trials given the } k^{th} \text{ 1 occurs}$$
$$\text{on the } x^{th} \text{ trial}) \times P(1 \text{ on the } x^{th} \text{ trial})$$

$$= \binom{x-1}{k-1}\pi^{k-1}(1-\pi)^{x-k}\pi = \binom{x-1}{k-1}\pi^k(1-\pi)^{x-k}.$$

Any random variable with such a distribution function is called negative binomial, with $x \geq k$ and integer. It is also called the **Pascal distribution** of the binomial waiting time distribution. The probabilities are the successive terms in the expansion of the second factor in powers of $1 - \pi$ of $\pi^k(1 - (1 - \pi))^{-k}$, giving it its name.

General results for the negative binomial distribution: The mean and variance of the negative binomial distribution are $\mu = k/\pi$ and $\sigma^2 = k(1/\pi - 1)/\pi$, respectively. The moment generating function of the negative binomial distribution is:

$$M(t) = \pi^k e^{kt}(1 - (1-p)e^t)^{-k}.$$

The following relationship between the pdf of a negative binomial distribution and a binomial distribution holds: $b^*(x; k, \pi) = \frac{k}{x} b(k; x, \pi)$.

The proofs of the general results for the negative binomial are based on the fact that:

$$\sum_{x=k}^{\infty} \binom{x-1}{k-1} \pi^k (1-\pi)^{x-k} = 1.$$

The last equation can be rearranged as:

$$\sum_{x=k}^{\infty} \binom{x-1}{k-1} \pi^k (1-\pi)^x = \left(\frac{1}{\pi} - 1\right)^k.$$

From this and the fact that $\exp(xt) = \exp((x-k)t)\exp(kt)$, we can derive the moment generating function. An exercise is to establish the formulae for the mean and variance using the mgf.

Example 8: Let the random variable X have a negative binomial distribution with parameters $\pi = 0.4$ and $k = 4$. Then $\mu = 10$ and $\sigma^2 = 15$. Thus, the average number of trials to get the 4th success is 10 trials when the probability of success is 0.4.

Example 9: A scientist needs 20 specimens of Drosophila grimshawi (a type of fruitfly) for a study on its wing structure. Many different types of fruitfly inhabit an open area, and the process of capturing the number of specimens needed is to set several traps and then identify those that are D. grimshawi. The trapping procedure yields, on average, 4 D. grimshawi for every 80 fruitflies captured. Let X be the random variable counting the number of fruitflies that must be captured to get the number needed for the study, and assume X has a negative binomial distribution. The process takes time, and the scientist can only afford to pay an assistant to capture up to 2 standard deviations more than the average number of trials needed to get the sample. Can the scientist get the necessary number of specimens if the assistant needs to capture 600 fruitflies?

Solution: Since X is assumed negative binomial with $\pi = 4/80 = 1/20$, then the average number of fruitflies that must be captured to get the necessary sample is $20/(1/20) = 400$. The variance of X is $20(20-1)/(1/20) = 7600$, and hence, the standard deviation is 87.2 (approximately). The mean plus two standard deviations is 574.4, which should be rounded down to 574. In this case, the scientist will not be able to get the sample necessary. However, the probability it will take no more than 574 trials to get the sample necessary is 0.967 (approximately).

Example 10: Using appropriate software, the probability X (of Example 9) falls within two standard deviations of its mean can be shown to be 0.957. In comparison, Chebyshev's Inequality only guarantees that this probability is at least 0.75.

The median number of trials necessary is the value u for which:

$$\sum_{x=20}^{u} \binom{x-1}{19}(0.05)^{20}(0.95)^{x-20} = 0.5.$$

By using an appropriate computer program, this is 393, approximately.

Example 11: It can be shown that for the example above, where X is the number of trials needed to acquire the sample necessary, $E((X - \mu)^3) = 296400$ and $E((X - \mu)^4) = 190615600$. Determine γ_1 and γ_2, and interpret these numbers in the context of this distribution.

4.2.5 The Geometric Distribution

Definition 6: The geometric distribution is $B^*(1, p)$, that is, it is the distribution of the random variable that counts the number of independent Bernoulli trials until the first success.

Example 12: Write out the conditions under which an experiment is geometric. If X is a geometric random variable, write out its pdf, mean, variance, and moment, generating function, using the general results for the negative binomial, and then by direct calculation.

Example 13: The probability it takes 5 rolls of a fair die to get the first 6 is $\frac{1}{6}(\frac{5}{6})^4$. It will take on average $1/(1/6) = 6$ rolls to get the first 6.

Example 14: One-tenth of all the cattle in a very large herd are infected with hoof-and-mouth disease. A government inspector will test up to 5 cattle for this disease, and stops and quarantines the herd if any of the 5 selected test positive for hoof-and-mouth. What is the probability the inspector does not find a single cow with the disease? How many cattle would she have to test to have a 75% probability of finding a cow with the disease?

Solution: Some assumptions have to be made: the presence of the disease in any cow is independent of its presence in any other cow; the cattle are randomly mixed together; the inspector makes her selection randomly; and the herd is sufficiently large so that the random variable X counting the number of trials necessary to find a cow with hoof-and-mouth is geometric.

Based on these assumptions, then $X \sim B^*(1, 1/10)$ and we want $P(X \geq 6) = 1 - P(X \leq 5) = 1 - \sum_{x=1}^{5} \binom{x-1}{0}(1/10)^1(9/10)^{x-1} \approx 1 - .41 = .59$. Thus, approximately 6 times out of 10 the inspector will not find one cow with the disease, under the given circumstances.

The last result may not be very reassuring, and a greater percentage of times an animal is found under the given conditions may be highly desirable. If the inspector wants to have an approximately 75% chance of finding a cow with hoof-and-mouth (under the given circumstances), she would have to be prepared to test at least 13 cows. If the inspector wanted to be 95% certain of finding a cow with the disease, she would have to be prepared to test up to 29 cows. Such results depend on the percentage infected. As an exercise, determine the number of cows the inspector should be prepared to test if 5% of the herd is infected and she wants to be 95% certain of finding an infected cow. Would you expect this number to be $2 \times 29 = 58$?

This last example illustrates some of the problems government inspectors face: To be 100% certain of finding an animal with hoof-and-mouth, it might be necessary to test a very large fraction of the herd, and this could be very costly in terms of money and time. In this scenario, the proportion with the disease is $1/10$. What if it were $1/100$ or even smaller? How does the inspector know if the cattle are uniformly mixed so that the selection process is random? A decision to quarantine the herd will be very hard on the rancher, so the inspector also must have confidence in the test for the disease. Although statistics cannot address the ethical behavior of people, it can at least provide information about how to proceed with such testing.

4.2.6 Hypergeometric Distribution

Definition 7: Let a population consisting of N experimental units be divided into two mutually exclusive and exhaustive categories. In category 1, there are K units, and in category 0 there are $N - K$ units. Take a simple random sample of n units without replacement from this population, and let the random variable X count the number of units in the sample in category 1. Then, X is said to have a **hypergeometric distribution**, with notation $X \sim H(N, K, n)$, and with pdf denoted $h(x; N, K, n)$.

Example 15: There is a population of N thrushes, K which have been tagged in the past. Consider the random variable X = number of tagged birds found in a random sample of n thrushes selected without replacement from the N. Thus the binomial distribution does not apply, because there is not constant probability of selection of a tagged bird from selection to selection.

Conditions for a hypergeometric random variable: A **hypergeometric experiment** consists of sampling without replacement from a finite population with:

1. N, the population size, is known;
2. there are two mutually exclusive and exhaustive categories, labeled 1 and 0, of experimental units in the population;
3. there are K experimental units that fall in category 1, and hence $N - K$ that fall in category 0; and
4. a random sample of size $n \leq N$ is taken without replacement from the population.

If the random variable X counts the number of experimental units in the sample that fall in category 1, then X is said to have a **hypergeometric distribution** or to be distributed as a **hypergeometric random variable**. The values N and K are the parameters of this distribution.

The pdf of the hypergeometric distribution: There are K units in the population falling in category 1 and $N - K$ in category 0. Let the value of the hypergeometric random variable X that fall in category 1 be x (with certain restrictions to be developed), and hence, $n - x$ is the number falling in category 0. We have to count the number of ways such a sample could have been chosen. It must be the case that $0 \leq x \leq \min\{n, K\} \leq n \leq N$. Now, $\binom{K}{x}$ is the number of ways of selecting x of the K in category 1, and $\binom{N-K}{n-x}$ is the number of ways of selecting $n - x$ in category 0. This means we will also require $n - x \leq N - K$. The multiplication rule applies, and so the total number of ways of selecting the sample with x in category 1 is $\binom{K}{x}\binom{N-K}{n-x}$. The total number of ways of selecting n units from the population is $\binom{N}{n}$. Thus:

$$P(X = x) = \frac{\binom{K}{x}\binom{N-K}{n-x}}{\binom{N}{n}}.$$

If any of the restrictions are violated, the probability is 0. Since $\binom{M}{a} = 0$ if $a > M$ or either M or a (or both) are negative, we can summarize the conditions on the variable and the parameters by saying $P(X = x)$ is given by the above formula, as long as all of the binomial coefficients are defined.

General results for the hypergeometric distribution: Let X have hypergeometric distribution $H(N, K, n)$. Then $\mu = \frac{nK}{N}$ and $\sigma^2 = \frac{nK(N-K)(N-n)}{N^2(N-1)}$.

Note: Hypergeometric random variables do have moment-generating functions. However, they are defined in terms of hypergeometric functions, which involve infinite series and will not be used in this book.

For a hypergeometric random variable $X \sim H(N, K, n)$, consider what happens as $N \to \infty$, with $\lim_{N \to \infty} K/N = \pi \in (0, 1)$, and as n stays fixed: $\lim_{N \to \infty} \mu = n\pi$ and $\lim_{N \to \infty} \sigma^2 = n\pi(1 - \pi)$, the mean and variance of a binomial variable with n trials. Because the hypergeometric distribution involves binomial coefficients, which grow very quickly, this observation plus some other theoretical considerations means we can approximate the hypergeometric distribution with a binomial distribution, as long as N and K are large compared to n. As a rule of thumb, if $20n \leq K$, this approximation can be used. If $X \sim H(N, K, n)$ and $n \leq 20K$, then use the variable $Y \sim B(n, \pi), \pi = K/N$, to approximate X.

Chapter 4 — Special Discrete Random Variables and their Distributions

Example 16: Let $X \sim H(10\,000, 3\,000, 20)$. Use the binomial approximation to the hypergeometric to estimate the probability $P(3 \leq X < 8)$.

Solution: Let $Y \sim B(20, 0.3)$. Then $P(3 \leq X < 8) \approx P(3 \leq Y < 8) = P(3 \leq Y \leq 7) = \sum_{y=3}^{7} \binom{20}{y} 0.3^y 0.7^{20-y} = 0.736788$. The exact value, to six decimals of accuracy is 0.737120.

If the hypergeometric is used, calculations would include evaluating $\binom{10000}{5}$, a very large number. There are ways to simplify some of the combinations of binomial coefficients, but the use of the binomial approximation is certainly adequate.

Example 17: From 150 patient records, 5 records are selected without replacement. There are 25 patients who are smokers, and that information is part of their record. What is the probability that all of the records selected will be of patients who do not smoke?

Solution: We define X to be the random variable that counts the number of records selected of patients who do not smoke. Thus:

$$P(X = 5) = \binom{125}{5}\binom{25}{0} \Big/ \binom{150}{5} = 0.3964$$

to four decimals of accuracy.

Suppose that we used a binomial distribution with $n = 5$ and $\pi = 25/150 = 1/6$. A numerical comparison of the probabilities from the hypergeometric and from the binomial yields:

$X = x$	0	1	2	3	4	5
$h(x; 150, 125, 5)$	.00009	.0027	.030	.161	.410	.396
$b(x; 5, 5/6)$	.00013	.003	.032	.161	.402	.402

There are already differences, especially in the tails, between the two distributions. As n gets larger, the differences become significant. It is essential in small-population problems to choose the correct distribution. What would be the best distribution if the experimental units were replaced in the population and then could be selected again?

4.2.6.1 The Multivariate Hypergeometric Distribution

The hypergeometric distribution has a natural multivariate extension:

Definition 8: Let a finite population have $k \geq 2$ mutually exclusive and exhaustive subpopulations labeled 1 to k. Let the population be of size N, and let the number in subpopulation i be N_i and so $\sum_{i=1}^{k} N_i = N$. Let a random sample of size n be taken from the population (without replacement), and the numbers falling into each of these subpopulations or categories be n_i, $1 \leq i \leq k$, $\sum_{i=1}^{k} n_i = n$. The random vector $(X_1, X_2, \ldots, X_k)$ of counts in each subpopulation is then said to have a **multivariate hypergeometric** distribution. The joint pdf of this vector is given by:

$$mh(n_1, n_2, \ldots, n_k) = P(X_1 = n_1, \ldots, X_k = n_k) = \frac{\binom{N_1}{n_1} \cdots \binom{N_k}{n_k}}{\binom{N}{n}}.$$

If we restrict attention to just one of the X_i random variables, then its properties are the same as an ordinary hypergeometric variable. Although we will not consider the multivariate hypergeometric distribution in any further detail, it is useful to see this distribution since there are instances where it occurs. To be able to use the multivariate hypergeometric distribution, we have to know the numbers in each of the subpopulations. An example would be a fish tank with known numbers of marine life of different types. A problem may be to sample a fixed number of marine creatures from the tank, and then determine probabilities associated with such selections. If the subpopulations are large and unknown, but the proportions of the different subpopulations in the tank or other large vessel are known, then we may be able to use the **multinomial** distribution as a good approximation.

4.2.7 The Poisson Distribution

We have encountered a distribution that had a countably infinite number of values for which the pdf was nonzero. Another important example of this is the Poisson.

Definition 9: Let X have distribution function $p(x; \lambda) = \frac{\lambda^x e^{-\lambda}}{x!}$, $x = 0, 1, 2, \ldots$ where $\lambda > 0$ is a given constant. Then we say X has a **Poisson distribution**, and write $X \sim Po(\lambda)$. The value λ is the sole parameter of this distribution.

This distribution is defined in terms of its pdf, rather than in terms of experimental conditions. The exact conditions are somewhat technical, but the general description is as follows:

Conditions for a Poisson random variable: In general, the requirements include counting rare events in a fixed period of time (or in a fixed distance or similar). In this case, an event is simply the occurrence of some outcome of interest — the use of event for an outcome in the case of the Poisson is standard. A **Poisson experiment** in an interval of time requires:

1. the probability of one occurrence of the event in a small interval of time Δt is (approximately) proportional to Δt;
2. the probability of two occurrences of the event in Δt is essentially 0;
3. the probability of the occurrence of the event is constant over intervals of the same length; and
4. the probability of the occurrence of the event is independent of the occurrence of the event in any other nonoverlapping interval of time.

If the random variable X counts the number of occurrences of the event in a fixed interval of time, then X is said to have a **Poisson distribution** or be distributed as a **Poisson random variable**. The pdf of the Poisson is given by $p(x; \lambda) = \exp(-\lambda)\lambda^x/x!$, $x = 0, 1, \ldots$.

General results for Poisson variables: Let $X \sim Po(\lambda)$. Then $E(X) = \lambda$, $\sigma^2 = \lambda$, and the moment generating function is $M(u) = \exp(\lambda(\exp(u) - 1))$.

These results follow from the Maclaurin expansion: $\exp(u) = \sum_{x=0}^{\infty} u^x/x!$. This result will be assumed true for all $u \in \mathbb{R}$.

Example 18: The formulae for the mean, variance, and moment generating function of $X \sim Po(\lambda)$ come from using the fact that the given Maclaurin expansion of $\exp(u)$ is valid regardless of the value of u, and from making changes in the index of summation to create a sum of the correct form. For example:

$$E(X) = \exp(-\lambda) \sum_{x=0}^{\infty} \frac{\lambda^x}{x!} x = \exp(-\lambda) \sum_{x=1}^{\infty} \frac{\lambda^x}{(x-1)!}$$

$$= \exp(-\lambda) \sum_{y=0}^{\infty} \frac{\lambda^{y+1}}{y!} = \exp(-\lambda) \lambda \sum_{y=0}^{\infty} \frac{\lambda^y}{y!}$$

$$= \exp(-\lambda) \lambda \exp(\lambda) = \lambda.$$

The calculations of the variance and the moment generating function are left as exercises.

A **Poisson random variable** counts usually rare and random events. Examples include: the number of dehydration cases at a large outdoor rally; the number of particles emitted in a given time in radioactive decay; the number of galaxies in a portion of space; or the number of aberrant cells in a cc sample from a healthy patient.

The interpretation of λ is that it represents the average number (usually small) of events per unit of time, or distance, or area, etc. For example, the number of radioactive particles emitted in a millisecond may be the random variable X, and then there may be 0, 1, 2, etc. particles emitted in any given millisecond. The average number of particles counted in a millisecond may be $\lambda = 4$.

Example 19: Suppose there are 3×10^9 basepairs in the human genome. See Dennis and Gallagher, Editors (2001): The Human Genome, New York: *Nature Palgrave* for an extensive examination of this project. The mutation rate per generation per basepair is 10^{-9}. If X counts the number of new mutations that a child will have, then X is assumed to follow a Poisson distribution with λ = mutation rate × number of basepairs = $10^{-9} \times 3 \times 10^9 = 3$.

Example 20: During the day, the average number of visits per hour to a particular flowering bush by hummingbirds is $\lambda = 2$. It is assumed the count of the visits per hour follows a Poisson distribution.

Once λ is known, then all standard calculations such as the mean, variance, probabilities of events, etc., are computed as in other examples where the pdf is known.

Note: For any Poisson random variable, the mean $\mu = \lambda$ and the variance $\sigma^2 = \lambda$, although the units of σ^2 are the square of those of the mean. Thus, if the mean is known, then that becomes λ. In trying to decide if a random variable is Poisson, we often take a large sample of observations per unit measurement (time, distance, etc.) and compute the sample mean and variance. If these values are close, then a Poisson distribution may be justified.

Example 21: Let $X \sim Po(\lambda) = Po(3)$, where $Po(3)$ means the Poisson distribution with $\lambda = 3$. [X may be the number of rabbits visiting a watering hole per hour.] Then:

$$E(X) = \lambda = 3 \text{ and } var(X) = \sigma^2 = 3.$$

Other sample calculations include:

$$P(X = 5) = e^{-3}3^5/5! \approx 0.10082, \quad P(X \leq 2) = \sum_{i=0}^{2} e^{-3}3^i/i! \approx 0.4232$$

$$P(X \geq 2) = 1 - P(X < 2) = 1 - [P(X = 0) + P(X = 1)] \approx 0.80085$$

$$E(3X + 2) = 3E(X) + 2 = 11, \quad var(2X + 1) = 4var(X) = 12 \text{ and}$$

$$var(3 - 2X) = 4var(X) = 12.$$

If we change the unit of measurement, so that, for example, we count the number of rabbits per half hour instead of per hour, then the average number of rabbits for this new unit will simply be half of the average in the original unit of time. If Y then is the number of rabbits per half hour, Y will still be Poisson, but with parameter 1.5. Questions about Y will be answered using this new value of the parameter λ.

One use of the Poisson distribution is to approximate the binomial for large n and π small. It can be shown that for $n \geq 20$ and $\pi \leq 0.05$ approximately that the Poisson distribution with $\lambda = np$ is a good approximation of the binomial. For $n \geq 100$ and $\pi n \leq 10$, the approximation is extremely good.

The approximation is relatively simple: If X is binomial, with n trials and π the probability of success, we simply let Y be the Poisson random variable with mean $\lambda = np$, the mean of the binomial. Generally, the approximation is best in the left tail rather than the right tail of the binomialdistribution.

Example 22: The probability that a bat has rabies is 0.005. What is the probability 15 of 2,000 bats captured randomly from a large population will have rabies?

Solution: The bat population is assumed so large that X = no. of bats with rabies in a random sample of size 20 is approximately binomial. In this case, the Poisson approximation to the binomial is valid, and $P(X = 15) \approx p(15; 10) = \frac{10^{15} e^{-10}}{15!} = 0.0347$, since $\lambda = 0.005 \times 2000 = 10$.

The assumption of a sufficiently large population so that the count of the number of objects with a required property is binomial is quite common. Strictly speaking, the bat population in Example 22 is finite, and the distribution of X is hypergeometric. However, under the circumstances where the population is very large compared to the sample size, we can use a binomial distribution as an approximation. It is unlikely we would know the population of bats in any case.

Example 23: In a large population, the probability of an individual having a particular mutation is 0.0004. What is the probability of at least 9 individuals having this mutation in a sample of size 10,000?

Solution: Let X = number of mutations counted in the sample of 10,000. Since the population is assumed very large compared to the sample, we can view X as having a binomial distribution (approximately), and since the value $0.0004 \times 10000 = 4 = \lambda$. Thus, $X \sim B(10000, 0.0004)$ and, in turn, we assume that we can approximate X by the random variable Y which is Poisson, with parameter $\lambda = 4$. $P(X \geq 9) = 1 - P(X \leq 8) \approx 1 - P(Y \leq 8)$. From tables or direct calculation, $P(X \geq 9) \approx 1 - .979 = .021$.

4.2.8 The Multinomial Distribution

The multinomial distribution is a generalization of the binomial distribution:

Definition 10: In a large (infinite) population, there are k mutually exclusive and exhaustive categories, labeled 1 to k. The probability an experimental unit falls in category j is $\pi_j > 0$, with $\pi_1 + \cdots + \pi_k = 1$. Assume these probabilities stay constant regardless of the number of units selected from the population. A random sample of size n is selected from the population, and the number of units falling into each of the different categories is recorded. Let the random vector $(X_1, X_2, \ldots, X_k)$ be the numbers of units from the sample that fall into the respective categories. We then say this vector has a **multinomial distribution**.

Conditions for a multinomial random vector: A **multinomial experiment** consists of sampling a fixed number of units n from a large (infinite) population with:

1. the population consists of k mutually exclusive and exhaustive categories labeled 1 to k;
2. the number of sampled units is n, a constant;
3. the probability a selected unit falls in category j stays constant from selection to selection;
4. the outcome of which category a unit falls in is independent of the outcome of the categories any other unit or units fall in; and
5. the outcome of the selection of the n units is the record of the number of selected units falling into the individual categories.

If $(X_1, \ldots, X_k)$ is the record of the counts of the number of sampled units falling in the respective categories then this random vector X is said to have a **multinomial distribution**. The values k, n and π_i, $1 \leq i \leq k$ are the parameters of the distribution. Since $X_1 + \cdots + X_k = n$, the variables are not independent.

Chapter 4 — Special Discrete Random Variables and their Distributions

The pdf of the multinomial: The multinomial distribution is a joint distribution of the random variables $X_1, \ldots, X_k$, where X_i counts the number of the n sampled units that fall in category i. The argument deriving the pdf of the multinomial is a generalization of that used to derive the binomial distribution: There are n independent trials and k mutually exclusive and exhaustive categories. In general, on any one trial, we assume the probability a randomly selected individual falls in category i is π_i. Let x_i be the number falling in category i in the n trials. Clearly, $\sum_{i=1}^{k} x_i = n$ and this is called a **constraint** of the problem. To find the value of $P(X_1 = x_1, X_2 = x_2, \ldots, X_k = x_k)$, we take one specific arrangement of the experimental units that produces this particular outcome, and then multiply by the number of combinations that would produce the result. Thus:

$$P(X_1 = x_1, X_2 = x_2, \ldots, X_k = x_k) = \binom{n}{x_1, \ldots, x_k} \pi_1^{x_1} \ldots \pi_k^{x_k},$$

where:

$$\binom{n}{x_1, \ldots, x_k} = \frac{n!}{x_1! x_2! \ldots x_k!}$$

are called the multinomial coefficients. As in previous arguments, we have x_i of the n in category i, and we do not care about the order we get them in the sample. Thus, we take all the arrangements of n distinct items ($n!$) and then divide by the number of ways to arrange nondistinct items. This means we have to divide $n!$ by each of the $x_i!$ values.

As in the case of the binomial distribution, as long as the population is large enough, the multinomial distribution can be used. The rule of thumb is that $20n$ should be less than the number in the smallest population category. Classifying individuals by blood types, eye colour, species of animals, etc., are examples where assuming a multinomial distribution for the counts in the categories is adequate.

Example 24: Suppose there are four species of wren (in a given locale) and 40 are captured, banded, and released. It is known that the relative frequency of species 1, 2, 3, 4 can be expressed as 3:2:2:1. What is the probability that there were exactly 14, 10, 10, 6, respectively, of each species captured and released?

Solution: Let X_i be the number of species i captured in the 40. We can show that $\pi_1 =$ probability of species 1 being captured in any one trial $= 3/8$, and then $\pi_2 = \pi_3 = 2/8$ and $\pi_4 = 1/8$. From this, we have:

$$P(X_1 = 14, X_2 = 10, X_3 = 10, X_4 = 6) = \binom{40}{14, 10, 10, 6} \left(\frac{3}{8}\right)^{14} \left(\frac{1}{4}\right)^{10} \left(\frac{1}{4}\right)^{10} \left(\frac{1}{8}\right)^{6} = 0.00372.$$

Exercises

Throughout the development of this chapter, several examples were exercises involving extensions of known results. Many of these problems are theoretical in nature, but do help develop skills and techniques that are essential in any quantitative analysis. They are fundamental to understanding the conditions under which results hold and should be completed, as well as the following exercises.

1. There are several other discrete distributions that are used in special circumstances. These include: the beta-binomial, Gibbs, Benford, logarithmic, and Zipf distributions. Choose at least one of these to investigate, and determine, if the values exist, the pdf, mean, variance, and mgf. Also determine the experimental conditions that lead to the selected discrete distribution, and write out an example using this distribution.

2. Let $X \sim B(10, 0.8)$. Determine:

    (a) $P(X = 2)$;

    (b) $P(X < 3)$;

(c) $P(X > 3)$;

(d) $E(X)$;

(e) $var(X)$.

3. Let the random variable X have a hypergeometric distribution with parameters $N = 52, K = 6$ and $n = 12$. Determine:

   (a) $P(X = 3)$;

   (b) $P(X \geq 10)$;

   (c) $P(X < 9)$;

   (d) $E(X)$;

   (e) $var(X)$.

4. Let $X \sim B^*(1, 0.8)$, that is, geometric with parameter 0.8. Determine:

   (a) $P(X = 2)$;

   (b) $P(X < 3)$;

   (c) $P(X > 3)$;

   (d) $E(X)$;

   (e) $var(X)$.

5. Let $X \sim B^*(10, 0.8)$. Determine:

   (a) $P(X = 2)$;

   (b) $P(X < 3)$;

   (c) $P(X > 3)$;

   (d) $E(X)$;

   (e) $var(X)$.

6. Let $X \sim Po(2)$. Determine:

   (a) $P(X = 2)$;

   (b) $P(X < 3)$;

   (c) $P(X > 3)$;

   (d) $E(X)$;

   (e) $var(X)$.

7. For each of the following situations, determine the distribution of the random variable X. Identify the parameters, the possible values of X, the pdf in each case, as well as the mean and variance.

   (a) From past experience, the staff in a doctor's office knows that 2 percent of all appointments do not show. Forty appointments are expected today. Let X represent the number that show.

   (b) In one version of a lottery, each player selects 6 different numbers from 1 to 54. The organizers randomly select 6 different numbers from 1 to 54, which are then deemed the "winning" numbers. Let X count the number of a given player's 6 selections that are winning numbers.

   (c) A standard deck of 52 playing cards is shuffled, and the cards are successively turned over. Let X denote the number of aces appearing in the first 10 cards turned over.

   (d) The number of births per year in a town in Southern Ontario is approximately 690. Let X represent the number of births per day.

   (e) A family has six children. Assume that the sex ratio is 1:1. Let X denote the number of males born.

(f) Assume again that the sex ratio in a family is 1:1. A couple would like to have a boy and plan to have children until the first male child is born. Let X denote number of females born before the first male.

(g) Assume the conditions in (f). Let X denote the number of females born before the third male is born.

8. (a) Given that there is an equal probability of a baby being a girl or a boy, determine the probability distribution of a family having 4 children and state the distribution in table form.

(b) Given that the first 4 children born in the family were girls, determine the probability that a fifth child born into the family will be a boy.

(c) Use appropriate notation when answering the following questions. What is the probability that the family has

   (i) at least two girls?
   (ii) at most one girl?
   (iii) two or three boys?

(d) State the cumulative distribution in table form and as a function.

9. In Mendel's experiments (see Freeman, (2006)) purple flowering pea plants were crossed with white flowering pea plants. The allele for purple flowers is dominant in pea plants, while the allele for white flowers is recessive. In a cross between two heterozygous purple flowering pea plants, the probability that the resulting offspring will have purple flowers is 0.75. Suppose that 10 seeds from such a cross are planted. Let X denote the number of plants resulting from this cross with purple flowers.

(a) Explain why X can be assumed to have a binomial distribution with parameters $n = 10$ and $\pi = 0.75$.

(b) Assume X is the random variable in (a). Calculate each of the following, and give a statement interpreting the value obtained.

   (i) $P(X = 3)$;
   (ii) $P(X \geq 4)$;
   (iii) $P(X < 8)$;
   (iv) $P(X < 3 \cup X > 8)$;
   (v) $E(X)$;
   (vi) $var(X)$.

10. The population of black bears in a particular region has been estimated to be 110. Since the last estimate of the population in that region, some human-made environmental changes have occurred. Researchers would like to investigate the population once again. They catch and tag 12 bears, release them back into the region, and then track these bears. A year later, 15 bears are caught. See Seber (2002).

(a) Assume that the population size is in fact 110. Let X represent the number of bears captured in the second sample that had been tagged a year earlier. Explain why X has a hypergeometric distribution. What assumptions are necessary about the conditions prior to taking the second sample to ensure X is hypergeometric? State the relevant parameters for the distribution.

(b) Determine the probability that of the 15 bears caught:

   (i) none have been previously tagged;
   (ii) three have been previously tagged;

(iii) less than 4 have been previously tagged;

(iv) at least 4 have been previously tagged.

(c) Determine the expected number of tagged bears in the second sample, and calculate the variance of X. What does the variance represent in this problem?

11. Consider a family in which one of the parents is a carrier for a particular disease. Suppose further that the probability a child inherits this particular disease is 5%. Assume the number of children has no upper bound.

(a) Determine the probability that:

(i) the fourth child born will be the first to inherit the disease;

(ii) at least 4 disease-free children will be born before a child is born with the disease;

(iii) the fifth child born will be the second to inherit the disease;

(iv) the fourth child born has the disease given that the third child born has the disease.

(b) Let X represent the number of children before the first child with the disease is born. Determine the expected value of X, and interpret the value in the context of the problem.

(c) Suppose the parents would like to have 5 children. Calculate the exact probability that at least 1 of 5 children will have the disease. Explain your choice of distribution.

12. Let X be a discrete random variable with a geometric distribution.

(a) Use the fact that for a random variable with a geometric distribution and probability of success π, $P(X \leq x) = 1 - (1 - \pi)^x$ to show that $P(X > n + k | X > n) = P(X > k)$.

(b) Given an interpretation of the result that $P(X > n + k | X > n) = P(X > k)$.

13. Haemophilia B is a congenital, recessive X, chromosome-linked disorder that arises from a deficiency of coagulation factor IX in the blood and it is estimated to occur in 1 out of 30,000 males worldwide (see, for example, the U.S. National Library of Medicine ®, under Hemophilia). Let X represent the number of male babies in the next 55,000 single births of male babies that have haemophilia B.

(a) Explain why X has a binomial distribution with $n = 55{,}000$ and $\pi = 1/30{,}000$.

(b) Calculate the probability that exactly 4 of the 55,000 single births of male babies have haemophilia B.

(c) Show that the Poisson distribution can be used to approximate the binomial distribution in this case.

(d) Use the Poisson approximation to the binomial to determine the probability, to 4 decimal places, that out of the 55,000 males born:

(i) exactly 4 males born have haemophilia B;

(ii) at least 3 males born have haemophilia B;

(iii) at most 2 males born have haemophilia B.

(e) How many of the 55,000 male babies would you expect to have haemophila B?

14. A student is asked the following problem on a test:

A tank of fish contains two types of small fish. Assume 30% of the fish in the tank are minnows. If there are 100 fish in the tank, what is the probability that 8 of 20 fish taken without replacement from the tank are minnows?

The student's solution is as follows:

Let X be the number of minnows captured. Since 30% of the fish are minnows, the probability of getting a minnow on one trial is 0.3.

$$\therefore P(X = 8) = \binom{20}{8} 0.3^8 (1 - 0.3)^{12} \approx 0.114$$

What is wrong with this solution? What change or changes to the experimental procedure would make this solution correct? Calculate the correct probability under the original conditions.

15. Consider the population of feral cats in Australia. They fall into three categories: those with large ears and large paws; those with small ears and small paws; and others. From past studies, the respective proportions of these categories in the population are 0.25, 0.30, and 0.45. A random sample of size 20 from the population of feral cats is selected and categorized. Let X_L, X_S, and X_O be the numbers in the sample falling into the different categories.

   (a) Determine the distribution of the random vector (X_L, X_S, X_O).
   (b) Find the mean and variance of X_L, X_S, and X_O.
   (c) Evaluate the probability that the sample contains fewer than two cats with small ears and small paws.
   (d) Find the covariance between X_L and X_S, and then determine the correlation between these two variables.

CHAPTER 5

SPECIAL CONTINUOUS RANDOM VARIABLES

Synopsis

Special continuous random variables and their distributions are introduced. These include: uniform; gamma; exponential; χ^2 (chi-square); beta; normal; log normal; Student's t; Snedecor's F; bivariate normal; and Weibull. Some of the conditions under which they arise are discussed, but they are defined primarily by the form of their pdf. Major properties are derived. Relationships between variables are discussed, and examples illustrate the use of the results.

5.1 INTRODUCTION

In many instances, the theoretical values of a random variable X fill out at least one interval of real numbers, either finite or infinite in length. For simplicity, it is assumed throughout that X's values fill out a single interval of nonzero length.

As noted earlier, the event $X = x$ for continuous variables is not of importance, since the probability of such an event must be 0. [The proof of this fact is beyond the scope of this text, and is assumed true.] The **probability density function (pdf)**, introduced in Chapter 3, generates probabilities for such variables and is the focus of the discussions in this chapter. Recall that the pdf f for random variable X satisfies: $\int_{-\infty}^{\infty} f(x)dx = 1$ and $f(x) \geq 0 \, \forall x \in \mathbb{R}$.

Certain characteristic values and relationships help describe the nature of a distribution. In Chapter 3, moments were used to quantify center or location (using the first moment or the mean, μ), dispersion or variance (using the second central moment, σ^2), skewness (using the standardized third central moment, γ_1) and kurtosis (using the standardized fourth central moment β_2, or $\gamma_2 = \beta_2 - 3$) of a distribution. [The non-central moments of the random variable X with pdf f are, in the continuous case, defined by $\mu'_k = \int_{-\infty}^{\infty} x^k f(x)dx$, while the central moments are given by $\mu_k = \int_{-\infty}^{\infty} (x - \mu)^k f(x)dx$.] As the different special distributions are discussed, these values and others (such as the median and the mode) will be evaluated and used to quantify properties of and distinguish between distributions.

There is no reason to always use μ for the mean or σ^2 for the variance of a variable. These are conventions only, but do occur frequently in the literature. As long as notation is fully defined, other letters can be used.

An important use of moments, when they exist, is given in Chebyshev's Inequality (see Chapter 3), which links the mean and the variance, and gives the minimum probability for events that are intervals centered on the mean:

Chebyshev's Theorem (Inequality): Let the random variable X have finite mean, μ, and variance, σ^2. Then for every $k > 0$, $P(|X - \mu| < k\sigma) \geq 1 - 1/k^2$ for any $k > 0$. This result is true for discrete distributions as well as continuous ones.

Example 1: Let $f(x) = 630x^4(1-x)^4$, $0 < x < 1$, and $f(x) = 0$ otherwise. Find the probability that X takes a value within 2 standard deviations of the mean, and then compare this to the Chebyshev lower bound.

Solution: Note that $\mu = 1/2$ and $\sigma^2 = 1/44$, so that $2\sigma \approx .30$. The approximate probability of the event in question is $P(0.20 < X < 0.80) = 630 \int_{0.20}^{0.80} x^4(1-x)^4 dx \approx 0.9610$ (the exact value, to four decimals, is 0.9621). The Chebyshev lower bound would be 0.75. The reason for the discrepancy is that this distribution is symmetric and very nearly bell-shaped, and hence, more closely follows the Empirical Rule.

In the next section, several of the most frequently used continuous distributions are defined, and their properties examined. Since many results depend on integrating functions, a review of Appendix B prior to starting this section would be helpful.

Previously, we defined certain quantiles, with the finest being the percentiles. We add to that list as follows:

Definition 1: Let a continuous random variable X take nonzero probability on an interval, with cdf F. Let α be such that $0 < \alpha < 1$. We call the value L_α, such that $F(L_\alpha) = \alpha$ the $\alpha \times 100$ **lower percentage point**, or just the $\alpha \times 100$ **percentage point**, of the distribution. We call the value U_α such that $1 - F(U_\alpha) = \alpha$ the $\alpha \times 100$ **upper percentage point** of the distribution. Upper percentage points are also called **critical values** of the distribution.

In particular, we will have need of 0.01 and 0.025 (and multiples of these) upper and lower percentage points in several distributions. Some of these numbers can be computed directly, but others need sophisticated computer programs to generate them. In this latter case, tables of values are available, so it will not be necessary to illustrate how this is done.

5.2 SPECIAL CONTINUOUS DISTRIBUTIONS

In this book, the distribution of a continuous random variable X is determined by the form of its density function. There are sound theoretical reasons for the form of the density (the normal is a limit of a sum of small random errors, for example); however, the arguments are more complicated than those for discrete distributions. As distributions are introduced, experimental conditions that give rise to them will be discussed, but full derivations of the densities will not be included. Interconnections between distributions are important in developing applications of the statistical results, and these relationships will be explored in section 5.4.

5.2.1 The Uniform Distribution

Definition 2: Let the random variable X have pdf f given by $f(x) = 1/(\beta - \alpha)$ on (α, β) where $\alpha < \beta$, and $f(x) = 0$ otherwise. We say X has a **uniform distribution** or is **distributed as a uniform random variable** or simply X is **uniform**. We write $X \sim U(\alpha, \beta)$, and α and β are the parameters of this distribution.

The uniform variable has many important theoretical uses, but is of limited application as a model for real phenomena. However, if there is no information available about a variable whose values fill out (α, β), sometimes the uniform distribution is assumed. This is revised as more information becomes available.

Because single numbers do not matter in terms of continuous random variables, the interval (α, β) could be $[\alpha, \beta]$, $[\alpha, \beta)$ or $(\alpha, \beta]$. For the purpose of exposition, we will use (α, β) as the interval where probability is nonzero for a uniform variable.

Conditions for a Uniform Random Variable: Experiments that give rise to a continuous uniformly distributed random variable are characterized by:

1. the outcomes fill out an interval of the form (α, β), where α and β are finite, and $\alpha < \beta$;
2. the probability of an event $I = (a, b) \subseteq (\alpha, \beta)$ is the same as the probability of $(c, d) \subseteq (\alpha, \beta)$ whenever both intervals have the same length; and
3. the experiment is to select an experimental unit and record the outcome of the experiment for this individual.

Chapter 5 — Special Continuous Random Variables

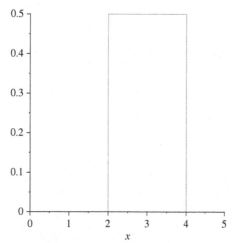

FIGURE 5-1. Uniform Distribution on [2,4].

General Results for the Uniform: Let the continuous random variable X have a uniform distribution on (α, β). Then:

$$E(X) = \mu = (\alpha + \beta)/2, \quad \sigma^2 = (\beta - \alpha)^2/12, \quad \text{and } M(t) = \frac{1}{t} \frac{e^{\beta t} - e^{\alpha t}}{\beta - \alpha}$$

with $M(0) = 1$. There is no mode (Why?), and the median is $(\alpha + \beta)/2$. If the density at x is denoted $u(x)$, and the cdf is denoted $U(x)$, then:

$$U(x) = \begin{cases} 0 & x \leq \alpha \\ (x - \alpha)/(\beta - \alpha) & \alpha < x < \beta \\ 1 & x \geq \beta \end{cases}$$

The proofs are left as exercises.

Example 2: Percentage points of the uniform. Let $X \sim U(0, 1)$. It is readily shown that the cdf is $F(x) = x$ for $0 \leq x \leq 1$, and $F(x) = 0$ if $x < 0$ and $F(X) = 1$ if $x > 1$. The percentage point corresponding to $0 < \alpha < 1$ would be the value x_α for which $F(x_\alpha) = \alpha$. Given F's formula, that means $x_\alpha = \alpha$. Upper percentage points, i.e., critical values, would be of the form $1 - \alpha$. Similar forms for percentage points can be worked out for $U(a, b)$ variables.

Example 3: The time T (in seconds) that a honeybee will "visit" a flower is uniformly distributed over [5, 20].

(a) Determine the probability that a honeybee will visit a flower for longer than 10 seconds.
(b) Determine the probability that a honey bee will visit a flower at least 15 seconds, given it stays at least 10 seconds.

Solutions:

(a) Since T is uniformly distributed, then its density is $f(t) = 1/15$, for $t \in [5, 20]$ and is 0 otherwise. Thus,

$$P(T > 10) = \int_{10}^{20} \frac{1}{15} dt = (20 - 10)/15 = 2/3.$$

(b) The probability required can be expressed as a conditional probability: $P(T \geq 15 \mid T \geq 10)$. This would be the same as:

$$P(T \geq 15 \mid T \geq 10) = P(T \geq 15 \cap T \geq 10)/P(T \geq 10).$$

However, $P(T \geq 15 \cap T \geq 10) = P(T \geq 15) = 5/15 = 1/3$, and $P(T \geq 10) = 2/3$, so the required probability is $1/2$.

5.2.2 The Gamma Distribution

Definition 3: Let the continuous random variable X have pdf given by:

$$f(x) = \begin{cases} kx^{\alpha-1}e^{-x/\beta}, & x > 0 \\ 0, & \text{otherwise} \end{cases}$$

where α and β are positive real numbers, and $k > 0$ is such that $\int_{-\infty}^{\infty} f(x)dx = 1$. Then we say X has a **gamma distribution**, or is **distributed as a gamma variable**. The parameters of the distribution are α and β.

To determine k, the change of variable $x/\beta = u$ yields:

$$\int_0^\infty x^{\alpha-1}e^{-x/\beta}dx = \beta^\alpha \int_0^\infty u^{\alpha-1}e^{-u}du.$$

The integral $\int_0^\infty x^{\alpha-1}e^{-x}dx$ occurs very frequently, and is defined to be the gamma function, denoted $\Gamma(\alpha)$, that is:

$$\Gamma(\alpha) = \int_0^\infty u^{\alpha-1}e^{-u}du.$$

From this, we have:

$$\int_0^\infty x^{\alpha-1}e^{-x/\beta}dx = \Gamma(\alpha)\beta^\alpha$$

and hence:

$$k = 1/[\beta^\alpha \Gamma(\alpha)].$$

The gamma function is a generalization of factorials for non-negative integers. It is left as an exercise to show that $\Gamma(a+1) = a\Gamma(a)$ for any $a > 0$ and to verify that $\Gamma(1) = 1$ and $\Gamma(2) = 1$. Then $\Gamma(3) = 2\Gamma(2) = 2 \times 1 = 2!$. An induction argument establishes the general result that $\Gamma(n+1) = n!$ for $n \in \mathbb{N}$.

The general observation that $\Gamma(a+1) = a\Gamma(a)$ for all $a > 0$ helps determine other values. For example, to compute the mean, we have to evaluate:

$$\mu = \frac{1}{\beta^\alpha \Gamma(\alpha)} \int_0^\infty x \, x^{\alpha-1}e^{-x/\beta}dx = \frac{1}{\beta^\alpha \Gamma(\alpha)} \int_0^\infty x^{(\alpha+1)-1}e^{-x/\beta}dx.$$

If we had:

$$\frac{1}{\beta^{\alpha+1}\Gamma(\alpha+1)} \int_0^\infty x^{(\alpha+1)-1}e^{-x/\beta}dx$$

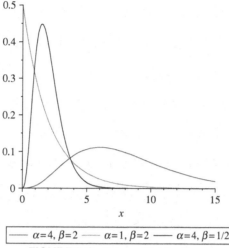

FIGURE 5-2. Gamma Distributions.

this would evaluate to 1, because:

$$g(x) = \frac{1}{\beta^{\alpha+1}\Gamma(\alpha+1)} x^{(\alpha+1)-1} e^{-x/\beta},$$

for $x > 0$, is just a gamma variable with parameters $\alpha + 1$ and β. By multiplying and dividing the integral defining the mean by $\beta^{\alpha+1}\Gamma(\alpha+1)$, μ is simply:

$$\mu = \beta^{\alpha+1}\Gamma(\alpha+1)/(\beta^{\alpha}\Gamma(\alpha)) = \beta\alpha.$$

This follows from the identity $\Gamma(a+1) = a\Gamma(a)$ and from simplifying the exponents on β.

General Results for the Gamma Distribution: If $X \sim \Gamma(\alpha, \beta)$, then:

1. The mean is $\alpha\beta$ and the variance is $\alpha\beta^2$.
2. The moment generating function is $(1 - \beta t)^{-\alpha}, t < 1/\beta$.
3. The skewness coefficient is $2/\sqrt{\alpha}$.
4. The kurtosis coefficient β_2 is $6/\alpha + 3$.
5. The mode, for $\alpha \geq 1$, is $(\alpha - 1)\beta$.

As α gets larger, the properties of the gamma are similar to those of the normal distribution (see 5.3).

The gamma variable is an effective model of a continuous random variable X that has nonzero probabilities for $X > 0$, and where the probability of small values is low, as is the probability of large values. Most of the distribution will be concentrated around a single point. There must be essentially no upper bound to the possible values of X, although in practical circumstances, this would be satisfied if there were a very large upper bound.

Example 4: The gamma distribution has been used to model daily rainfall in a specific region. See, for example, Wilks (2006). Let the random variable X measure the amount of rainfall in a 24-hour period in a given city on days it rains. The amount of rainfall observed in a given rainy day is x and is the average amount recorded at different stations around the city. [If other forms of precipitation occur, such as sleet, hail, or snow, the equivalent amount of water is included.] Over several years, a model of the amount of rain on rainy days was developed using a gamma distribution. If the parameters are $\alpha = 1.5$ and $\beta = 2.2$ when X is in inches, determine the probability (to three decimals of accuracy) that at least 4 inches of rain will fall on a given rainy day.

Solution: The density for X is

$$f(x) = \frac{1}{\beta^\alpha \Gamma(\alpha)} x^{\alpha-1} e^{-x/\beta} = \frac{1}{2.2^{1.5}\Gamma(1.5)} x^{0.5} e^{-x/2.2}$$

for $x > 0$ and $f(x) = 0$ otherwise. Also, $\Gamma(1.5) \approx 0.88623$ and $2.2^{1.5} \approx 3.26313$ from which the required probability is:

$$P(X \geq 4) = \int_4^\infty \frac{1}{2.2^{1.5}\Gamma(1.5)} x^{0.5} e^{-x/2.2} dx \approx 0.304.$$

Example 5: The health of an individual in a large population is often measured on a continuous scale increasing from 0, corresponding to perfectly healthy, to very large numbers, corresponding to increasing health problems. See, for example, Streiner and Norman (2008). If X is the health value (as measured on this scale), it is assumed to follow a gamma distribution with mean 0.6 and variance 1.5. This would correspond to $\alpha = 0.24$ and $\beta = 2.5$ since:

$$\alpha\beta = 0.6 \text{ and } \alpha\beta^2 = 1.5. \text{ Thus, } 0.6\beta = 1.5$$

so $\beta = 2.5$ and $\alpha = 0.6/2.5 = 0.24$. The probability that a randomly selected individual would have a health rating of 1.0 or less would be:

$$\int_0^1 \frac{1}{(2.5)^{0.24}\Gamma(0.24)} x^{0.24-1} \exp(-x/2.5) \, dx = 0.823$$

and 95% of the population would have a health index of 2.94 or less, since:

$$\int_0^{2.94} \frac{1}{(2.5)^{0.24}\Gamma(0.24)} x^{0.24-1} \exp(-x/2.5) \, dx = 0.9500$$

to four decimals of accuracy.

Example 6: The gamma distribution is used to model the time until the kth occurrence of an event, when the rate r of the occurrence of such an event is known, and essentially, the occurrence is random. The pdf is most frequently written in the form:

$$f(t) = \frac{r^k}{\Gamma(k)} t^{k-1} \exp(-rt), \; t > 0.$$

Note that this is just a relabeling of the parameters in our original form of the pdf. Expressed in this way, the mean of this distribution is k/r and the variance is k/r^2.

Thus, if we want to find the probability that it will take no more than 30 milliseconds until the 50th new strand of DNA is generated from a PCR reaction, knowing that the rate of production is 2 strands per millisecond (with the production of such strands independent of each other), then this probability is:

$$\int_0^{30} \frac{2^{50}}{\Gamma(50)} \exp(-2t) t^{50-1} \, dt = 0.9156.$$

For details of such models, see, for example, Miklos and Freyer (1990).

Chapter 5 — Special Continuous Random Variables

Here, we assume a gamma distribution, with the rate of the event (production of DNA strands) = $r = 2$ per millisecond, and the number of events needed $k = 50$. The mean of the gamma distribution of time in this case is just $50/2 = 25$ milliseconds.

Finding upper or lower percentage points for the general gamma distribution is not straightforward. To find them requires finding the cdf in a reasonable form, and that is not readily available. Fortunately, we do not need them in this book.

The gamma distribution is a family of distributions indexed by two parameters, α and β. Two subfamilies based on specific choices of these parameters occur often enough to merit some individual attention. These are the exponential and the chi square distributions.

5.2.2.1 The Exponential Distribution

Definition 4: (Exponential Distribution). If we take $\alpha = 1$ and relabel $\beta = \theta$ (to match standard notation), the resulting distribution is called the exponential distribution and so its density is:

$$f(x) = \begin{cases} \dfrac{1}{\theta} e^{-x/\theta}, & x > 0 \\ 0, & \text{otherwise} \end{cases}.$$

We write $X \sim E(\theta)$.

The exponential distribution arises very naturally in many situations. It is involved in lifetime models in which the length of life X of an experimental unit frequently is assumed to follow such a distribution. It is often used as a model of the time between occurrences of a random and relatively rare event, where that number of those events in a fixed time follows a Poisson distribution. The standard example is the time between emissions of particles of a radioactive material.

General Results for the Exponential Distribution: If $X \sim E(\theta)$, then:

1. The mean is θ and the variance is θ^2.
2. The moment generating function is $(1 - \theta t)^{-1}$, $t < 1/\theta$.
3. The skewness coefficient is 2.
4. The kurtosis coefficient is 9.
5. The mode is $x = 0$.

Even though the exponential has parameter θ, its shape is unaffected by changes in this parameter.

Example 7: The exponential distribution is used as a model of the time between impulses along nerve fiber. For example, if the average time between impulses along a particular type of nerve fiber is 0.5 milliseconds, and the distribution of T, the time between impulses (in milliseconds) is assumed exponential; then we may want to determine the probability that there is at least a 1.0 millisecond gap between impulses. Since we are assuming T is exponentially distributed, then its pdf is:

$$f(t) = \begin{cases} \dfrac{1}{\theta} e^{-x/\theta}, & t > 0 \\ 0, & \text{otherwise} \end{cases}$$

Recall that $E(T) = \theta$, so $\theta = 0.5$ mS here. We want $P(T \geq 1)$, which is given by:

$$\int_1^\infty \frac{1}{0.5} \exp(-t/0.5)\, dt = 0.2030$$

that is, about 20% of the time the gap between impulses along this type of fiber will be at least one millisecond.

Example 8: The exponential distribution has been used as a model of the time from the introduction of a childhood disease in a small community (with essentially no contact outside the community) until the extinction of that disease within the community. In large communities, this model is generally seen as not valid, and childhood diseases persist. See Keeling and Grenfell (1997) for fundamentals of modeling such diseases.

In a remote, small community, a particular childhood disease has entered, and it has been observed that, in other such communities, the time to extinction of this disease is exponential, with mean 50 years from onset to extinction of the disease. Then, the probability disease will become extinct within 25 years is:

$$\int_0^{25} \frac{1}{50} \exp(-t/50)\, dt = 0.3934$$

so the probability the disease will disappear, or become extinct, in this community within 25 years is 40%.

Example 9: Percentage points of the exponential. Let X be exponential, with parameter θ. It is easily shown that X's cdf is $F(x) = 1 - \exp(-x/\theta)$ so the $\alpha \times 100$ percentage point L_α is $L_\alpha = -\theta \ln(1-\alpha)$, while the upper percentage point is $U_\alpha = -\theta \ln \alpha$. In particular, if $\theta = 2$, the lower 1% point is $x_{0.01} = -2\ln(1-0.01) \approx 0.020$ and the lower 97.5% point is 7.38 (approximately). This also means the upper 2.5% percentage or critical point is 7.38.

5.2.2.2 The Chi-Square (χ^2) Distribution

Definition 5: (Chi-square (or χ^2) Distribution). If we take $\beta = 2$ and replace α by $\nu/2$, the resulting density:

$$f(x) = \begin{cases} \dfrac{1}{2^{\nu/2}\Gamma(\nu/2)} x^{(\nu-2)/2} e^{-x/2}, & x > 0 \\ 0, & \text{otherwise} \end{cases}$$

is the density of a Chi-square distribution. This distribution is characterized by ν, which is called the degrees of freedom, and we write $X \sim \chi^2(\nu)$. We will discuss the meaning of this term when we get to applications in later chapters.

General Results for the χ^2 Distribution: Let $X \sim \chi^2(\nu)$. Then:

1. The mean is ν and the variance is 2ν.
2. If $\nu = 2$, the distribution is exponential.
3. The mode is 0 if $\nu \leq 2$ and is $\nu - 2$ when $\nu > 2$.
4. The skewness coefficient has value $\sqrt{8/\nu}$.
5. The kurtosis coefficient is $12/\nu + 3$.
6. The moment generating function is $(1 - 2t)^{-\nu/2}$.

Percentage Points: There is no simple formula for finding percentage points, so many values have been tabulated. These values depend on ν, the degrees of freedom, and so the creation of tabulated values are typically restricted to positive integer values of ν. This will be discussed further in section 5.4.1.

As can be seen from the general results, the skewness tends to 0 and the kurtosis tends to 3 as $\nu \to \infty$. In fact, for ν large enough, we can approximate the χ^2 distribution by a normal distribution with mean ν and variance 2ν.

The chi-square distribution is linked to the normal distribution and is frequently used in **testing hypotheses** and finding **confidence intervals**, developed in Chapter 8. This distribution is most often used in analyzing variances, something we will use extensively in later chapters.

5.2.3 The Beta Distribution

In the case of gamma variables, the pdf is nonzero on an infinitely long interval, specifically from 0 to ∞. This is one of the conditions in an experiment that would suggest a gamma variable would be appropriate. In a number of important problems, the range will be finite but the density will not be constant, so a uniform distribution is not appropriate. A class of densities that are nonzero on finite intervals is the class of beta distributions:

Definition 6: We say the random variable X has a beta distribution if its pdf is given by:

$$f(x) = \begin{cases} kx^{\alpha-1}(1-x)^{\beta-1}, & 0 < x < 1 \\ 0, & \text{otherwise} \end{cases}$$

where α and β are positive real numbers. The constant $k > 0$ is the value that makes the integral of the density equal 1. We write $X \sim \text{Beta}(\alpha, \beta)$.

Definition 7: The two variable function $B(\alpha, \beta)$ is defined by:

$$B(\alpha, \beta) = \int_0^1 x^{\alpha-1}(1-x)^{\beta-1}dx.$$

It can be shown that $B(\alpha, \beta) = \Gamma(\alpha)\Gamma(\beta)/\Gamma(\alpha + \beta)$. The proof of this result is not obvious, and we will assume it is true.

Thus, the density function of the beta distribution is:

$$f(x) = \begin{cases} \dfrac{\Gamma(\alpha + \beta)}{\Gamma(\alpha)\Gamma(\beta)}x^{\alpha-1}(1-x)^{\beta-1}, & 0 < x < 1 \\ 0, & \text{otherwise} \end{cases}.$$

General Results for the Beta Distribution: Let the random variable $X \sim \text{Beta}(\alpha, \beta)$. Then:

1. $E(X) = \frac{\alpha}{\alpha + \beta}$ and $var(X) = \frac{\alpha\beta}{(\alpha + \beta)^2(\alpha + \beta + 1)}$.
2. The moment generating function exists, but has a complicated formula.
3. The skewness coefficient is $2(\beta - \alpha)\sqrt{\alpha + \beta + 1}/((\alpha + \beta + 2)\sqrt{\alpha\beta})$.

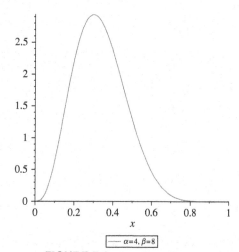

FIGURE 5-3. Beta Distribution.

4. The kurtosis coefficient is complicated.
5. The mode exists for $\alpha, \beta > 1$, and is equal to $(\alpha - 1)/(\alpha + \beta - 2)$.

One of the most common uses of a beta distribution is in modeling the probability of an event, or the proportion in a population:

Example 10: Consider a test for detecting the presence of anabolic steroids in an athlete's system. Let X be the probability that the test does not detect steroids in those that have in fact used them. Note that $0 < X < 1$ and similar problems suggests X follows a beta distribution, with parameters $\alpha = 4$ and $\beta = 30$. We want to determine a value p_0 of X so that $P(X \leq p_0) = 0.95$, that is, a 95% upper bound to the probability of non-detection. This is also the 95% point. Given our assumptions, this means solving:

$$\int_0^{p_0} \frac{\Gamma(\alpha + \beta)}{\Gamma(\alpha)\Gamma(\beta)} x^{\alpha-1}(1-x)^{\beta-1} dx = \int_0^{p_0} f(x) dx = \int_0^{p_0} \frac{\Gamma(34)}{\Gamma(4)\Gamma(30)} x^{4-1}(1-x)^{30-1} dx = 0.95.$$

This value is $p_0 = 0.2185$ (to four decimals of accuracy). This type of problem generally needs the use of a computer to solve.

The average value μ of X and its variance σ^2 are, approximately:

$$\mu = \frac{4}{34} = .1176 \text{ and } \sigma^2 = \frac{120}{(34)^2(35)} = 0.00296$$

and so the standard deviation is 0.05446.

The beta distribution is used in a method of parameter estimation, called **Bayesian Analysis**. In the next chapter, we will consider in detail how it is used to estimate the proportion in a population, making use of Bernoulli variables.

Percentage Points: The main application of the beta distribution in this text is as a model of proportions, and percentage points are not needed. Clearly, the percentage points now depend on two parameters, and tables can be created as needed.

5.3 THE NORMAL DISTRIBUTION

The **Normal Distribution** is perhaps the most important distribution that we will consider in this text. It is used as a model of errors in a vast array of problems and can be shown to be the limiting distribution of other distributions. Generally, the normal is used when a variable is symmetric about a point, with most values falling near the point and progressively fewer are farther away. There are other characteristics, but we will not go into them in this text.

Definition 8: Let the random variable X have pdf given by:

$$f(x; \mu, \sigma^2) \equiv f(x) = \frac{1}{\sqrt{2\pi}\sigma} \exp(-(x-\mu)^2/(2\sigma^2)),$$

$-\infty < x < \infty$, where $\mu \in \mathbb{R}$ and $\sigma > 0$. Then we say X has a normal distribution and write $X \sim N(\mu, \sigma^2)$, with parameters μ and σ^2.

One fundamental result, which we won't prove, is: $\sqrt{\pi} = \int_{-\infty}^{\infty} \exp(-u^2) du$. Based on this, we can show the parameters μ and σ^2 are in fact the mean and variance, respectively of the normal distribution. Note that the median is also μ as is the mode (which, for continuous variables, is the maximum of the pdf). The effect on

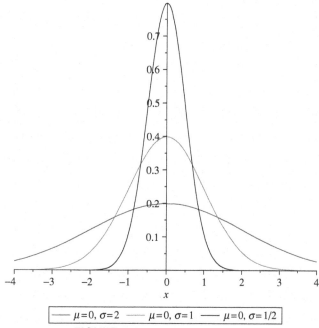

FIGURE 5-4. Normal Distributions.

changing μ but holding σ^2 fixed is to simply shift the graph of the pdf along the horizontal axis. On changing σ^2 but holding μ fixed, the density function stays centered on μ and symmetric about the mean, but the tails look "fatter" and the height of the distribution at the mode decreases as σ^2 increases. If σ^2 decreases, the tails look thinner, and the height at the mode increases. However, all normal distributions fit the description of being bell-shaped, because they all have the same kurtosis value of 3.

General Results for the Normal Distribution: Let the random variable $X \sim N(\mu, \sigma^2)$. Then:

1. $E(X) = \mu$ and $var(X) = \sigma^2$.
2. The moment generating function is $\exp(\mu t + \sigma^2 t^2/2)$.
3. The skewness coefficient is 0.
4. The kurtosis coefficient is 3.
5. The mode and median both occur at μ.

The normal distribution is the model for bell-shaped distributions and is the basis of the Empirical Rule. Often, distributions are compared to the normal, especially with respect to the kurtosis. Sometimes, the **excess kurtosis** γ_2 of a variable is reported, rather than the kurtosis coefficient in this text. The excess kurtosis is the kurtosis of a given distribution, minus 3, so that it indicates thicker tails than the normal when this value is positive, and thinner tails when it is negative.

The normal distribution is effective in modeling a vast array of phenomena. In many situations, we can think of the random variable X as having the form $X = \mu + \epsilon$, where μ is the average value of X, and ϵ is the error in measuring this mean or average value. For example, X may be the measurement of an individual's temperature (in degrees Fahrenheit). The average body temperature of an individual has been stated to be 98.6°, but it is now clear than an individual's average body temperature may be higher or lower than this value. For an individual, we will not know their mean body temperature, but we can take several readings of it, and the different temperature readings is what X represents. When we write $X \sim N(\mu, \sigma^2)$, we are saying that the overall average value of all body temperatures for this individual is μ, but for a specific reading, the value will be x, and it will almost assuredly not be μ exactly. The difference between the measured value x and the true value μ, $x - \mu$ is then called the error in the measurement and is called ϵ. If many measurements of body temperature for this individual were taken and then averaged, this average should be very close to μ. Another way to say this is that the errors $\epsilon = x - \mu$ should average (close) to 0.

The value σ^2 is, by definition, the variance, or second moment. Chebyshev's Inequality shows that as σ^2 gets smaller, the greater the fraction of the distribution of the random variable is within a given distance of the mean. In the context of the errors in temperature measurements, this means the smaller σ^2 is, the greater the proportion of temperatures taken of an individual will be near the true mean temperature.

It is clear that we don't know the mean temperature of the individual, and we don't know the true variability in the errors, i.e. σ^2. These need to be estimated, and how we go about this is dealt with in the next chapter.

In general, let X be a random variable measuring some population quantity with average μ and call the difference between an observed value x and μ the error. If the errors satisfy:

1. the errors can be positive or negative and average to 0;
2. most errors are small in absolute terms, but there is no theoretical upper or lower bound to them;
3. large errors are increasingly rare as their absolute size increases; and
4. the errors have the same variance σ^2 from measurement to measurement (that is, the conditions under which the measurements are taken are fixed).

then a reasonable model for the errors is a normal distribution with mean 0 and variance σ^2.

An alternate scenario leading to the normal as a reasonable choice is the situation where the errors can be thought of as the result of many small, uncontrollable additive factors. Again, the normal may be a reasonable choice for modeling the errors. On the other hand, the normal would not be effective in modeling errors if there are frequent large errors, or more errors are positive than negative (or viceversa).

A special case of the normal has $\mu = 0$ and $\sigma^2 = 1$. We then use the letter Z for the variable, and say Z has a standard normal distribution, and write $Z \sim N(0, 1)$. To distinguish this special case from other normal variables, the pdf is often denoted ϕ and the cdf Φ.

Standardization and Probability Calculations: If $X \sim N(\mu, \sigma^2)$ and we want $P(a < X < b)$, we can find it by referencing the standard normal through the following calculations. If $X \sim N(\mu, \sigma^2)$, then the standardized variable $Z = (X - \mu)/\sigma$ has a standard normal distribution. This important result means that percentage points and probabilities for any normal variable can be computed using values from the standard normal. If we need $P(a < X < b)$, we proceed as follows:

$$P(a < X < b) = P((a - \mu)/\sigma < (X - \mu)/\sigma < (b - \mu)/\sigma) = P((a - \mu)/\sigma < Z < (b - \mu)/\sigma)$$

where Z has a standard normal distribution. Set $z_a = (a - \mu)/\sigma$ and $z_b = (b - \mu)/\sigma$, so the calculation of probabilities of any normal variable can be found by converting to the standard normal and finding the value of $P(z_a < Z < z_b)$. The process of taking a value, say a, from a normal variable's distribution, or, in fact, from any distribution with a finite mean μ and variance σ^2, and transforming it to $(a - \mu)/\sigma$ is called **standardization**. If this is done to the random variable Y, that is, we form $(Y - \mu_Y)/\sigma_Y \equiv W$ then $E(W) = 0$ and $var(W) = 1$. In fact, W has no units. This process makes it possible to compare two different variables by standardizing each.

Percentage Points: Since all normal distributions can be converted to the standard normal, comprehensive tables of values of the cumulative distribution function $P(-\infty < Z \leq z) \equiv \Phi(z)$ of the standard normal can be found in any introductory statistics text, or on the web. Typically, the z values range from 0 to perhaps 4 or larger, and usually in increments of 0.01, so that, for example, the value of $P(-\infty < Z \leq 2.31)$ is included. These tables are sufficiently comprehensive that upper or lower percentage points (or values very close to them) can be found in the body of the table. For example, if we need the lower 10% point, we find $\Phi(-1.28) = 0.1003$ and so take the lower 10% point to be -1.28. Of course, this means the upper 10% point is 1.28. The upper 2.5% point is taken to be 1.96, because $\Phi(1.96) = 0.975$ or $1 - \Phi(1.96) = 0.025$, approximately. Since $1 - \Phi(1.64) = 0.0505$ and $1 - \Phi(1.65) = 0.0495$ then we take the upper 5% point to be 1.645.

Since the standard normal density is symmetric about 0, then, for example,

$$\Phi(z) = 1 - \Phi(-z), \text{ and } P(z < Z < 0) = P(0 < Z < -z).$$

These observations mean that if the table of values includes only positive values of z, we can always find probabilities involving negative values of z.

Example 11: These illustrative examples retain four decimals of accuracy.

(a) $P(-1 < Z < -0.55) = \Phi(-0.55) - \Phi(-1) = 0.2912 - 0.1587 = 0.1325$. This is the same value as $P(0.55 < Z < 1)$.

(b) $P(Z > -2.22) = P(-2.22 < Z < 0) + 0.5 = 0.5 - 0.0132 + 0.5 = 0.9868$.

(c) $P(Z < -1.96) = \Phi(-1.96) = 0.02500 = P(Z > 1.96)$.

(d) $P(Z > 1.645) = 1 - P(Z < 1.645) = 0.04998 = P(Z < -1.645)$.

Outliers: It is often the case that past history suggests a particular distribution for a random variable, and often, $X \sim N(\mu_0, \sigma^2)$ is assumed, where μ_0 is the expected value of X from this past history. However, when we run the experiment again and observe $X = x$, x is "very far" away from μ_0. Such a value can be considered unusual and may merit special attention. We call such values **outliers**.

The normal, or other variables that are very nearly normal, give us a way to quantify what might be considered an outlier in an experiment. If we run an experiment and observe a value of X, say x (from virtually any reasonable distribution), then if $|(x - \mu_0)/\sigma| \geq 3$, we would suspect it of being an outlier. This is because for the standard normal variable, $P(Z > 3) = P(Z < -3) \approx 0.00135$. Thus, if $X \sim N(3, 16)$, then an unusual X value would be any value greater than 15 or less than -9. If we see such a value, we would report that it is a suspected outlier. It is not impossible to see such values under the assumptions about the distribution of X; it is only rare that such values would occur. True outliers are often the result of cross-contamination; that is, we really have two separate populations, but we think we only have one. Another possibility is that conditions in the experiment have changed, and this may suggest returning to the experimental design and the assumptions about what we are observing. For example, the distribution of X may not be normal and so we may appeal to Chebyshev's Inequality. If that is the case, we expect to see up to $1/9^{th}$ of the distribution further than 3 standard deviations from the mean, and that is no longer so unusual.

Example 12: A healthy adult man has body temperature T that is assumed to be normally distributed, with mean μ and constant standard deviation $\sigma = 0.75$ (in degrees Fahrenheit). Suppose one such man is chosen at random, and his mean healthy temperature is 98.3. What is the probability that his temperature is 101 or higher, if he is healthy?

Solution: If we let T be the man's temperature when healthy, then $T \sim N(98.3, 0.75^2)$. Thus, we want $P(T \geq 101) = P(Z = (T - 98.3)/0.75 \geq (101 - 98.3)/0.75) = P(Z \geq 3.6) \approx 0.00016$. Although it is possible, the probability that this is the temperature of a healthy man is extremely low.

Normal Approximation to the Binomial: A specific application of the normal distribution is as an approximation to the binomial when non-rare events are being counted and the number of trials is reasonably large. If the probability of a 1 (or success) in the Bernoulli trials is π and $(\pi - 3\sqrt{\pi(1-\pi)/n}, \pi + 3\sqrt{\pi(1-\pi)/n}) \subset (0, 1)$, then we can use a normal distribution to approximate the binomial. A different criterion is that n be greater than 20, with $n\pi \geq 5$ and $n(1-\pi) \geq 5$. The first criterion guarantees a very good approximation, while the second provides conditions for a good approximation.

We will use the criterion that $(\pi - 3\sqrt{\pi(1-\pi)/n}, \pi + 3\sqrt{\pi(1-\pi)/n}) \subset (0, 1)$. The binomial with n trials and π as the probability of success is well-approximated by the normal distribution with mean $\mu = n\pi$ and variance $\sigma^2 = n\pi(1-\pi)$. If the population proportion π is not known, then replace π by $p = \sum_{i=1}^{n} x_i/n$

in the criterion, where the x_is are the outcomes of the n Bernoulli trials run in the binomial experiment, where $x_i = 0$ or 1.

The normal approximation uses a continuous distribution to approximate a discrete one. In order to increase the accuracy of the approximation, we use a **continuity correction factor** in our calculations. For example, suppose we want $P(2 < X \leq 5)$ for a binomial random variable X. We want to approximate the discrete distribution of X by a continuous normal one as closely as possible. Thus, we want to cover (in this case) the possible values $X = 3, 4, 5$. If we consider the structure of the histogram for a binomial, then we use bars centered at each possible value of X, with base length 1 and height the probability of that value. Thus, to make sure as much of the bar is covered as possible by a continuous curve, we should have our continuous variable start at 2.5 to exclude 2 in the binomial distribution and go to 5.5 to cover all of 5 in the binomial. Hence, $P(2 < X \leq 5) = P(2.5 < X < 5.5) \approx P(2.5 < Y < 5.5)$, where Y is the normal variable with the appropriate mean and variance. Once in this form, all we need do now is standardize the endpoints and then use the standard normal probabilities.

Example 13: Suppose 200 Bernoulli trials, with probability of success $p = 0.08$, are run and X counts the number of successes. Then $P(X \geq 20) = P(X > 19.5) \approx P(Y > 19.5) = P(Z > (19.5 - 16)/3.837) = P(Z > 0.912) = 0.1814$ where we have used $Y \sim N(16, 14.72)$ and then standardized it. Note that $(p - 3\sqrt{p(1-p)/n}, p + 3\sqrt{p(1-p)/n}) = (0.0225, 0.138) \subset (0, 1)$, so the normal approximation is justified. The true value of this probability is, to three decimals of accuracy, 0.179.

One final observation is that if $Z \sim N(0, 1)$, then $Z^2 \sim \chi^2(1)$, or, in words, the square of a standard normal variable is chi-square, with 1 degree of freedom. This relationship plays an important role in quantifying results of experiments, as we shall see in later chapters on hypothesis testing, confidence intervals, and regression.

5.3.1 The Log Normal (Lognormal, Log-normal) Distribution

The log normal (lognormal, log-normal) distribution is used as a model of survival times (see, for example, Royston (2001)), and in situations where the random variable is itself the product of many other independent random variables. This is in contrast to the normal, since the normal can be thought of as the sum of other independent random variables.

Definition 9: The random variable X is said to have a log normal distribution if its pdf is given by:

$$f(x) = \frac{1}{x\sigma\sqrt{2\pi}} \exp(-(\ln x - \mu)^2/(2\sigma^2)), \quad x > 0$$

and $f(x) = 0$ otherwise. We write $X \sim LN(\mu, \sigma^2)$.

If we set $Y = \ln X$, then the density of Y is $g(y) = \frac{1}{\sigma\sqrt{2\pi}} \exp(-(y - \mu)/(2\sigma^2))$, $y \in \mathbb{R}$, that is, Y has a normal distribution with mean μ and variance σ^2. Thus, we can restate the definition as X is $LN(\mu, \sigma^2)$, if its (natural) logarithm is normally distributed as $N(\mu, \sigma^2)$. Recall that if $Y = \ln X$, then on changing variables in X's density, $dy = dx/x$.

The moment generating function does not exist for any positive value of its variable. On the other hand, this distribution will have finite moments of all orders.

General Properties of Log Normal Variables: Let $X \sim LN(\mu, \sigma^2)$. Then:

1. $E(X) = \exp(\mu + \sigma^2)$ and $var(X) = (\exp(\sigma^2) - 1)\exp(2\mu + \sigma^2)$.
2. The moment generating function does not exist except at 0, but the distribution has finite moments of all order.
3. The mode is $\exp(\mu - \sigma^2)$ and the median is $\exp(\mu)$.

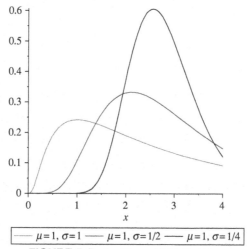
FIGURE 5-5. Log Normal Distributions.

4. The skewness coefficient is $(\exp(\sigma^2) + 2)\sqrt{\exp(\sigma^2) - 1}$.
5. The kurtosis coefficient is $\exp(4\sigma^2) + 2\exp(3\sigma^2) + 3\exp(2\sigma^2) - 3$.

From these values, it is clear the log normal is very different from the normal. No choice of μ and σ^2 (except $\sigma^2 = 0$) will generate the same kurtosis and skewness coefficients of the normal.

Example 14: Let $X \sim LN(1, 1)$. Determine the mean and variance of X, and the coefficients of skewness and kurtosis. Compare the coefficient values to those of the normal.

Solution: From the general results above, $E(X) = \exp(\mu + \sigma^2) = \exp(2)$ and $var(X) = (\exp(1) - 1)\exp(3)$.
The coefficient of skewness is $(\exp(\sigma^2) + 2)\sqrt{\exp(\sigma^2) - 1} = (\exp(1) + 2)\sqrt{\exp(1) - 1} \approx 6.18$ while the coefficient of kurtosis is $\exp(4\sigma^2) + 2\exp(3\sigma^2) + 3\exp(2\sigma^2) - 3 = \exp(4) + 2\exp(3) + 3\exp(2) - 3 \approx 113.94$. Clearly, the distribution is much different from that of any normally distributed random variable.

Percentage Points and Probability Calculations: Because of the relationship with the normal, calculations of probabilities of events for the log normal only require referring to the normal. For example, if $X \sim LN(1, 1/4)$ and we want $P(0.8 < X < 3.2)$, then we take logarithms appropriately, and use the fact that $Y = \ln X$ is distributed as $N(1, 1/4)$ as follows:

$$P(0.8 < X < 0.9) = P(\ln(0.8) < \ln X < \ln(3.2))$$

$$= P((\ln(0.8) - 1)2 < Z < (\ln(3.2) - 1)2) = 0.6207.$$

5.4 RANDOM VARIABLES RELATED TO THE NORMAL

There are several distributions that we will find useful: in particular, the χ^2, **Student's t-distribution** and **Snedecor's F-distribution**, or simply the F **distribution**. Each of these has its own pdf, which we will discuss in brief. They arise from important relationships between normal variables and are important in the quantitative analysis of a wide range of problems. These distributions occur so frequently that tables of percentage points are available in virtually any introductory text. The forms of the tables are similar to the standard normal but are indexed by the degrees of freedom.

5.4.1 The χ^2 Distribution

Recall that a random variable X has a $\chi^2(\nu)$ distribution if it has pdf:

$$f(x) = \begin{cases} \dfrac{1}{2^{\nu/2}\Gamma(\nu/2)} x^{(\nu-2)/2} e^{-x/2}, & x > 0 \\ 0, & \text{otherwise} \end{cases}.$$

It is also a fact that the sum of independent χ^2 variables is again χ^2, with degrees of freedom the sum of the degrees of the individual summands.

Also recall the observation that if $Z \sim N(0,1)$, then $Z^2 \sim \chi^2(1)$, or, in words, the square of a standard normal variable is chi-square, with 1 degree of freedom. Thus, if each random variable X_i, $1 \le i \le n$, is such that $X_i \sim N(\mu, \sigma^2)$ and these random variables are independently distributed, then each $(X_i - \mu)^2/\sigma^2$ is $\chi^2(1)$. Thus, $\sum (X_i - \mu)^2/\sigma^2 \sim \chi^2(n)$.

Cochran's Theorem (Paraphrased): Let $Z_i \sim N(0,1)$, $1 \le i \le n$, and assume they are independently distributed. Set $Q = \sum_{i=1}^n Z_i^2$. For each j with $1 \le j \le k$ ($k < n$ a given natural number), let Q_j be a sum of squares of linear (but not **affine**) combinations of the Z_is, with $Q = \sum_{j=1}^k Q_j$. Let r_j be the number of **linearly independent** combinations of the Z_is making up Q_j and call r_j the **rank** of Q_j. Assume $\sum_{j=1}^k r_j = n$. Then the Q_js are independent, and $Q_j \sim \chi^2(r_j)$.

See Hogg, Craig, and Allen (2005) for a complete statement and proof of Cochran's Theorem. It is better expressed in terms of matrices and quadratic forms, but this is not essential for the purposes of this text. This is the reverse of the theorem that states the sum of independent χ^2 variables is also χ^2, with degrees of freedom equal to the sum of the degrees of freedom of the summands.

An affine combination of, say, x and y is an expression of the form $ax + by + c$, where a, b, c are constants, and $c \ne 0$. The expression $ax + by$ is linear. The real variables $x_1, \ldots, x_n$ are said to be **linearly independent** if and only if the only set of constants $a_1, \ldots, a_n$ for which $\sum_{i=1}^n a_i x_i \equiv 0$ for all choices of the x_is are $a_1 = a_2 = \cdots = a_n = 0$. Linear independence and statistical independence are different concepts.

Essentially, Cochran's Theorem allows us to decompose a sum of squares of the Z_is into other sums of squares that are also χ^2 variables and are independent of each other. Each Q_j is itself a sum of squares. It is a linear combination of the Z_is that is being squared in each summand in the Q_js. The total number of linearly independent combinations in all of the Q_js must be exactly the same as the number of independent variables that we start with, which is why $r_1 + \cdots + r_k = n$ is required. This would occur, for example, if each Q_j was made up of distinct subsets of the Z_is.

Example 15: Under the conditions of Cochran's Theorem, with $n = 6$, take $Q_1 = Z_1^2$, $Q_2 = Z_2^2 + Z_3^2$ and $Q_3 = Z_4^2 + Z_5^2 + Z_6^2$. For real variables with values in $\mathbb{R}$ and no restrictions, the only way to make a linear combination of them zero identically is to have the coefficients all zero. Thus, $r_1 = 1$, $r_2 = 2$ and $r_3 = 3$. Cochran's Theorem says the Q_js are independent and, in this case, $Q_j \sim \chi^2(j)$, $j = 1, 2, 3$.

Example 16: The main use of Cochran's theorem is in decomposing sums of squares into certain other sums of squares. For example, let the conditions of Cochran's Theorem hold, with $Q = \sum_{i=1}^n Z_i^2$. It is an algebraic exercise to show:

$$Q = \sum_{i=1}^n (Z_i - \bar{Z})^2 + n(\bar{Z})^2 \equiv Q_1 + Q_2.$$

$\bar{Z} = \sum_{i=1}^n Z_i/n$. It is not difficult to show the covariance between $\bar{Z}$ and each $Z_i - \bar{Z}$ is 0, and hence, they are independent (why?). But what is the distribution? We know that only one linear combination of the Z_is is making up $\bar{Z}$. In the case of $\sum_{i=1}^n (Z_i - \bar{Z})^2$ there are only $n - 1$ linearly independent combinations, because if we form $\sum_{i=1}^n (Z_i - \bar{Z})$, this will be zero. Thus, $r_1 = n - 1$, $r_2 = 1$ so $r_1 + r_2 = n$. Based on this, we can conclude that $Q_1 \sim \chi^2(n-1)$, $Q_2 \sim \chi^2(1)$ and Q_1 and Q_2 are independent.

Chapter 5 — Special Continuous Random Variables

Linear Algebra and Rank (Optional): The simplest way to determine the number of linearly independent combinations of expressions we have is to use the rule from linear algebra: Let expressions $u_1, \ldots, u_m$ be linear combinations of the variables $x_1, \ldots, x_n$. In a matrix, called the coefficient matrix, where each column represents the coefficient of x_i in each u_j, list the coefficients of the x_is for each u_j in order in row j. Use row reduction methods to put the matrix in row echelon form. The number of nonzero entries on the main diagonal is the rank of the matrix and is the number of linearly independent u_js.

Example 17: Consider $Z_1 - \overline{Z}$, and $Z_2 - \overline{Z}$, with $n = 2$ in the last example. We can rewrite these as $\frac{1}{2}Z_1 - \frac{1}{2}Z_2$ and $-(\frac{1}{2}Z_1 - \frac{1}{2}Z_2)$, respectively. Clearly, the second is just the negative of the first, so once we know, say, the value of the first expression, the value of the second must be its negative. There is only one linearly independent expression here. In matrix terms, the coefficient matrix would be:

$$\begin{pmatrix} 1/2 & -1/2 \\ -1/2 & 1/2 \end{pmatrix} \to \begin{pmatrix} 1/2 & -1/2 \\ 0 & 0 \end{pmatrix}$$

where the first row has been added to the second to produce the second matrix. This is row echelon form, and there is only one nonzero entry on the main diagonal.

Consider $Z_1 - \overline{Z}, Z_2 - \overline{Z}$ and $Z_3 - \overline{Z}$, with $n = 3$. The coefficient matrix is:

$$\begin{pmatrix} 2/3 & -1/3 & -1/3 \\ -1/3 & 2/3 & -1/3 \\ -1/3 & -1/3 & 2/3 \end{pmatrix} \to \begin{pmatrix} 2/3 & -1/3 & -1/3 \\ 0 & 1/2 & -1/2 \\ 0 & -1/2 & 1/2 \end{pmatrix} \to \begin{pmatrix} 2/3 & -1/3 & -1/3 \\ 0 & 1/2 & -1/2 \\ 0 & 0 & 0 \end{pmatrix}.$$

The first row in the first matrix is added to its second and third rows to produce the second matrix, and then the second row of the second matrix has been added to the third row of the second matrix to produce the third matrix. Now there are two nonzero elements in the row-echelon form.

It is straightforward to generalize these two calculations to show that the rank for Q_1 in the previous example is $n - 1$.

A heuristic argument providing the rank is to take the total number of expressions and subtract one for each condition imposed. If we know the sample mean, $\overline{Z}$, then once $Z_1, \ldots, Z_{n-1}$ are given, Z_n is known. It cannot be arbitrary. Beyond this, there are no restrictions on the choices of the rest of the Z_is.

Definition 10: Let X_i, $1 \leq i \leq n$ be n random variables. Define the random variable $\overline{X}$ by the formula $\overline{X} = \sum_{i=1}^{n} X_i/n$, and call it the **sample mean**. Define the random variable S^2 by the formula $S^2 = \sum_{i=1}^{n}(X_i - \overline{X})^2/(n-1)$: and call it the **sample variance**.

Corollary to Cochran's Theorem: Let the n random variables X_i be independently and identically distributed as $N(\mu, \sigma^2)$. Then the sample mean and the sample variance are independently distributed.

Proof: Define $Z_i = (X_i - \mu)/\sigma$. These new variables are independent and identically distributed as standard normal variables. We can then apply the results in the last example. Note that adding constants to variables or multiplying variables by constants does not affect whether the variables are independent or not.

Cochran's Theorem is important because it allows sums of squares of normal variables to be decomposed into smaller pieces or summands that are independent and are related to χ^2 variables. Particular summands can be associated with the different factors in an experiment. This is the basis of much of the Analysis of Variance and Regression procedures.

Upper Percentage Points/Critical Values of the χ^2: In most applications, percentage points are the most useful values to which to have access. We can refer every normal distribution to the standard normal, and then a table of the cdf for a wide range of Z values is all that is needed. Sufficiently detailed and accurate tables will include percentage points, at least approximately. For the χ^2, we would have to have a table for every different choice of the degrees of freedom. Formally, although the definition of pdf of a χ^2 does not require ν, the degrees of freedom, to be whole numbers, this value can be assumed integer for our purposes.

For the χ^2, we determine specific percentage points for different degrees of freedom. Recall that if $0 < a < 1$ is specified, then the $a \times 100$ percentage point is the value x_a such that for the cdf F, $F(x_a) = a$. Most tables of values include the **upper percentage points** instead. Upper percentage points of a distribution are also called **critical values** of the distribution. By this, we mean for a given $0 < \alpha < 1$ we find the value χ^2_α for which $1 - F(\chi^2_\alpha) = \int_{\chi^2_\alpha}^{\infty} f(x)dx = \alpha$. This is the area under the right tail of the distribution from χ^2_α to infinity. A typical table of values for the χ^2 distribution lists the chosen α values along the top, has left most column the degrees of freedom, and, where a row meets column, the value in the table is χ^2_α. A brief table of upper percentage points would look like:

		\multicolumn{6}{c}{α}
		0.01	0.025	0.05	0.95	0.975	0.99
	5	15.1	12.8	11.1	1.15	0.831	0.554
DOF	6	16.8	14.4	12.6	1.64	1.24	0.872
	7	18.5	16.0	14.1	2.17	1.69	1.24

So, for example, $P(X \geq 2.17) = 0.95$ when $X \sim \chi^2(7)$. This also means $P(X \leq 2.17) = 0.05$. The χ^2 variables are not symmetric about any point, so upper and lower tail values must be tabulated. The order of the values of α is reversed in some tables. Most tables include several more upper percentage points, and the number of significant digits vary from table to table.

5.4.2 Student's t Distribution

Definition 11: We say a random variable T has a **Student's t distribution** with ν **degrees of freedom**, and write $T \sim t(\nu)$, if its density function is given by:

$$g(t) = \frac{\Gamma\left(\frac{\nu+1}{2}\right)}{\sqrt{\pi\nu}\,\Gamma(\nu/2)} \left(1 + \frac{t^2}{\nu}\right)^{-(\nu+1)/2}$$

where $t \in \mathbb{R}$ and $\nu > 0$.

The reason $\nu > 0$ is to ensure this is a proper density function. Also, ν does not have to be an integer, but for our purposes, we will assume it is.

It is shown in more advanced courses that for independently distributed random variables $X_i \sim N(\mu, \sigma^2)$, $1 \leq i \leq n$ then $T = \sqrt{n}(\overline{X} - \mu)/\sqrt{S^2}$, where $\overline{X}$ is the sample mean and S^2 is the sample variance as defined earlier, has a t distribution with $n - 1$ degrees of freedom. Note that the distribution is free of unknown parameters.

The result in the last paragraph is based on the following theorem:

Theorem: Let $Z \sim N(0, 1)$ and $W \sim \chi^2(\nu)$ be independent. Then the random variable T defined by:

$$T = Z/\sqrt{W/\nu}$$

is distributed as a Student's t variable, with ν degrees of freedom.

This theorem does not reference the sample variance, but from our previous discussion of the properties of χ^2 variables, includes the result quoted above concerning the sample mean and variance.

General Results for Student's t: Let $T \sim t(\nu)$. Then:

1. The mode and median are always 0. The mean is 0 if $\nu > 1$.
2. The variance is $\nu/(\nu - 2)$ when $\nu > 2$.
3. The moment generating function does not exist.

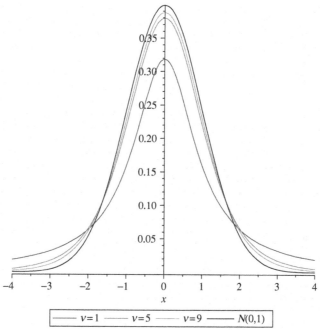
FIGURE 5-6. Student's t Distributions.

4. The distribution is symmetric.
5. The kurtosis is $6/(\nu - 4) + 3$ for $\nu > 4$.
6. Student's t tends to $N(0, 1)$ as $\nu \to \infty$.

Student's t is to some extent bell shaped, but is definitely different than the normal, especially for small values of ν, as can be seen from the kurtosis coefficient (recall that the normal has kurtosis coefficient value of 3). It is symmetric about $X = 0$, and for each fixed ν, it looks somewhat like the standard normal distribution, but has thicker tails and a lower mode. It can be shown that as $\nu \to \infty$, Student's t does tend to the standard normal distribution, and they are virtually identical once the degrees of freedom exceed 30. However, no Student's t with $\nu < \infty$ has a finite moment generating function, except at 0, which contrasts with the normal, which has a moment generating defined everywhere.

Since the Student's t distribution plays a crucial role in statistical analysis, values of its cumulative distribution function are tabulated in most introductory texts. It should be noted that these values change as the degrees of freedom change, and it is not possible to include values for each different value of ν. Typically, we need such values when ν is a positive integer, and for most applications, only certain percentage points. In this case, as in the case of the χ^2 already discussed and the F distribution to come, the table of values is the reverse of the normal. In the case of the standard normal distribution, the value of Z is along the outside of the table, and the corresponding value of the cdf is inside. In the case of Student's t, since it is symmetric about 0, upper percentage points only are needed. Different values of α are written along the top row, and the degrees of freedom of the particular distribution are written in the left-hand column. Where a row and column meet is the T value u_α for the given degrees of freedom that has $1 - cdf(u_\alpha) = \alpha$. This is the same format used in the χ^2 distribution. The F distribution is similar, but there are two parameters, each called degrees of freedom, and different authors use a modification of the format for t and χ^2 distributions. It should also be noted that most statistical packages and spreadsheets have built-in procedures to derive the necessary values.

Upper Percentage Points/Critcal Values of the t Distribution: Summarizing the previous paragraph: Just as in the case of the χ^2 distribution, Student's t depends on the parameter called the degrees of freedom. As is done for the χ^2, tables of upper percentage points or critical values for different degrees of freedom are presented, usually with the upper tail area specified by $0 < \alpha < 1$ and t_α the value for which the area under the given t pdf from t_α to infinity is α. The values of α are given along the top row, and the degrees of freedom along the left side of the table. Student's t is always symmetric, so lower tail values are readily computed from upper tail values. A part of a typical table looks like:

		α					
		0.10	0.05	0.025	0.01	0.005	0.0005
DOF	8	1.40	1.86	2.31	2.90	3.36	5.041
	9	1.38	1.83	2.26	2.82	3.25	4.78
	10	1.37	1.81	2.23	2.76	3.17	4.59

For example, if $X \sim t(9)$, then $P(X > 2.26) = 0.025$. This also means that $P(X < -2.26) = 0.025$ by symmetry. Note that the α values were listed in descending order. Again, different texts have different conventions, but each will have either explanatory diagrams or equations to help make sure mistakes are infrequent.

5.4.3 Snedecor's F Distribution

The t family of distributions provide pivotal statistics for the mean of a normal variable (see Definition 13, Chapter 6 for the definition of pivot and pivotal statistic). The F distribution is a two parameter family that will provide pivotal statistics in analyzing variances.

Definition 12: Let the continuous random variable X have pdf given by:

$$f(x) = \frac{(m/n)^{m/2}}{B(m/2, n/2)} \frac{x^{m/2-1}}{(1 + mx/n)^{(m+n)/2}}, \quad x > 0$$

and $f(x) = 0$ otherwise, where $m, n > 0$. Then we say X has an F distribution and write $X \sim F(m, n)$. The value m is called the numerator degrees of freedom, and n is called the denominator degrees of freedom.

Theorem 1: Let $Y \sim \chi^2(m)$ and $W \sim \chi^2(n)$ be independent. Then:

$$(Y/m)/(W/n) = (nY)/(mW) \sim F(m, n).$$

In words, the last theorem states that the ratio of two independent χ^2 variables divided by their degrees of freedom has an F distribution. The order of the degrees of freedom is essential, and this result explains the nomenclature for the different degrees of freedom.

Example 18: If $X_i \sim N(\mu, \sigma^2)$, $1 \le i \le n$ are independently distributed, then $\sqrt{n}(\overline{X} - \mu)/\sigma \sim N(0, 1)$ and $(n-1)S^2/\sigma^2 \sim \chi^2(n-1)$. But the square of a standard normal variable is $\chi^2(1)$, so $n(\overline{X} - \mu)^2/\sigma^2$ and $(n-1)S^2/\sigma^2$ are independent χ^2 variables (by Cochran's Theorem). Hence,

$$(\sqrt{n}(\overline{X} - \mu)/\sigma)^2 / [(n-1)S^2/((n-1)\sigma^2)] = \frac{n(\overline{X} - \mu)}{S^2} \sim F(1, n-1).$$

Since the F distribution depends only on the parameters $m = 1$ and n, and no unknown parameters, then this is called a pivotal statistic for μ.

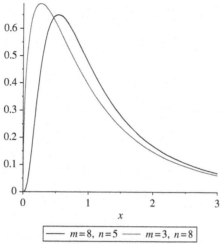

FIGURE 5-7. The F Distribution.

Since we will not be using the F distribution as a model, but as a pivot for certain means and variances, we will not consider its properties, with one exception:

Theorem 2: Let X be distributed as $F(m, n)$, and define $Y = 1/X$. Then $Y \sim F(n, m)$.

Proof: A formal proof would make use of a change of variable in the density of X. However, the result must be true because the numerator and denominator χ^2 variables in X simply change places in Y.

This last observation is useful in developing upper percentage points or critical values of the F distribution because we can relate the critical numbers of $F(m, n)$ to those of $F(n, m)$. Generally, tables of critical values for the $F(m, n)$ are listed by first fixing a value of m, and then determining critical values for a range of values of n. F tables can be extensive in texts, especially if a wide range of combinations of numerator and denominator degrees of freedom are incorporated. Other, usually shorter, tables are based on fixing a small number of upper percentage values, and then determining the upper percentage points for various combinations of m and n. Typically, the numerator degrees of freedom are listed along the top row, and the denominator degrees of freedom are listed in the left column. For example, let $F_{\alpha,m,n}$ be the upper tail percentage point, that is, the area under the density of the $F(m, n)$ density from this value to infinity is α. Fix $\alpha = 0.05$. A part of the table of upper percentage points would look like:

		$\alpha = 0.05$ Numerator DOF		
		10	12	15
Denominator	10	2.98	2.91	2.85
Degrees of	11	2.85	2.79	2.72
Freedom	12	2.75	2.69	2.62

Thus, for example, $F_{0.05,10,12} = 2.75$. Not all consecutive numerator degrees of freedom are given explicitly in this snapshot. Suppose we need $F_{0.05,14,11}$. The values change slowly, and we get a reasonable approximation by taking the average of the closest two values in the table. Thus, $F_{0.05,14,11} \approx (2.79 + 2.72)/2 = 2.76$.

The observation that if $X \sim F(m, n)$ then $Y = 1/X$ is distributed as $F(n, m)$ allows us to compute lower percentage points as follows: Suppose $X \sim F(m, n)$, and we want the value x_α so that $P(X \leq x_\alpha) = \alpha$. This is the same as:

$$P(X \leq x_\alpha) = P(1/x_\alpha \leq 1/X) = P(1/x_\alpha \leq Y) = \alpha.$$

We now find the upper percentage point of Y from tables or otherwise, call it $F_{\alpha,n,m}$, set it equal to $1/x_\alpha$, and then solve for x_α.

5.5 THE BIVARIATE NORMAL DISTRIBUTION

Definition 13: The continuous and jointly distributed random variables X and Y have a joint bivariate normal distribution iff their joint pdf is given by:

$$f(x,y) = \frac{1}{2\pi\sigma_1\sigma_2\sqrt{1-\rho^2}} \exp\left(-\frac{1}{2(1-\rho^2)}Q\right)$$

where

$$Q = \frac{(x-\mu_1)^2}{\sigma_1^2} - 2\rho\left(\frac{x-\mu_1}{\sigma_1}\right)\left(\frac{y-\mu_2}{\sigma_2}\right) + \frac{(y-\mu_2)^2}{\sigma_2^2}.$$

Observations of X and Y are recorded simultaneously for each experimental unit, so the ordered pair (X, Y) represents a random observation from the joint distribution, and (x, y) an observed value. The domain of the jpdf is $\mathbb{R}^2$.

The parameters are μ_1, μ_2, σ_1^2, σ_2^2 and ρ. These values must satisfy $\mu_1, \mu_2 \in \mathbb{R}$, $\sigma_1, \sigma_2 > 0$ and $-1 < \rho < 1$. We call ρ the correlation coefficient, and later, we discuss how we can interpret and estimate this value. It plays an important role in fitting lines to data. Instead of subscripting the appropriate parameters with 1 or 2, we often use X and Y as the subscript.

It is not immediately obvious, but $\int_{-\infty}^{\infty}\int_{-\infty}^{\infty} f(x,y)dxdy = 1$. Further, the marginal distributions of X and Y are $X \sim N(\mu_1, \sigma_1^2)$ and $Y \sim N(\mu_2, \sigma_2^2)$. We can also show that $cov(X, Y) = \rho\sigma_1\sigma_2$. Further, the conditional distributions can be evaluated: $X \mid Y = y \sim N(\mu_1 + \rho\frac{\sigma_1}{\sigma_2}(y - \mu_2), \sigma_1^2(1-\rho^2))$ and $Y \mid X = x \sim N(\mu_2 + \rho\frac{\sigma_2}{\sigma_1}(x - \mu_1), \sigma_2^2(1-\rho^2))$. A notation that is sometimes used for the conditional variances is $\sigma_{X|y}^2 = \sigma_1^2(1-\rho^2)$ and $\sigma_{Y|x}^2 = \sigma_2^2(1-\rho^2)$, with similar notation for the means, i.e., $\mu_{X|y}$ and $\mu_{Y|x}$. One very important theoretical result is that if X and Y are bivariate normal, then X and Y are independent iff $\rho = 0$. Note what happens to the conditional distributions when $\rho = 0$: the dependence on the "conditioning" variable vanishes. These forms and ideas will again occur when we discuss regression in the last chapter.

5.6 OTHER IMPORTANT RESULTS

A major result is that if X and Y have normal distributions, then so does any linear combination of them. This in fact generalizes: if $X_1, \ldots, X_n$ each have normal distributions, then so does any linear combination of them. This is true regardless of whether the variables are dependent or independent. If they are independent, and we set $Y = a_1X_1 + \cdots + a_nX_n$, then have:

$$Y \sim N\left(a_1\mu_1 + \cdots + a_n\mu_n, a_1^2\sigma_1^2 + \cdots + a_n^2\sigma_n^2\right) = N\left(\sum_{i=1}^{n} a_i\mu_i, \sum_{i=1}^{n} a_i^2\sigma_i^2\right).$$

In particular, if all the X_i have identical distributions, and we take $a_i = 1/n$, then Y is just the average of the Xs, and $Y \sim N(\mu, \sigma^2/n)$. When the variables are not independent, we have to account for the interdependence of the variables, and this is done through the covariances:

$$Y \sim N\left(\sum_{i=1}^{n} a_i\mu_i, \sum_{i=1}^{n} a_i^2\sigma_i^2 + 2\sum\sum_{i<j} a_ia_j cov(X_i, X_j)\right).$$

Recall that $cov(X, Y) = E((X - \mu_X)(Y - \mu_Y)) = E(XY) - E(X)E(Y)$.

These facts and formulae will be used when we discuss regression analysis in Chapter 11.

The results in this section will be used without proof, but are fundamental to many of the procedures developed in the remainder of this work.

Chapter 5 — Special Continuous Random Variables

5.7 FINAL OBSERVATION

The material in this chapter is not meant to be a compendium of all distributions one might encounter in the life sciences. For example, consider the **Weibull distribution**, with density given by:

$$f(x) = \begin{cases} Cx^{\beta-1}\exp(-\alpha x^{\beta}) & x > 0 \\ 0 & \text{elsewhere.} \end{cases}$$

It generalizes the gamma distribution and has been used in studies related to health and smoking, wind speed, and many other areas of research. Experimental conditions may require new distributions, or modifications of old ones.

This chapter does introduce many of the most commonly used distributions in the life and other sciences, and develops some of the tools needed to perform high-quality quantitative analysis. The results make use of standard manipulations in algebra and calculus, and carrying out the calculations indicated in the text will help develop skills in reading research papers, and in exposition.

Exercises

1. There are many other continuous, random variables with important applications. These include: logistic, Rayleigh, Weibull, extreme value, Cauchy, Laplace, and hyperbolic secant. Select one of these distributions, determine the fundamental properties including moments, modes, etc., and find and express in your own words a problem using the distribution and the characteristics of the experiment that led to this model.

2. State whether the following random variables are discrete or continuous, giving reasons for your answers in each case.

    (a) X: the number of white blood cells found in 1 cubic millimeter of blood.
    (b) X: the weight gain of a woman during the first three months of pregnancy.
    (c) X: the amount of rainfall per day in Waterloo Region.
    (d) X: the number of times that a pollinating insect visits a flower.
    (e) X: the amount of time that elapses between each visit of a pollinating insect to a flower.
    (f) X: the body length of a female nurse shark.
    (g) X: the blood pressure of a 25 year-old male professional athlete.

    In problems 3 to 10 inclusive, determine the distribution of the random variable, as well as the requested information. For example, if $f(x) = 2\exp(-2x)$, $x \geq 0$ and $f(x) = 0$ otherwise, then f is exponential. This can also be identified as a gamma variable. In each case, determine the parameters of the distribution.

3. Recall that if $f(x) \geq 0$ for all $x \in \mathbb{R}$ and $\int_{-\infty}^{\infty} f(x)dx = 1$, then f can be a probability density function (pdf).

    (a) Determine the constant k in each of the following that makes the given function f a pdf:

    (i) $f(x) = kx^2(1-x)$, $0 \leq x \leq 1$, and $f(x) = 0$ otherwise.
    (ii) $f(x) = kx(1-x)^2$, $0 \leq x \leq 1$ and $f(x) = 0$ otherwise.
    (iii) $f(x) = k\exp(-3x)$, $x \geq 0$ and $f(x) = 0$ otherwise.

    (b) For each of the pdfs in (a), determine the cdf, and use it to evaluate $P(1/2 < X \leq 3/4)$ in each case.

4. Let $f(x) = k$, $a \leq x \leq b$ and $f(x) = 0$ otherwise. Here, $k > 0$ and $a < b$. Determine (in terms of a and b) the value of k that makes f a pdf. From this, determine the mean and variance of this distribution.

5. Let $f(x) = 1/2$, $0 \leq x \leq 2$ and $f(x) = 0$ otherwise. Determine the corresponding cumulative distribution function and then graph it on the interval $[-2, 5]$. Use the cdf to evaluate $P(1/2 < X \leq 3/2)$.

6. Let $f(x) = 5\exp(-5x)$, $x \geq 0$ and $f(x) = 0$ otherwise.

    (a) Determine f's cdf, and then graph it on the interval $[-2, 5]$.

    (b) Determine $E(X)$ and $var(X)$. Hint: you will need to use integration by parts.

7. Let $g(x) = 20x^3(1-x)$, $0 \leq x \leq 1$ and $g(x) = 0$ otherwise. Determine μ_X and σ_X.

8. Let $h(x) = k/x$, $1 \leq x \leq 5$ with $h(x) = 0$ otherwise.

    (a) Determine the value of k that makes h a pdf.

    (b) If the random variable X has the pdf in (a), determine its mean and its variance.

    (c) For the random variable X in (b), use integration to determine the probability $P(2 < X < 3)$, to two decimals of accuracy.

9. Let T be a continuous random variable with pdf $f(t) = 4t^3$, $0 \leq t \leq 1$, and $f(t) = 0$ otherwise.

    (a) Determine the median of this distribution to three decimal places.

    (b) Determine the 80th percentile of this distribution.

    (c) Determine the first quartile of this distribution.

10. Let the random variable X have pdf $f(x) = 5\exp(-5x)$, $x > 0$ and $f(x) = 0$ otherwise.

    (a) Determine the median of this distribution.

    (b) Determine the first and third quartiles of this distribution.

    (c) Comment, with reasons, on the skewness of this distribution.

11. For each of the following pdfs of continuous random variables, determine the moment generating function.

    (a) Use integration by parts: $f(x) = 2x$, $0 \leq x \leq 1$ and $f(x) = 0$ otherwise.

    (b) Combine the exponents and recognize the integral is again exponential except not multiplied by the correct constant: $f(x) = 3\exp(-3x)$, $x > 0$ and $f(x) = 0$ otherwise.

12. Let X be a continuous random variable uniformly distributed over $[2, 5]$.

    (a) Write out the probability density function of X.

    (b) Use integration to determine the following probabilities:

        (i) $P(X = 3)$
        (ii) $P(1 < X \leq 4)$
        (iii) $P(X > 4)$

    (c) Determine the cumulative distribution function of X.

    (d) Use the CDF obtained in (c) to compute the probabilities in (b).

        (i) $P(X = 3)$
        (ii) $P(1 < X < 4)$
        (iii) $P(X > 4)$

13. Let X be a continuous random variable with exponential distribution given by $f(x) = 0.18\exp(-0.18x)$, $x \geq 0$, and $f(x) = 0$ otherwise.

    (a) Verify that f(x) is in fact a probability density function.

    (b) Find the cdf, and use it to determine the following probabilities:

(i) $P(X \leq 3)$

(ii) $P(X < 3)$

(iii) $P(4 \leq X < 8)$

(iv) $P(-1 < X < 2)$

(v) $P(X \geq 4)$

(c) Verify your results in part (b) by evaluating the probabilities using the cumulative distribution function.

(d) Use integration by parts to find the mean of X.

14. Let $X \sim N(0, 1)$, i.e., the standard normal. Determine the following probabilities using the table of values in the text. In each case, sketch the area under the curve that corresponds to the given probability.

    (a) $P(X \leq 1.55)$
    (b) $P(X < 1.55)$
    (c) $P(-1 \leq X < 0.75)$
    (d) $P(X \geq 0.657)$

15. Repeat Problem 14, using $X \sim N(0.5, 1.44)$ instead.

16. Find the first, second, and third quartiles of the standard normal distribution.

17. Consider a random variable X with a beta distribution with $\alpha = 3$ and $\beta = 4$.

    (a) Use integration to determine the cumulative distribution function of X.
    (b) Use the cumulative distribution function found in (a) to find:

        (i) $P(0.5 < X < 0.8)$
        (ii) $P(X \geq 0.3)$

18. The amount of rainfall during the summer in a particular region has a gamma distribution with $\alpha = 2$ and $\beta = 5/4$. Let X represent the amount of rainfall (in millimeters) in this region during the period of interest.

    (a) State the probability density function for the amount of rainfall in simplified form.
    (b) Use integration by parts to determine $P(X < 2)$ and $P(X \geq 3.5)$, and interpret the values in the context of the problem.
    (c) Determine the average amount of rainfall over this period of time.

19. Suppose that the weight of adult female manatees is normally distributed with an average weight of 210 kg and a variance of 47 kg². A marine biologist catches an adult female manatee off the coast of Cuba. Determine the probability that the weight of the manatee is:

    (a) at least 220 kg;
    (b) no more than 180 kg;
    (c) within 1.5 standard deviations of the mean. Compare this to the probability of such an event guaranteed by Chebyshev's inequality.

20. Refer to question 19. Suppose that 100 adult female manatees were caught and their weights measured. How many of the manatees caught would we expect have weights between 185 kg and 205 kg?

21. During the daylight hours in the month of June, the number of visits of a pollinating insect to a particular type of flower in an hour is Poisson, with $\lambda = 3$. Let X represent the time between visits to this type of flower.

(a) What is an appropriate distribution for X based on the material in this chapter?

(b) Determine that probability that it takes less than an hour from the most recent visit of a pollinating insect until the next one arrives at these flowers.

(c) Determine the probability that it takes 10 minutes or less between visits by such insects to these flowers.

22. Let $X \sim B(140, 0.15)$, $X = 0, 1, \ldots, 140$.

   (a) Show that the normal approximation to the binomial can be applied.

   (b) Using continuity correction factor(s) estimate the following probabilities. Keep three decimal places of accuracy in your calculations.

   (i) $P(X = 15)$
   (ii) $P(X \geq 22)$
   (iii) $P(X \leq 25)$
   (iv) $P(X < 25)$
   (v) $P(34 \leq X \leq 46)$

CHAPTER 6

PARAMETER ESTIMATION, SAMPLING DISTRIBUTIONS AND INFERENCE

Synopsis

Different methods for estimating unknown parameters in distributions, including maximum likelihood, least squares, moment estimation, and Bayesian estimation, are used to generate point estimates. The estimators are functions of sample data, and so have distributions, called sampling distributions, and where possible, these are identified. The basic problems in inference are discussed. A summary of sampling distributions for different parameters and of distributional results is provided at the end of the chapter.

6.1 INTRODUCTION

In the previous chapters, a number of families of distributions used in modeling and analyzing data in the life sciences were introduced. In particular, an array of specific discrete variables and continuous variables were considered in detail. In all cases, specific family members were characterized by certain constants, called parameters, in each family. For a given experiment, some of these values will be known and some will not: In a binomial model, the number of trials, n, is known, but often the probability π of seeing a 1 on any of the Bernoulli trials is not. In the case of a normal distribution, the mean μ or the variance σ^2 or both may be unknown. In the exponential distribution, the parameter λ may be unknown. Thus, assumed properties of the random variable of interest indicates which particular family of distributions should be considered, but the specific family member may depend on one or more unknown parameter values.

Definition 1: Let a probability distribution be determined by the pdf $f(x, \theta)$, where θ is the mean, variance, standard deviation, proportion or correlation of the population. Then we call θ the **true (population) mean, variance, standard deviation, proportion,** or **correlation** of the population.

The definition extends to other named characteristics of a population, such as the true coefficient of skewness or kurtosis, but generally, these suffice. In such cases, we are then looking for ways to estimate, say, the true mean of a population.

Definition 2: Let θ be a parameter to be estimated. A numerical value, based on data collected in an experiment, used to estimate this parameter is called a **point estimate** of the parameter. The formula for generating the estimate of θ is called the **estimator** of θ.

Note that an estimator is a function of the data. As such, it is a random variable, and hence has a distribution. In this chapter, we will consider two problems:

Problem 1: Given a population with some unknown parameters, estimate those parameters.

Problem 2: For an estimator of a parameter, determine the distribution of the estimator.

Where will the estimated value of a parameter come from? For a random variable X with pdf $f(x; \theta)$ and θ unknown, we need data from the distribution itself. This means we need to take a **sample**, i.e., collect data, of appropriate size and with particular characteristics from this distribution, and then use this information to infer a reasonable estimate of θ. This estimate will be a function of the data (and known parameters), and this function will be the **estimator**. However we decide on what the specific numerical value to use to estimate θ, there is every chance that the estimated value is not the true population value. The only way to know for certain would be to include the entire population in the sample, and that is impractical. What we need, then, is a way to quantify how certain we are that our estimate is close to the true value. That is Problem 2. This will require taking a sample of size n sufficiently large to get the necessary level of confidence, but it also means we have to take the sample in a way that allows us to develop a distribution. Typically, this means taking a simple random sample, but in more complicated problems, other sampling schemes may be better.

Definition 3: Let X be a random variable defined on a sample space of outcomes for given population of experimental units and a given experiment. Let n experimental units be selected from the population and X observed on each. We call the resulting X values a **random sample** if every experimental unit in the population has a known probability of being selected. [Some authors define random sampling as the selection of experimental units such that there is a random element to the selection.]

Example 1: Suppose we want to sample a population of adults, in which 45% are male and 55% are female. A random sample of 60 would be a selection of individuals that conform to these probabilities. We would expect about 27 males to be selected and 33 females to be selected, if every individual had the same probability of being selected. However, a specific sample may be composed of 35 males and 25 females. An alternate approach would be to randomly choose 27 males and 33 females to ensure proportionate representation of the groups, and overall, this would still be a random sample. This second approach is called **stratified random sampling**.

Definition 4: Let the random variables $X_1, \ldots, X_n$ be such that they are independently and identically distributed, each with the same distribution as the random variable X. Then we say these variables form a **simple random sample** from the distribution of X. We abbreviate "independently and identically distributed" as *iid*. The terms "simple random sample" and *iid* are used interchangeably.

A **simple random sample** is generated when each time the experiment is run, the exact same conditions hold as for the previous runs of the experiment. If we have a finite population, this means that we sample with replacement. Basically, all experimental units must have the same chance of being selected throughout the n runs of the experiment.

In the case of very large populations and relatively small sample sizes, the amount of correlation among the random variables representing the selection of the sample values is sufficiently small that it can be ignored. Thus, if we take a random sample without replacement in large populations, we can assume the variables are in fact independent and with the same distribution of the original variable. In the case of continuous random variables, the selection of any finite number of values has no impact on the distribution of the remaining values.

How is a simple random sample drawn if a population is relatively small? Each member can be assigned a number from 1 to N, where N is the population size. If N has three digits, then all numbers are of the form abc; so, for example, the first experimental unit would be assigned 001. Then a random number generator or table of random numbers is used to get numbers in blocks of three, and any such block that corresponds to a number assigned to an experimental unit causes that unit to be selected. If $N = 876$, say, and a block of the form 923 is generated from the random number generator or table, it is ignored, and the next block of three is selected. One can then choose the sample of given size with or without replacement.

The procedure above works well if all experimental units can be assigned a number. Suppose, instead, a simple random sample of pine trees from a large forest is required. It would be impractical to first number all pine trees in the forest. One approach would be to determine the boundaries of the forest, and then generate random latitude and longitude values in the forest boundaries, and then randomly select a set of such numbers, and then select the tree closest to the given coordinates. If a simple random sample of cod is required, again, it is not sensible to think of numbering all of them first. Instead, a selection procedure that minimizes selection bias is used. Some cod captured at different times of the day and from different locations in the sea could be used to form the sample. Essentially, the sampling procedure itself must be fully considered before the sample is actually taken. Much of the statistical analysis done in experiments is based on assuming the sample is a simple random one, or conforms to some other scheme that generates random samples. If this is violated, the analysis may not be valid.

Throughout the remainder of this book, we will be assuming the samples are simple random samples, unless otherwise specified. Getting such samples requires great attention to the details of the experimental procedure to try to keep sampling bias to a minimum. Such bias occurs when experimental units with particular characteristics are chosen more frequently than they should be. This occurs if, for example, some segments of the population are harder to reach than others, and the ones easier to get to are deliberately chosen more often. One cautionary note: Just because it appears there is sampling bias doesn't mean there is. Suppose in a population of 5,000 undergraduates at a particular university, a sample of 50 are chosen and they are all males, even though the population is equally divided along gender lines. This does not immediately say there was a problem in the design. It simply means a rare event has happened, and rare events do not have 0 probability of occurring. Should such an event occur, it is a good idea to check the procedure put in place to generate the sample to determine if some unknown factor has caused this result.

Sampling without replacement has some advantages. In particular, it means you do not repeat information. Also, it is not necessary that independent and identically distributed variables be created to generate important theoretical results. It is enough that the variables satisfy a condition called **exchangeability**. This will not be explored further, but interested readers can find information on exchangeability in many sources.

There are different approaches to generating the estimator and the estimate of a parameter, including maximum likelihood, moment estimation, and Bayesian estimation. In this text, we concentrate on the maximum likelihood approach, and discuss briefly some aspects of moment and Bayesian estimation.

6.1.1 Principles Used in Parameter Estimation

Let the parameter to be estimated be denoted θ. Its estimator is a function of the data and is often denoted $\tilde{\theta}$ in general. Thus, we can express $\tilde{\theta}$ as $\tilde{\theta}(X_1, X_2, \ldots, X_n)$, that is, it is a function of the random sample $X_1, X_2, \ldots, X_n$ from the distribution of X. (This explicit dependence on the data values is not written as a general rule, but it is important to understand this connection to the data.) A **realization** of the sample is the actual observed numbers $X_1 = x_1, X_2 = x_2, \ldots, X_n = x_n$, and the numbers $x_1, \ldots, x_n$ substituted in the formula for $\tilde{\theta}$ provides the estimate of θ for the actual sample.

There can be many different estimators for the same parameter, but some have properties that make them better suited for analysis than others. In general, the data is assumed to form a simple random sample, which makes the estimator itself a random variable on the space of n-tuples $(X_1, X_2, \ldots, X_n)$. It is this fact that allows us to analyze just how confident we are that our specific point estimate is close to the true population value θ.

Unless otherwise specified, our samples will be simple random samples from the distribution of X, the random variable of interest. Each X_i is a random variable in its own right, and in the case of simple random samples, each X_i has exactly the same distribution as the original variable X. This holds if sampling is with replacement or we are sampling from a continuous distribution. If sampling is from a finite population, and without replacement, as long as the population is much larger than the sample size, we can assume the sample is a simple random one.

Definition 5: Let X be a random variable and let $X_1, X_2, \ldots, X_n$ represent a simple random sample from X's distribution. We call X's distribution the **parent distribution**. In conjunction with Definition 4, we say $X_1, \ldots, X_n$ are independently and identically distributed or *iid* as X.

Other important properties of estimators in considering which to use include:

Definition 6: (Properties of Estimators).

1. **Unbiased**: An estimator is **unbiased** for a parameter if its expected value is the value of the parameter. Otherwise, it is called **biased**.
2. **Consistent**: An estimator is **consistent** if its variance decreases to 0 as the sample size increases to infinity.
3. **Efficient**: An estimate is **efficient** if it has the least variance of all estimators under consideration.
4. **Sufficient**: An estimator is **sufficient** for a parameter if it exhausts all the information about the parameter in the given sample.
5. **Tractability**: An estimator is **tractable** if its distribution can be related to well-known distributions (for which procedures are already available) or is easily analyzed mathematically.

Let θ be the parameter to be estimated, and let $\tilde{\theta}$ be an estimator of θ. As noted earler, $\tilde{\theta}$ has a distribution, so we can discuss its mean, variance, and other moments and properties. Some comments and observations about the properties introduced above include:

1. The estimator is **unbiased** for θ means that $E(\tilde{\theta}) = \theta$. If we repeat the sampling process a large number of times for the same fixed sample size, and ensure the experimental conditions remain constant, then the average of all the estimated values of θ using $\tilde{\theta}$ should be θ. Although the actual value of $\tilde{\theta}$ for a given sample is not likely to be θ exactly, the average value of $\tilde{\theta}$ should be.
2. For any given parameter to be estimated, there can be many unbiased estimators. Each estimator has its own distribution, making direct comparison difficult. However, we can compare variances: As seen from either Chebyshev's Inequality or the Empirical Rule, the smaller the variance, the greater the concentration of the distribution about the mean. Thus, if we consider estimators that are unbiased for θ, then the one with the smallest variance would mean its values are more frequently close to θ than the others, at least from the application of Chebyshev's Inequality. It would seem reasonable that the larger the sample, the better the information about the distribution of the estimator. Generally, the distribution of the estimator will depend on n, the sample size, and we would expect more of the values of the estimator to be closer to θ as n gets larger. A **consistent** estimator is one for which the variance goes to 0 as the sample size tends to infinity, meaning for larger samples we get more precise information about θ.
3. In general, we will concentrate on unbiased estimators. Again, there can be several different unbiased estimators, each dependent on the sample size n. The preferred choice would be the estimator with the least variance, which is then said to be **efficient**, because for the same information, the estimate should be closer to the population value θ on average, and hence, gives the best information for the same data.
4. **Sufficiency** is a more technical term than the previous three. Basically, an estimator is sufficient for θ if the conditional distribution of X, given $\tilde{\theta} = u$, is free of θ. This means that once we know $\tilde{\theta}$, there is no more information about θ in the sample. It is also the case that any monotone function of a sufficient statistic is again a sufficient statistic.
5. There can be many different estimators with good properties, but if the distribution is difficult to work with, it may not be as useful as an estimator with a well-known distribution. Many distributions are not explicitly defined, and so finding moments may require computer simulations or other algorithms. An explicit form of a pdf makes it easier to compute characteristic values and to verify distribution assumptions are valid. A **tractable** distribution is highly desirable, but computer power has made it less necessary for analysis.

It would be best to have an estimator of θ that satisfies all of these properties, but it isn't always possible. We need to choose the specific estimator that best reflects these considerations. We also need a method to determine the actual estimator. In the next section, the principle of **maximum likelihood** and its consequences is studied. For the more advanced reader, the books by Kendall, Stewart and Ord (1987), Hogg, McKean and Craig (2005) and Larsen and Marx (2001) are recommended reading.

6.2 MAXIMUM LIKELIHOOD ESTIMATION

Let the random variable X have pdf $f(x, \theta)$, where θ is an unknown parameter, and f is the pdf of a general family of distributions determined up to the value of θ. We will also consider the problem of estimating two or more parameters, once the general procedure for one unknown parameter has been considered. Throughout, we assume $X_1, X_2, \ldots, X_n$ is a simple random sample from the parent distribution, or equivalently, that the variables X_i, $1 \leq i \leq n$ are *iid* with common pdf $f(x, \theta)$.

Definition 7: Let $X_1, X_2, \ldots, X_n$ be *iid* with common pdf $f(x, \theta)$. Let $x_1, x_2, \ldots, x_n$ be the observed values of $X_1, X_2, \ldots, X_n$, respectively. We call $x_1, x_2, \ldots, x_n$ a **realization** of the random variables $X_1, X_2, \ldots, X_n$. The **likelihood function** L of θ is defined to be:

$$L \equiv L(\theta) \equiv \prod_{i=1}^{n} f(x_i, \theta).$$

Since the random variables X_i, $1 \leq i \leq n$, are independent, the likelihood function looks precisely the same as the joint pdf of these random variables. However, L is a function of θ alone, since the values x_i are taken as known, and hence constant.

Definition 8: (The Principle of Maximum Likelihood). Let X_i, $1 \leq i \leq n$, be *iid* random variables, with common pdf $f(x, \theta)$, and such that parameter θ is unknown. Let $\theta \in \Omega$, where Ω is a subset of real numbers, usually an interval. Ω is called the **parameter space**. If there is a function $\tilde{\theta}$ of the random variables such that when it is evaluated at $x_1, x_2, \ldots, x_n$, $L(\tilde{\theta}) \geq L(\theta)$ for all $\theta \in \Omega$, then we call $\tilde{\theta}$ a **maximum likelihood estimator**, or **MLE**, of θ. The **Principle of Maximum Likelihood** is: Select as estimator of θ a maximum likelihood estimator. The **method of maximum likelihood** is the process of determining MLEs.

It is possible to have a set of independently distributed random variables with pdfs that are different, and the method of maximum likelihood could still be applied. If X_i has pdf $f_i(x_i, \theta)$, $1 \leq i \leq n$, and the variables are independent, then the likelihood function would take the form $L(\theta) = \prod_{i=1}^{n} f_i(x_i, \theta)$ and the MLE of θ would be the value that maximizes L. The main difference is the level of computational complexity. Since we are concerned with simple random samples, and hence variables that are *iid*, we will not consider this more general situation in this text.

It is frequently the case that a likelihood function has a unique maximum likelihood estimator of θ, and in this case, the Principle of Maximum Likelihood is: Select as estimator of θ the maximum likelihood estimator. The problem of finding the MLE is most often, but not always, a problem in differential calculus.

A way to interpret this principle is as follows: If we think of $\prod_{i=1}^{n} f(x_i, \theta)$ as the jpdf of $X_1, X_2, \ldots, X_n$, then by multiplying by $dx_1 dx_2 \cdots dx_n$, the result can be thought of as the probability of a sample falling in a small region of n dimensional space, precisely in the same way we can think of $f(x, \theta) dx$ as the probability that X falls in a small interval around x. The maximum likelihood principle then means to choose our estimate of θ that maximizes this probability. Thus, the MLE of θ maximizes the chances of seeing the specific sample.

Definition 9: (The Principle of Maximum Likelihood for two or more unknown parameters). Let X_i, $1 \leq i \leq n$, be *iid* random variables, with common pdf $f(x, \boldsymbol{\theta})$, and such that the vector of parameters $\boldsymbol{\theta} = (\theta_1, \theta_2, \ldots, \theta_k)$, $k \geq 2$, consists of k unknown parameters. Let $\boldsymbol{\theta} \in \Omega$, where Ω is a subset of $\mathbb{R}^k$. Ω is called the **parameter space** of the vector of unknown parameters. If there is a vector function $\tilde{\boldsymbol{\theta}}$ of the random variables such that when it is evaluated at $x_1, x_2, \ldots, x_n$, $L(\tilde{\boldsymbol{\theta}}) \geq L(\boldsymbol{\theta})$ for all $\boldsymbol{\theta} \in \Omega$, then we call $\tilde{\boldsymbol{\theta}}$ a **maximum likelihood estimator**, or **MLE**, of $\boldsymbol{\theta}$. The **Principle of Maximum Likelihood** is: Select as estimator of $\boldsymbol{\theta}$ a maximum likelihood estimator.

A vector function of the data is just a vector where each entry is a function of the data. In the case of two or more unknowns, partial derivatives are often, but not always, used in optimizing functions of more than one variable.

The method of maximum likelihood is dependent on knowing the form of the pdf. Information concerning a random variable from past experiments may suggest a particular form, or the experimental procedure may lead to a specific family of distributions. In this chapter, we will assume the form of the pdf is known, but supporting evidence should be presented in any analysis. For example, suppose a random variable is assumed to be normally distributed. One check would be to see if the Empirical Rule holds for the sample drawn: within one standard deviation of the mean should be about 67% of the data; within two standard deviations about 95%; and within three standard deviations, we should see nearly all, or in excess of 99%, of the data. If the problem is one involving counting, then it must be clear that the experimental conditions give rise to the binomial, or Poisson, or negative binomial, etc., distribution.

The principle of maximum likelihood is highly dependent on having a simple random sample so that the joint pdf is just the product of the individual pdfs. If the variables are not *iid*, the procedure is doubtful. In designing an experiment, it is very important to verify as best one can that the appropriate experimental procedures are carried out properly to ensure the distributional results are valid.

Example 2: Let the random variable X be Bernoulli, with unknown parameter π. Although the distribution is discrete, the unknown parameter can take any value in $[0,1] = \Omega$. Let $X_1, X_2, \ldots, X_n$ be a simple random sample from X's distribution, or, in other words, let them be *iid* $B(1, \pi)$, so the common pdf is $f(x) = \binom{1}{x}\pi^x(1-\pi)^{1-x}$, with $x = 0, 1$. The likelihood function will be:

$$L(\pi) = \prod_{i=1}^{n}\binom{1}{x_i}\pi^{x_i}(1-\pi)^{1-x_i} = \pi^{\sum x_i}(1-\pi)^{n-\sum x_i}.$$

Our goal is to find a function $\tilde{\theta}$ of $x_1, \ldots, x_n$ that always produces a maximum of L, if such exists. To do this, we differentiate L with respect to π (in this case) and set this derivative to 0. We solve the resulting equation for θ, and call this value $\tilde{\theta}$. In general, there can be more than one critical number, and sometimes, we have to consider values that make L undefined, within the parameter space. The value or values should be proven to yield a maximum by considering second derivatives and comparing values to values at endpoints, just as typically done in calculus.

To solve this problem, we need to use the product rule. The resulting equation is not difficult to solve, but is messier than necessary. The standard alternative is to recognize that if $f(x)$ is maximized at x_0, then so is any monotone increasing differentiable function of $f(x)$. In particular, if we take natural logarithms (any log base can be used) of L, then L is maximized exactly where $\ln(L)$ is, and vice versa. We call $\ln(L)$ the **log likelihood function**, and the derivative of the log likelihood set to 0 is the likelihood equation. Thus:

$$\ln(L) = \left(\sum x_i\right)\ln(\pi) + \left(n - \sum x_i\right)\ln(1-\pi), \text{ so that } \frac{d\ln(L)}{d\pi} = \frac{\sum x_i}{\pi} - \frac{n - \sum x_i}{1 - \pi}.$$

The likelihood equation is

$$\frac{\sum x_i}{\pi} - \frac{n - \sum x_i}{1 - \pi} = 0.$$

The solution to this equation is denoted $\hat{\pi}$, and by straightforward algebra (left as an exercise) $\hat{\pi} = \sum_{i=1}^{n} x_i/n$.

Our solution is denoted $\hat{\pi}$ to distinguish it from other estimators, and we will use the hat notation for the maximum likelihood estimator, when it exists. Its value is always in the parameter space. If we replace each x_i by the corresponding random variable X_i, the result is a random variable written as $\hat{\pi} = \hat{\pi}(X_1, \ldots, X_n)$; it can be shown that this random variable is unbiased, consistent, and efficient for θ. For example:

$$E(\hat{\theta}) = E\left(\sum_{i=1}^{n} X_i/n\right) = \sum_{i=1}^{n} E(X_i/n) = nE(X)/n = \pi$$

while

$$\text{var}(\hat{\theta}) = \frac{1}{n^2} \sum_{i=1}^{n} \text{var}(X_i) = \frac{1}{n^2} n \, \text{var}(X) = \frac{\pi(1-\pi)}{n}.$$

The estimator is unbiased, and it is clearly consistent. It is also true that among all unbiased estimators, it has the minimum variance possible, and so it is efficient.

It is also important to note that our estimator is a sum of Bernoulli random variables, divided by n. This means that $n\hat{\pi}$ must have a binomial distribution $B(n, \pi)$. We can use this fact as an alternate way to derive the formula for the variance. Since $n\hat{\pi} \sim B(n, \pi)$, then $\text{var}(n\hat{\pi}) = n\pi(1-\pi)$, as this is the variance of a binomial variable. Because n is a constant, $\text{var}(n\hat{\pi}) = n^2 \text{var}(\hat{\pi})$, and dividing by n^2 yields the result. This illustrates the value of knowing relationships between variables.

The fact that our estimator in the previous example has the least variance of all unbiased estimators is true and is based on:

Theorem 1: The Rao-Cramér Inequality for Unbiased Estimators. Under regularity conditions, the least variance an unbiased estimator based on random samples of size n can have is:

$$1/(nE((\partial \ln f(X, \theta)/\partial \theta)^2)) \equiv -1/(nE(\partial^2 \ln f(X, \theta)/\partial \theta^2)).$$

The proof is beyond the scope of this text, and the interested reader is referred to Hogg, Craig and Allen (2005), and Kendall, Stuart and Ord (1987).

In Example 2, $\partial^2 \ln f(X, \theta)/\partial \theta^2 = -X/\pi^2 - (1-X)/(1-\pi)^2$, which has expected value $-(1/\pi + 1/(1-\pi)) = -1/(\pi(1-\pi))$. The Rao-Cramér lower bound is $\pi(1-\pi)/n$, which is the variance of the MLE.

This example also illustrates the importance of using the logarithm of the likelihood function, rather than the likelihood function itself, in searching for a maximum. It works so well because the likelihood function is a product and the logarithm of a product is a sum of logarithms.

Summary of the Method of Maximum Likelihood: The Bernoulli example illustrates the procedure to use in calculating the MLE of an unknown parameter:

1. Form the likelihood function $L(\theta) = L = \prod_{i=1}^{n} f(x_i, \theta)$.
2. Form the log likelihood function $\ln L = \sum_{i=1}^{n} \ln f(x_i, \theta)$, and simplify as much as possible.
3. Differentiate $\ln L$ with respect to θ.
4. Set $d \ln L/d\theta$ to 0, and call the resulting equation the likelihood equation.
5. Solve the likelihood equation for θ, and denote the result by $\hat{\theta}$.
6. Verify $\hat{\theta}$ does yield a maximum.

Once the MLE of θ is determined, properties such as unbiasedness, consistency, and efficiency can be included in the analysis.

Example 3: Let the random variable $X \sim N(\mu, \sigma^2)$. Determine the MLE(s) in each of the following cases: (i) μ is unknown, but σ^2 is known; (ii) μ is known but σ^2 is unknown; and (iii) both μ and σ^2 are both unknown.

Solution: The pdf of a normal is:

$$f(x) = \frac{1}{\sigma\sqrt{2\pi}} \exp\left(-\frac{(x-\mu)^2}{2\sigma^2}\right)$$

so the likelihood function of a random sample of size n is:

$$L = \prod_{i=1}^{n} \frac{1}{\sigma\sqrt{2\pi}} \exp\left(-\frac{(x_i - \mu)^2}{2\sigma^2}\right) = (2\pi)^{-n/2} \sigma^{-n} \exp\left(-\sum_{i=1}^{n} \frac{(x_i - \mu)^2}{2\sigma^2}\right)$$

regardless of the parameters that are known or unknown. The log likelihood function is:

$$\ln L = \text{Const.} - n \ln \sigma - \sum_{i=1}^{n} \frac{(x_i - \mu)^2}{2\sigma^2}.$$

All summands not involving μ and/or σ are collected into the term "Const."

(i) On differentiating with respect to μ, we have:

$$\frac{d \ln L}{d\mu} = \frac{2}{2\sigma^2} \sum_{i=1}^{n} (x_i - \mu) = \frac{1}{\sigma^2} \sum_{i=1}^{n} (x_i - \mu).$$

Setting this to 0 and simplifying leads to:

$$\sum_{i=1}^{n} x_i - n\mu = 0, \text{ so that } \hat{\mu} = \frac{1}{n} \sum_{i=1}^{n} x_i = \bar{x},$$

where $\bar{x}$ is the sample mean, $\sum_{i=1}^{n} x_i / n$.

(ii) Before differentiating $\ln L$ with respect to σ^2, rewrite the log likelihood function as:

$$\ln L = \text{Const.} - \frac{n}{2} \ln \sigma^2 - \sum_{i=1}^{n} \frac{(x_i - \mu)^2}{2\sigma^2}.$$

This makes the differentiation with respect to σ^2 easier. Now the derivative with respect to σ^2 yields:

$$d \ln L / d\sigma^2 = -\frac{n}{2\sigma^2} + \frac{1}{2\sigma^4} \sum_{i=1}^{n} (x_i - \mu)^2.$$

On setting this to 0 and simplifying,

$$\widehat{\sigma^2} = \frac{1}{n} \sum_{i=1}^{n} (x_i - \mu)^2.$$

Note carefully that the differentiation was with respect to σ^2, not σ.

(iii) Critical points for functions of two or more variables are found by partially differentiating with respect to the individual variables, setting these to 0, and then simultaneously solving the resulting equations for the variables (if possible). There can be several solutions, but we will concentrate on situations such as this where there will be a unique solution. We will also assume this unique solution produces a maximum without further verification.

Thus,

$$\frac{\partial \ln L}{\partial \mu} = \frac{1}{\sigma^2} \sum_{i=1}^{n} (x_i - \mu) \text{ and } \frac{\partial \ln L}{\partial \sigma^2} = -\frac{n}{2\sigma^2} + \frac{1}{2\sigma^4} \sum_{i=1}^{n} (x_i - \mu)^2.$$

From the first equation: we again find $\hat{\mu} = \bar{x}$ while in the second, $\widehat{\sigma^2} = \sum_{i=1}^{n} (x_i - \bar{x})^2 / n$.

Chapter 6 — Parameter Estimation, Sampling Distributions and Inference

Since a simple random sample from the normal distribution in Example 3 consists of n iid random variables with the same distribution as $X \sim N(\mu, \sigma^2)$, it is easy to show in cases (i) and (ii) that the resulting MLEs are unbiased. Each has variance that is the minimum possible for unbiased estimators, and in each case, the variance tends to 0 as n tends to ∞.

In the third case, when both μ and σ^2 must be estimated, the results for $\hat{\mu}$ are unchanged, but the estimate of the variance is different than in case (ii). In particular, we know that $E(\widehat{\sigma^2}) = (n-1)\sigma^2/n$, which means this is a biased estimator. Because the bias is a multiplicative constant, we can readily create an unbiased estimate, known as the sample variance: $s^2 = \sum_{i=1}^{n}(x_i - \bar{x})^2/(n-1)$. When we replace x_i by X_i and $\bar{x}$ with $\bar{X}$, the resulting random variable is denoted S^2.

Theorem 2: The sample variance $S^2 = \sum_{i=1}^{n}(X_i - \bar{X})^2/(n-1)$ is unbiased for the population variance σ^2, where $X_1, \ldots, X_n$ are *iid* with parent distribution having finite variance σ^2 and mean μ.

Proof: It is important to observe that this result is always true and is not based on any assumptions about the distribution involved, except that it has finite mean and variance. The proof is based on rearranging certain sums of squares, recognizing a term is actually zero all the time, and using $var(X) = E((X - \mu)^2) = \sigma^2$ and $var(\bar{X}) = E((\bar{X} - \mu)^2) = \sigma^2/n$. The variance of the sample mean is found by using the fact that the sample mean is just a sum of *iid* random variables.

Consider the sum $\sum_{i=1}^{n}(X_i - \mu)^2$. If we add in and subtract out the same value in an expression, the value of the expression doesn't change, and so:

$$\sum_{i=1}^{n}(X_i - \mu)^2 = \sum_{i=1}^{n}(X_i - \bar{X} + \bar{X} - \mu)^2 = \sum_{i=1}^{n}((X_i - \bar{X}) + (\bar{X} - \mu))^2.$$

We square out the terms in the last sum and then rewrite the result as three sums:

$$\sum_{i=1}^{n}((X_i - \bar{X}) + (\bar{X} - \mu))^2 = \sum_{i=1}^{n}((X_i - \bar{X})^2 + 2(X_i - \bar{X})(\bar{X} - \mu) + (\bar{X} - \mu)^2)$$

$$= \sum_{i=1}^{n}(X_i - \bar{X})^2 + 2(\bar{X} - \mu)\sum_{i=1}^{n}(X_i - \bar{X}) + n(\bar{X} - \mu)^2).$$

As an exercise, show that $\sum_{i=1}^{n}(X_i - \bar{X}) = 0$. This result is simply an identity about numbers and is not a statistical or probability result. Assuming this is true,

$$\sum_{i=1}^{n}(X_i - \mu)^2 = \sum_{i=1}^{n}(X_i - \bar{X})^2 + n(\bar{X} - \mu)^2).$$

On taking expectations, it follows that:

$$n\sigma^2 = E\left(\sum_{i=1}^{n}(X_i - \bar{X})^2\right) + n\sigma^2/n = E\left(\sum_{i=1}^{n}(X_i - \bar{X})^2\right) + \sigma^2.$$

On rearranging, we have:

$$E\left(\sum_{i=1}^{n}(X_i - \bar{X})^2\right) = (n-1)\sigma^2$$

or $E(S^2) = \sigma^2$.

Definition 10: Let the unbiased estimator U_n of θ be a function of random samples of size n from some parent distribution with unknown parameter θ, and let V_n be the Rao-Cramér lower bound for unbiased estimators of that parameter. If:

$$\lim_{n \to \infty} var(U_n)/V_n = 1$$

then we say U_n is **asymptotically efficient**.

In Example 2, the MLE of σ^2 was in fact biased, but an unbiased estimator could be created that has the same asymptotic variance as the MLE. Since unbiasedness is a key feature of a good estimator, and little is lost by making the MLE unbiased, then this related variable is used instead of the MLE.

Example 4: Let $Y \sim B(n, \pi)$, where $\pi \in [0, 1]$. Find the maximum likelihood estimator of π.

Solution: This will generalize our results for the Bernoulli in Example 1, since the Bernoulli distribution is just $B(1, \pi)$. For a general binomial problem of estimating π, we repeat the binomial experiment of n Bernoulli trials m times. Thus, the variables $X_1, \ldots, X_m$ are assumed iid with parent distribution $B(n, \pi)$. It is important to keep the two numbers m and n straight: n is the number of Bernoulli trials that lead to random variable X, and m is the number of complete replications of the binomial experiment with n trials. The likelihood function for a realization of $X_1, \ldots, X_m$ is:

$$L = \prod_{j=1}^{m} \binom{n}{x_j} \pi^{x_j}(1-\pi)^{n-x_j}.$$

The log likelihood function takes the form:

$$\ln L = \text{Const.} + \ln \pi \sum_{j=1}^{m} x_j + \ln(1-\pi) \sum_{j=1}^{m}(n - x_j).$$

Setting the derivative with respect to π equal to 0 and solving, the MLE for π yields:

$$\hat{\pi} = \frac{1}{mn} \sum_{j=1}^{m} x_j.$$

Another way to consider the estimator is as follows: Each random variable X_j is $B(n, \pi)$ and can be viewed as the sum of n independently and identically distributed Bernoullis, and so we have $X_j = \sum_{i=1}^{n} Y_{ij}$, where the Y_{ij}s for fixed j and $1 \le i \le n$ are the iid Bernoullis making up X_j. The random variables X_j are independently and identically distributed, so $\sum_{j=1}^{m} X_j = \sum_{j=1}^{m} \sum_{i=1}^{n} Y_{ij}$ is the sum of mn iid Bernoullis. This is a random sample of size mn from the parent distribution $B(1, \pi)$, and the MLE of π in this setup is exactly the MLE of π derived for the binomial experiment.

It is easy to show that $\hat{\pi}$ is unbiased, and it attains the Rao-Cramér lower bound, so it is efficient. The variance is $\pi(1-\pi)/(mn)$, and this establishes it is consistent.

Example 5: Let $X \sim Beta(\alpha, \beta)$, that is, let the pdf of X be given by $f(x) = x^{\alpha-1}(1-x)^{\beta-1}/B(\alpha, \beta)$ for $0 \le x \le 1$ and $f(x) = 0$ otherwise. The expression:

$$B(\alpha, \beta) = \Gamma(\alpha)\Gamma(\beta)/\Gamma(\alpha + \beta)$$

is called the beta function. Recall that $\Gamma(u) = \int_0^\infty x^{u-1} \exp(-x)dx$ and generalizes factorials.

Because the beta function is defined in terms of gamma functions, the process of maximum likelihood when both parameters are unknown is non-trivial. The derivative of the gamma function is not just a power or other simple function. However, there are some cases that do yield reasonable values. For example, suppose α is unknown

Chapter 6 — Parameter Estimation, Sampling Distributions and Inference

and $\beta = 2$. As usual, we take a random sample of size n from $Beta(\alpha, 2)$ and denote the realization as $x_1, \ldots, x_n$, and hence, the likelihood function of α is:

$$L = L(\alpha) = B(\alpha, 2)^{-n} \prod_{i=1}^{n} x_i^{\alpha-1}(1-x_i) = (\alpha+1)^n \alpha^n \prod_{i=1}^{n} x_i^{\alpha-1}(1-x_i).$$

In the simplification of the formula for L, we used:

$$B(\alpha, 2) = \Gamma(\alpha)\Gamma(2)/\Gamma(\alpha+2) = \Gamma(\alpha)\Gamma(2)/((\alpha+1)\alpha\Gamma(\alpha)) = 1/((\alpha+1)\alpha).$$

The log likelihood function is:

$$\ln L = n \ln(\alpha+1) + n\alpha + (\alpha-1) \sum_{i=1}^{n} \ln x_i + \sum_{i=1}^{n} \ln(1-x_i)$$

and so:

$$d \ln L / d\alpha = \frac{n}{\alpha+1} + \frac{n}{\alpha} + \sum_{i=1}^{n} \ln x_i.$$

The likelihood equation results from setting $d \ln L / d\alpha = 0$, and the MLE of α is the solution of this equation and is denoted $\hat{\alpha}$. In this case,

$$\frac{n}{\hat{\alpha}+1} + \frac{n}{\hat{\alpha}} + \sum_{i=1}^{n} \ln x_i = 0.$$

To simplify matters, let $Q = -\sum_{i=1}^{n} \ln x_i / n$. Then the maximum likelihood estimate of α is:

$$\hat{\alpha} = (2 - Q + \sqrt{Q^2 + 4})/(2Q).$$

This follows from rearranging the likelihood equation as $Qx^2 + (Q-2)x - 1 = 0$, applying the quadratic formula, and eliminating the extraneous solution.

For example, the following values are a random sample of size 20 from $Beta(\alpha, 2)$:

0.718 0.584 0.793 0.737 0.756 0.704 0.885 0.619 0.793 0.714
0.433 0.706 0.746 0.683 0.560 0.840 0.674 0.590 0.521 0.523

The value of Q is 0.402, and hence, the MLE of α is 4.521. The actual population value of α is 4.

Example 5 illustrates that the solution of the likelihood equation(s) may be non-trivial and numerical methods such as the Newton-Raphson method for finding roots of equations may be necessary. In this case, the likelihood equation can be solved by converting it into a quadratic.

Example 6: Let X have a log normal distribution, $LN(\mu, \sigma^2)$. The relationship with the normal makes finding the maximum likelihood estimates of the parameters μ and σ^2 straightforward. In effect, we can view $\ln X$ as a normal variable, and so the MLEs must be:

$$\hat{\mu} = \frac{1}{n} \sum \ln x_i, \quad \text{and} \quad \hat{\sigma^2} = \frac{1}{n} \sum (\ln x_i - \hat{\mu})^2.$$

Remember that these are not estimates of the mean and variance of X, but of the specified parameters. Recall that for the log normal,

$$E(X) = \exp(\mu + \sigma^2) \quad \text{and} \quad var(X) = (\exp(\sigma^2) - 1) \exp(2\mu + \sigma^2).$$

Example 7: Let the random variable X denote the length of time, in months, from diagnosis of a fatal disease until death. Past experience with the disease suggests X will follow a log normal distribution. Twenty patients are followed from diagnosis until death, with the following results:

12.17 335.32 240.53 17.67 70.30 279.62 13.06 36.10 58.44 131.91
47.32 61.79 109.22 53.36 21.90 40.90 25.27 469.45 41.04 49.88

For this data, determine the maximum likelihood estimates of the parameters μ and σ^2.

Solution: Summary values include: $\sum \ln x_i = 82.115$ and $\sum (\ln x_i)^2 = 358.42$. From these values, we have $\hat{\mu} = 4.106$ and $\hat{\sigma}^2 = 1.063$. This example was computer generated from the LN(4,1.21) distribution.

Example 8: Consider the problem of estimating an onset value μ in an exponential distribution. Let the random variable X have pdf given by:

$$f(x) = \begin{cases} \exp(-(x-\mu)) & x > \mu \\ 0 & \text{otherwise} \end{cases}.$$

Such a model may reflect experimental conditions in which there is a time lag from the start of the experiment (administering a drug, for example) and when the random variable X is measurable. What is the maximum likelihood estimator of μ?

Solution: We proceed as in all the other examples, and so the likelihood function and log likelihood function of a random sample of size n from X's distribution are:

$$L = \prod_{i=1}^{n} \exp(-(x_i - \mu)), \text{ and } \ln L = -\sum_{i=1}^{n}(x_i - \mu).$$

Now

$$d\ln L/d\mu = n$$

which is strictly positive, and does not depend on μ and so the usual procedure of setting this derivative to 0 and solving will not work. Does this mean the method fails? No — it means we maximize L by different considerations.

Since $d\ln L/d\mu = n$, then $\ln L$ is monotone increasing, and then so is L. That means we make L as large as possible by taking μ as large as possible. But μ is an onset value, meaning μ has to be less than or equal to any value of X we observe. Thus, for the given sample, $\mu \leq \min\{x_1, \ldots, x_n\}$. We call the minimum of a sample the first-order statistic and denote it $x_{(1)}$, and so the value $\hat{\mu}$ that maximizes L based on the sample and the restriction (that μ is an onset value) is $\hat{\mu} = x_{(1)}$.

Example 8 illustrates that important problems require methods other than finding solutions to likelihood equations. Another problem that has not been addressed fully is the fact that sometimes the values of the variable X depend on the parameter being estimated. We will not examine these further, but the interested reader can find relevant information about this problem in Kendall, Stuart, and Ord (1987).

6.2.1 Moment Estimation

Parameter estimation by moments is the process of equating non-central population moments, which are expressed in terms of the unknown parameters, to non-central sample moments. If possible, the resulting equations are solved to generate the individual parameter estimates. More formally:

Let $\theta_1, \ldots, \theta_k$ be the unknown parameters, and let μ'_j, $1 \leq j \leq k$ be the first k non-central moments of the distribution. For a sample of size n, let $\tilde{\mu}'_j$, $1 \leq j \leq k$, be the first k non-central moments of the sample, that is, $\tilde{\mu}'_j = \sum_{i=1}^{n} X_i^j/n$. Set $\mu'_j = \tilde{\mu}'_j$, and solve for the θ_js.

In the case of the normal distribution, $\mu'_1 = \mu$ and $\mu'_2 = \sigma^2 + \mu^2$. Thus, we set $\mu = \bar{X}$ and $\sigma^2 + \mu^2 = \sum_{i=1}^{n} X_i^2/n$, yielding $m_1 = \bar{X}$ and $m_2 = \sum_{i=1}^{n}(X_i - \bar{X})^2/n$ as the moment estimates of μ and σ^2, respectively.

Let $W \sim \beta(\alpha, \beta)$. Then we know $E(W) = \alpha\beta$ and $E(W^2) = \alpha(\alpha + 1)\beta^2$. Equating these to the first two non-central sample moments and solving (using the same notation as above), we find $m_1 = \bar{X}^2/(\sum_{i=1}^{n}(X_i - \bar{X})^2/n)$ and $m_2 = (\sum_{i=1}^{n}(X_i - \bar{X})^2/n)/\bar{X}$ as the moment estimates of α and β, respectively.

As mentioned earlier, MLE for the β distribution when both parameters are unknown will be messy. However, there is no guarantee the moment estimates actually fall in the parameter space (it is guaranteed for MLE). Complicated MLEs use computer methods and we often use the moment estimates as a starting point in an iterative numerical procedure to generate the MLEs. It is also clear that the estimators from moments can be very complicated and determining distributional results is non-trivial. Some researchers use moment estimators, depending on the distribution and assumptions made, and so they are presented here to help with the interpretation of published results.

6.3 LEAST SQUARES ESTIMATION

Estimation of parameters by the method of Least Squares, due to Carl Friedrich Gauss, is perhaps the oldest procedure still in use. It is used primarily to estimate parameters in a linear model, and it will be revisited in Linear Regression. The general procedure is as follows:

Least Squares Estimation Procedure: Let Y be a random variable with mean μ, and assume $Y = \mu + \epsilon$, where ϵ is a random variable with mean 0 and finite variance σ^2. In other words, ϵ is the random variable that represents the difference between our observed value of Y and a constant μ. Let Y_i, $1 \leq i \leq n$, be a random sample from Y's distribution and $y_1, \ldots, y_n$ a realization of the random sample. Then $Y_i - \mu = \epsilon_i$ is the error or residual after "fitting" the mean. We then form $L = \sum_{i=1}^{n}(y_i - \mu)^2$ and call this the sum of the squared errors or **error sum of squares**. The main reasons for the square is that we don't care if the error value is positive or negative, and also, we can think of this as a generalization of the Pythagorean distance. We now choose our estimate of μ so that L is minimized. From basic calculus, we know that we first differentiate L with respect to μ, determine any critical values, and then verify this produces a minimum. Specifically:

$$\frac{d}{d\mu}L = -2\sum_{i=1}^{n}(y_i - \mu) = -2[\sum_{i=1}^{n}y_i - n\mu].$$

On setting this derivative to 0 and solving for μ, we get the least squares estimator, which is labeled $\tilde{\mu}$. Here, $\tilde{\mu} = \sum_{i=1}^{n} y_i/n = \bar{y}$, the sample mean. This is the point estimate, and replacing the y_i values with Y_i produces the least squares estimator.

The least squares method can be generalized to include several unknown parameters, and to the case where $E(Y) = h(\boldsymbol{\theta})$, that is, the expected value Y is some function of an unknown parameter.

The distribution of Y, or, equivalently, of ϵ does not have to be normal. However, if this is the case, the least squares estimate of the mean of Y is the same as the maximum likelihood estimate. As an exercise, determine why this should be the case.

A variant on least squares is weighted least squares. We will not be examining it in this text, but it does have a number of applications, and the interested reader can refer to Bickell and Doksum (2006).

The least squares method does not appear to include a way to estimate the unknown variance σ^2. However, Gauss' method included the estimate of the variance of the error distribution as well, at least in one important case. Let $h(\theta) = \sum_{j=1}^{k} x_j \theta_j$, that is, the expected value of Y is a linear combination of the unknowns. Let the least squares estimates of the parameters be $\tilde{\theta}_1, \ldots, \tilde{\theta}_k$, then the fitted mean will be $\sum_{i=1}^{k} x_i \tilde{\theta}_i$. Set $d_i = y_i - (\sum_{i=1}^{k} x_i \tilde{\theta}_i)$. It can be shown that $\sum_{i=1}^{n}(Y_i - \sum_{j=1}^{k} x_j \tilde{\theta}_j)^2/(n-k)$ is an unbiased estimator of σ^2. In the case of a single parameter, and with $x_1 = 1$, this reduces to the sample variance.

Numerical examples will be examined in Linear Regression.

6.4 BAYESIAN ANALYSIS

We will outline a simple version of Bayesian Analysis. Interested readers should consult Kendall, Stuart and Ord, (1987), Chapter 8, and Larsen and Marx, (2001), Chapter 3.2, for more details.

An alternative approach to maximum likelihood is Bayesian analysis. It is based on Bayes' Theorem (from the chapter on probability):

Bayes' Theorem/Rule: Let A and B be events in a sample space with probability function P, and assume $P(A) \neq 0 \neq P(\bar{A})$. Then:

$$P(A \mid B) = P(B \mid A)P(A)/[P(B \mid A)P(A) + P(B \mid \bar{A})P(\bar{A})].$$

More generally, if $A_1, \ldots, A_k, k \geq 2$ are mutually exclusive and exhaustive, and $P(A_i) > 0$ for $1 \leq i \leq k$ then:

$$P(A_i \mid B) = P(B \mid A_i)P(A_i)/[\sum_{j=1}^{k} P(B \mid A_j)P(A_j)].$$

As noted earlier, this allows us to revise the probability of A_i once we know B has occurred, using the information about B from the mutually exclusive and exhaustive events. The proof follows from the Theorem of Total Probability.

The general approach in Bayesian analysis is a change in perspective: In the previous methods, it has been assumed that the parameter of interest, θ say, was a constant, and we are using the data to estimate the constant. The Bayesian approach is to think of θ as a random variable with a distribution of its own, with values in a parameter space Ω. The random variable is denoted Θ, and its pdf h. The function h is assumed nonzero on Ω and 0 otherwise. This distribution is determined in advance of gathering data, and so it is called the **prior distribution**.

It is clear that one problem will be determining h and even Ω, but in maximum likelihood, the pdf of the random variable X must be known in advance. In the absence of any information about Θ, a **flat prior** is assumed. This just means h is the pdf of a uniform variable defined on Ω. This, too, can be a problem when Ω is an interval that is infinite in length, for example. There are ways to deal with this, but the details are left for more advanced texts.

Regardless of which method is used, a random sample of size n, denoted $X_1, \ldots, X_n$, must be selected to get information about the distribution. Again, this can be viewed as n iid random variables.

What is the goal? We have a prior distribution for Θ, before any data is gathered. We now have information in the form of the sample, so according to Bayes' Theorem, we should be able to update our knowledge about the probability distribution of Θ, given the sample values. Thus, we want the conditional distribution of Θ, given the sample values $X_1, \ldots, X_n$. We will denote this $K(\theta \mid \mathbf{X})$, where $\mathbf{X}$ is a vector (or n-tuple) containing the sample values.

Suppose we fix Θ, that is, let $\Theta = \theta$. As in the case of maximum likelihood, the form of the pdf for X is assumed known, so knowing θ means the conditional distribution of $\mathbf{X}$, given $\Theta = \theta$ is some function $f(\mathbf{x} \mid \Theta = \theta)$. The distribution function of Θ is h, so the joint distribution of $\mathbf{X}$ and Θ is k, where $k(\mathbf{x}, \theta) = h(\theta)f(\mathbf{x} \mid \Theta = \theta)$. In form, $f(\mathbf{x} \mid \theta = \theta)$ is the likelihood function.

The variable Θ could be continuous or discrete, but for this introduction to Bayesian analysis, we will assume it is continuous. Since it is a continuous random variable, we can determine the marginal distribution of $\mathbf{X}$, say g from:

$$g(\mathbf{x}) = \int_{-\infty}^{\infty} h(\theta)f(\mathbf{x} \mid \Theta = \theta)d\theta.$$

(If Θ's distribution is discrete, the integral would be replaced by a sum.) The conditional distribution of θ, given $\mathbf{X} = \mathbf{x}$ is:

$$K(\theta \mid \mathbf{X}) = k(\mathbf{x}, \theta)/g(\mathbf{x}) = h(\theta)f(\mathbf{x} \mid \Theta = \theta)/g(\mathbf{x})$$

where $g(\mathbf{x}) > 0$ is assumed. This is called the **posterior distribution** of Θ.

Chapter 6 — Parameter Estimation, Sampling Distributions and Inference

Where does h come from? Often, it is from past experience with similar problems, or simply a matter of personal choice. There is a certain amount of subjectivity in Bayesian analysis, but many prefer it because it allows a researcher to use his or her knowledge of the experiment under consideration. There are circumstances under which maximum likelihood does not apply, and Bayesian methods provide a good alternative.

Once we have all of the functions necessary, how do we actually find a point estimate of Θ? We can use the posterior distribution to find probabilities of seeing values of Θ in different intervals (assuming Θ is a continuous variable), but we still need other values such as means and variances to help understand what the posterior distribution indicates about Θ. We need a **decision rule**, and a general discussion can be found in Bickel and Doksum (2007). For our purposes, we will take as the estimate the mean of the posterior distribution.

Example 9: Let X be Bernoulli with parameter π. We know the maximum likelihood estimate for π, based on a random sample of size n from X's distribution is $\hat{\pi} = \sum_{i=1}^{n} x_i/n$, and the estimator is $\sum_{i=1}^{n} X_i/n$. We now compute the Bayes' estimate, using the mean of the posterior distribution as our estimate.

In this case, if $\pi = p$, then the likelihood function, or conditional distribution of the X's, given $\pi = p$, is:

$$f(\mathbf{x} \mid \theta) = \prod_{i=1}^{n} p^{x_i}(1-p)^{1-x_i} = p^{\sum x_i}(1-p)^{n-\sum x_i}.$$

Note that this is just the pdf of the binomial, except for a constant factor. What is a reasonable prior distribution for π? Since this has to be a value between 0 and 1, a good choice is the beta distribution, $h(p) = p^{\alpha-1}(1-p)^{\beta-1}/B(\alpha, \beta)$. The joint distribution of π and $\mathbf{X}$ is:

$$k(\mathbf{x}, p) = h(p)f(\mathbf{x} \mid \pi = p).$$

To find the marginal distribution of $\mathbf{X}$, we integrate k with respect to p:

$$g(\mathbf{x}) = \int_0^1 h(p)f(\mathbf{x} \mid \pi = p)dp = \int_0^1 p^{\alpha-1}(1-p)^{B-1}p^{\sum x_i}(1-p)^{n-\sum x_i}dp/B(\alpha, \beta)$$

$$= \int_0^1 p^{\alpha+n\bar{x}-1}(1-p)^{\beta+n-n\bar{x}-1}dp/B(\alpha, \beta)$$

$$= B(\alpha + n\bar{x}, \beta + n - n\bar{x})/B(\alpha, \beta)$$

where $\bar{x}$ is the sample mean.

Finally, we compute the posterior distribution of π:

$$K(p \mid \mathbf{X}) = k(\mathbf{x}, p)/g(\mathbf{x}) = h(p)f(\mathbf{x} \mid \pi = p)/g(\mathbf{x})$$

$$= p^{\alpha+n\bar{x}-1}(1-p)^{\beta+n-n\bar{x}-1}/B(\alpha + n\bar{x}, \beta + n - n\bar{x}).$$

This is the pdf of $Beta(\alpha + n\bar{x}, \beta + n - n\bar{x})$, so, using as our estimate of π the mean of the posterior distribution, we have:

$$\tilde{\pi} = (\alpha + n\bar{x})/(\alpha + n\bar{x} + \beta + n - n\bar{x}) = (\alpha + n\bar{x})/(\alpha + \beta + n).$$

This value is between 0 and 1 and makes use of the initial information in the prior distribution and the information in the sample. The maximum likelihood estimate is just $\bar{x}$, whereas the Bayes estimate can be written $\bar{x} + (\alpha - \bar{x}(\alpha + \beta))/(\alpha + \beta + n)$. As n increases, the Bayes estimate and the maximum likelihood estimate will become essentially the same.

This procedure has more technical challenges than maximum likelihood, but it does allow judgment to be used in a direct way.

6.5 SAMPLING DISTRIBUTIONS

In the previous sections, we addressed the first of two major problems in estimation, namely to find point estimates of unknown parameters in a distribution. In this section, we discuss the second problem: Find the distribution of the estimator of a parameter. One reason this is important is that our estimator generates a single value, called a point estimate, of the parameter. In the problems we have been considering, the possible values of the parameter fill out an interval, so the probability our estimate is exactly the population value of the parameter has to be zero. In some sense, we are always wrong. However, we do expect that, given the approaches that generate the estimate, the estimate should be close to the true population value. Solutions to this second problem will help us quantify how close we think we are, with a way to measure how often this belief is wrong.

In general, let the random variable be X, with pdf $f(x, \theta)$, where θ is a parameter to be estimated. Assume we have determined the point estimate, $\hat{\theta}$. The point estimate is a function of the observed values $x_1, \ldots, x_n$ of the random sample $X_1, \ldots, X_n$, or, equivalently, the n iid random variables $X_1, \ldots, X_n$ with common pdf $f(x, \theta)$. When we replace the sample values in the formula for $\hat{\theta}$ by the corresponding random variables, this becomes the estimator, a random variable in its own right, of θ. Let this be denoted $\hat{\theta}$. Since it is a random variable, it has a distribution, and the goal in this section is to identify the nature of that distribution.

Based on previous observations, we have developed estimators based on principles such as maximum likelihood, minimum variance, least squares, Bayesian and moment estimation. This illustrates that several different estimators can be available, so how do we decide which to use? For example, suppose we want to estimate the mean of a distribution. Since our standard method of estimation is to take a simple random sample, each X_i has the same distribution as the parent, so why would we choose the sample mean over a single observation? The main reason is in fact the variance of the estimator. The variance of a single observation is just σ^2, the variance of the parent distribution, while it is σ^2/n for the sample mean. This means that as n gets larger, more of the distribution of the sample mean is concentrated around the mean μ of the population than in the case of a single observation. We can also note that if the estimator is a single observation, it is neither consistent nor efficient, although it will be unbiased.

In this section and in the remaining chapters, we will assume that the estimator has been selected to reflect our main principles, and, if possible, is the maximum likelihood estimator (recall that the sample variance in the case of estimating σ^2 when $X \sim N(\mu, \sigma^2)$ where both μ and σ^2 are unknown is not the MLE, but is unbiased and nearly minimum variance). It is the distribution of these estimators that we explore.

Definition 11: The distribution of the random variable $\hat{\theta}$, the estimator of the parameter θ, is called the **sampling distribution** of the parameter. In particular, if the parameter is the mean of the parent distribution, then we call the distribution of its estimator the **sampling distribution of the mean**; if the parameter is the variance of the parent distribution we call the distribution of its estimator the **sampling distribution of the variance**; and if the parameter is the population proportion, we call the distribution of its estimator the **sampling distribution of the proportion**.

Definition 12: The standard deviation of the estimator of the parameter θ is called the **standard error** of the estimate.

Often, an estimate will be given, along with reference to the standard error. The standard error gives a measure of how close we expect the estimate to be to the true value. For example, in the absence of more information, we can always appeal to Chebyshev's inequality to get a reasonable measure of the distance from our estimate of a mean to the population mean. If we know the distribution is bell shaped, then we have more precise information about the difference between the population mean and our estimate.

The sampling distribution of a parameter may include other parameters that are unknown, so finding percentage points, for example, becomes a difficult problem. However, sometimes there is a relatively simple function of the estimator that has a distribution free of unknown parameters, making it a random variable we can work with.

Chapter 6 — Parameter Estimation, Sampling Distributions and Inference

Definition 13: Let the random variable X have unknown parameter θ, and let $\hat{\theta}$ be an estimator of θ. If Y is a function of $\hat{\theta}$, θ and the sample values such that its distribution is free of unknown parameters, including θ, then we call Y a **pivot**, or a **pivotal statistic**, for θ.

Example 10: Let X be $N(\mu, \sigma^2)$, with both μ and σ^2 unknown. We want to estimate μ, and we know the best point estimate is $\overline{X}$. Since $\overline{X}$ is a linear combination of normal variables, we also know $\overline{X} \sim N(\mu, \sigma^2/n)$. Clearly, by itself, $\overline{X}$ is not a pivot for μ because its distribution depends on σ^2. However, $(\overline{X} - \mu)/S$ (where S is the sample standard deviation) is unitless, and hence, free of σ^2. The pdf of this random variable is in fact Student's t with $n-1$ degrees of freedom.

Example 11: If $X \sim N(\mu, \sigma^2)$ with μ known, but σ^2 unknown, then the MLE of σ^2 is $\hat{\theta} = \frac{1}{n} \sum (X_i - \mu)^2$. Define $Y = \frac{n}{\sigma^2} \frac{1}{n} \sum (X_i - \mu)^2 = \frac{1}{\sigma^2} \sum (X_i - \mu)^2 \sim \chi^2(n)$. The distribution of Y is free of unknown parameters, so it is a pivot for σ^2.

Why are pivots important? Suppose we have the situation in the previous example, and now we want to know the probability that σ^2 is between 1 and 4, given a random sample from the distribution of the random variable. We want $P(1 \leq \sigma^2 \leq 4)$ based on the information in a random sample of size n from a normal population. We can introduce the pivotal statistic as follows:

$$P(1 \leq \sigma^2 \leq 4) = P(n\hat{\sigma}^2/4 \leq n\hat{\sigma}^2/\sigma^2 \leq n\hat{\sigma}^2) = P(n\hat{\sigma}^2/4 \leq \chi^2(n) \leq n\hat{\sigma}^2).$$

The inequalities come from breaking the original into two pieces, $1 \leq \sigma^2$ and $\sigma^2 \leq 4$, multiplying each by $n\hat{\sigma}^2/\sigma^2$ and then dividing by appropriate constants. Note that one of the inequalities gives the upper bound to the pivot, and the other gives the lower bound, since both inequalities must hold. Once the sample is taken, this is just a probability of a known random variable.

Example 12: Let $\hat{\sigma}^2 = 3$ from a random sample of size 10 under the conditions in the last paragraph. Then, given this information, the probability σ^2 is between 1 and 4 is:

$$P(1 \leq \sigma^2 \leq 4) = P(n\hat{\sigma}^2/4 \leq \chi^2(n) \leq n\hat{\sigma}^2) = P(30/4 \leq \chi^2(n) \leq 30) \approx 0.68.$$

Thus, under the given conditions, and for the sample values, the probability σ^2 is between 1 and 4 is about 0.68. If, instead, $\hat{\sigma}^2 = 6$, the required probability is now approximately 0.13. Such calculations can support or bring into question assumptions about the size of variability assumed in a distribution. This methodology will be used extensively in hypothesis testing and confidence intervals.

Some distributional form such as binomial, Poisson, normal, or Student's t is assumed for our random variable X. We can use **descriptive statistics**, including summary values such as the sample mean, variance, and other sample moments, or visual displays such as histograms and bar charts, to help decide if this assumption is valid, at least in a qualitative way. The goal, however, is to have a quantitative measure of the accuracy of our estimates of parameters. It cannot be stressed enough that independence of the observations is critical to being able to make such quantitative statements. In most problems, we must have a simple random sample of size n, or n iid random variables, to make sure the assumed distribution of the estimator is valid.

6.5.1 Summary of Sampling Distributions

In the previous sections of this chapter, we considered ways to generate point estimates of important parameters. As noted earlier, point estimates by themselves are not very useful, as they are almost always wrong.

However, if we know the distribution of the estimator and how it relates to the parameter of interest, then at least we can quantify how far away from the true parameter value we are likely to be. This provides a level of **confidence** in our estimate generated from a random sample. In particular, we have:

1. Let $X \sim N(\mu, \sigma^2)$.

    (a) Estimating μ when σ^2 is known.

    The best point estimate of μ is $\hat{\mu} = \overline{X}$, the sample mean. The pivot, or pivotal statistic, is $(\overline{X} - \mu)/(\sigma/\sqrt{n}) = \sqrt{n}(\overline{X} - \mu)/\sigma \sim N(0, 1)$.

    The distribution of $\overline{X}$ is $N(\mu, \sigma^2/n)$, and this is the sampling distribution of the mean. The value $\sigma/\sqrt{n}$ is the standard error.

    (b) Estimating μ when σ^2 is unknown.

    The best point estimate of μ is $\hat{\mu} = \overline{X}$. The pivot is $(\overline{X} - \mu)/(S/\sqrt{n}) = \sqrt{n}(\overline{X} - \mu)/S \sim t(n-1)$, where S^2 is the sample variance.

    The distribution of $\overline{X}$ is $N(\mu, \sigma^2/n)$: and this is the sampling distribution of the mean. The value $\sigma/\sqrt{n}$ is the standard error. Since σ^2 is unknown, then $S/\sqrt{n}$ is the **approximate standard error**.

    (c) Estimating σ^2 when μ is known.

    The best point estimate of σ^2 is $\widehat{\sigma^2} = \sum_{i=1}^{n}(X_i - \mu)^2/n$. The pivot is $n\hat{\sigma}^2/\sigma^2 \sim \chi^2(n)$. Although it is possible to write out the density for the sampling distribution of the variance, it does not fall into one of the standard distributions. In this case, the pivotal statistic is more useful and contains the same information.

    The standard error is $2\sigma^2/n$. Since σ^2 is unknown, we use $2\hat{\sigma}^2/n$ as the approximate standard error.

    (d) Estimating σ^2 when μ is unknown.

    The best point estimate of σ^2 is $\hat{\sigma}^2 = \sum_{i=1}^{n}(X_i - \overline{X})^2/(n-1) = S^2$. The pivot is $(n-1)S^2/\sigma^2 \sim \chi^2(n-1)$. Again, we do not identify the sampling distribution of the variance (it is just the distribution of the estimator), but do identify the pivot.

    The standard error is $2\sigma^2/(n-1)$, and since σ^2 is unknown, we can report the approximate standard error is $2S^2/(n-1)$.

2. Let X be a Bernoulli variable.

    The best point estimate of the proportion π is $\hat{\pi} = \overline{X}$. If the normal approximation to the binomial can be used, then:

    $$(\overline{X} - \pi)/(\sqrt{\hat{\pi}(1-\hat{\pi})/n}) \sim N(0, 1)$$

    approximately. We could use:

    $$(\overline{X} - \pi)/(\sqrt{\pi(1-\pi)/n}) \sim N(0, 1),$$

    approximately, as the pivot, but it is technically more complicated to use. The given approximations are good as long as $n \geq 30$ and $\hat{\pi}$ satisfies $(\hat{\pi} - 3\sqrt{\hat{\pi}(1-\hat{\pi})/n}, \hat{\pi} + 3\sqrt{\hat{\pi}(1-\hat{\pi})/n}) \subset (0, 1)$.

    The standard error is $\sqrt{\pi(1-\pi)/n}$. Since π is unknown, we report the approximate standard error as $\sqrt{\hat{\pi}(1-\hat{\pi})/n}$.

3. Let X have an exponential distribution with parameter θ.

    The best point estimate of θ is $\hat{\theta} = \sum_{i=1}^{n} X_i/n$. The pivot is $n\hat{\theta} \sim \Gamma(n, \theta)$.

    The standard error is $\theta/\sqrt{n}$, and since θ is unknown, then we report the approximate standard error as $\hat{\theta}/\sqrt{n}$.

As we introduce other parameters and their estimates, we will add to this list of best point estimates, pivots, sampling distributions, and standard errors or approximate standard errors.

6.5.2 Summary of Fundamental Rules about Distributions

The results in this and previous chapters depend on a relatively small number of observations. The following list is a compendium of major results to this point.

1. **Central Limit Theorem**: Let $X_1, \ldots, X_n$ be a random sample of size n from the distribution of the random variable X, where X has finite mean μ and variance σ^2. Then $Z = (\bar{X} - \mu)/(\sigma/\sqrt{n})$ has, for large n, a distribution that is approximately standard normal. If X is exactly normal, so is Z for any n. In general, $n \geq 30$ is sufficient for the normal approximation to hold. The asymptotic result holds if we replace the standard deviation of X with the sample standard deviation.
2. **Theorem**: Let $X_1, \ldots, X_n$ be a random sample from an infinite population with mean μ and variance σ^2. Then $\bar{X} = (\sum_{i=1}^{n} X_i)/n$ has mean μ and variance σ^2/n.
3. Let $X \sim N(\mu, \sigma^2)$. Then $Z = (X - \mu)/\sigma \sim N(0, 1)$.
4. Any linear combination of normal variables is normal.
5. If X and Y have a bivariate normal distribution, then X and Y are independent iff $cov(X, Y) = 0$.
6. If X and Y are independent, then $cov(X, Y) = 0$. If $cov(X, Y) = 0$, this does not necessarily imply X and Y are independent.
7. Let $X_1, \ldots, X_n$ be n independent normal variables. Then $Y = a_1 X_1 + \ldots + a_n X_n$ is distributed as:

$$N(a_1\mu_1 + \ldots + a_n\mu_n, a_1^2\sigma_1^2 + \ldots + a_n^2\sigma_n^2) = N\left(\sum_{i=1}^{n} a_i\mu_i, \sum_{i=1}^{n} a_i^2\sigma_i^2\right).$$

8. If $Z \sim N(0, 1)$ then $Z^2 \sim \chi^2(1)$.
9. If $Z \sim N(0, 1)$ and $U \sim \chi^2(r)$ are independent, then $T = Z/\sqrt{U/r} \sim t(r)$.
10. If $U_1 \sim \chi^2(r_1)$ and $U_2 \sim \chi^2(r_2)$ are independent, then $U_1 + U_2 \sim \chi^2(r_1 + r_2)$.
    Let $\bar{X}$ be the sample mean of n independent observations of the normal variable X, and let S^2 be the sample variance. Then:
11. $\bar{X}$ and S^2 are independent, and hence $corr(\bar{X}, S^2) = 0$.
12. $E(S^2) = \sigma^2$.
13. $(n-1)S^2/\sigma^2 \sim \chi^2(n-1)$.

6.6 INFERENCE

Probability and statistics are used to draw conclusions about characteristics of a population, including parameters and even the nature of the distribution of probabilities of the values of the random variables involved.

Definition 14: Inferential statistics is the drawing of conclusions about a population or several populations based on samples, and using properties of estimators and their distributions to quantify the confidence we have in these conclusions.

A point to be emphasized here is that this process does not prove a conclusion is true: it gives evidence, and the strength of evidence, for the conclusion.

Qualitative results are available through descriptive statistics. For example, if we have assumed some distributional form, a histogram of the actual data may help us decide if we have made a good choice. However, a formal **test of hypothesis** concerning the fit of the assumed distribution to the histogram may be necessary. If we compare a sample mean to a population mean, it may appear to be close, but closeness depends on the variance. Again, we would need a more formal procedure to determine what closeness means, and then the strength of the evidence that the true population value is close to the sample mean.

Inference problems are encapsulated in the following example.

Example 13: Suppose that the average yield per acre of a strain of corn in a region with a dry climate is 100 bushels per acre. A hybrid strain has been developed, and it is claimed that it has an average yield of at least 8 bushels more per acre than the previous strain. Twenty-five one-acre plots were selected at random from a very large field in the dry climate zone and planted with the new strain. Let Y be the yield per acre, and assume it is normally distributed with known variance $\sigma^2 = 16$. From the data, we find the (sample) average yield was 105 bushels. Does this mean the claim for the new strain is wrong?

We want to infer from the data whether the new strain of corn has a yield, on average, of 108 bushels. The variable Y, the yield per acre, is not a constant, but consists of a mean yield, and then plus or minus some amount due to different conditions, such as quality of soil, terrain (hill side versus flat, for example), amount of shade, etc. Thus, on any one acre, we may see a result that is much different from the overall average value. Because of earlier statements, we know the sample mean, $\overline{Y}$, is also normal, and in particular, $\overline{Y} \sim N(\mu_Y, 16/25)$, and again, one time running the experiment may yield a sample mean very much different from the population mean. Then again, it may appear a sample mean is not too far away from the supposed population mean, but that will be based on the scale that quantifies deviations from the mean.

In our problem, is 105 sufficiently close to 108 to say it is simply by chance that we have such a sample mean when the true mean is 108? Can we quantify the chances of seeing this value when the true mean is 108? In our example, the value 105 is actually 3.75 standard deviations below the assumed mean of 108, and since we are dealing with the normal distribution, we know this is a significant difference.

We will explore the use of the results concerning continuous and discrete variables in the remainder of the book. The applications are wide-ranging and concentrate on problems arising in the life sciences and related disciplines.

Exercises

1. There are other methods for generating samples, including: stratified sampling, cluster sampling, systematic sampling, multistage sampling, and inverse sampling. Choose at least one of these sampling plans, and from the web, text sources, or otherwise, determine the conditions under which such a sampling plan would be used, its advantages and disadvantages, and an example of an actual experiment using this procedure.

2. (a) There are many other methods for estimating parameters, including: BLUE (best linear unbiased estimation), MVUE (minimum variance unbiased estimation), and MMSE (minimum mean squared error). Choose one of these estimation procedures, determine the conditions under which it is used, and find an application.

   (b) The method of moments equates the non-central moments of the population to the non-central sample moments. Enough moments are computed so that there are as many equations as there are unknown parameters. These equations are then solved simultaneously for the various parameters.

      (i) Determine the moment estimators of the mean and variance in a normal distribution.

      (ii) Use the method of moments to determine the moment estimators of the parameters of the Gamma distribution.

3. Let $U \sim \Gamma(\alpha, \beta)$ and assume α is known. Let $U_1, \ldots, U_n$ be a random sample from U's distribution.

   (a) Determine the maximum likelihood estimator of β.

   (b) Show that the maximum likelihood estimator of β is unbiased for β.

   (c) Determine if the MLE of β attains the minimum variance bound for unbiased estimators here.

   (d) Determine the moment estimator of β, and compare it to the MLE.

Chapter 6 — Parameter Estimation, Sampling Distributions and Inference

4. Let $Y \sim \chi^2(\nu)$.

    (a) Determine $E(\sqrt{Y})$.

    (b) Use the result in (a) to show that for the normal distribution $N(0, \sigma^2)$, the square root of the sample variance is biased for σ.

5. Let W have a geometric distribution with unknown probability π of observing a 1 (or getting a success) on each Bernoulli trial.

    (a) Determine the maximum likelihood estimate of π.

    (b) Determine the moment estimator of π, and compare it to the MLE.

6. Let $X \sim N(\mu, \sigma^2)$ and assume σ^2 is known. Determine the Bayes' estimate of μ using as prior pdf of μ, $N(m, a^2), a > 0$.

7. Let X be a normal random variable, with mean $\mu = 30$ and variance $\sigma^2 = 100$. Let $\overline{X}$ and S^2 be the sample mean and sample variance, respectively, of a random sample of size n from X's distribution.

    (a) State the distribution of $\overline{X}$ when $n = 16$.

    (b) State the distribution of $(\overline{X} - 30)/(S/\sqrt{n})$ when $n = 16$.

    (c) Determine the smallest sample size n such that $P(28 < \overline{X} < 32) > 0.95$.

8. Let X have an exponential distribution with $\lambda = 2$. A sample of size 60 is taken from the population.

    (a) Determine $E(\overline{X})$, the expected value of the sample mean, and interpret this value.

    (b) Determine $var(X)$ and the standard error of the sample mean.

    (c) What is the sampling distribution of $\overline{X}$, assuming the Central Limit Theorem applies?

    (d) Determine the distribution of $Y = X_1 + \cdots + X_{60}$, where $X_1, \ldots, X_{60}$ is the random sample from X's distribution. From this, write out the density of $\overline{X}$. Hint: Consider the moment generating function of a single exponential variable, and use the fact a simple random sample consists of *iid* random variables.

9. Let X be a random variable with mean 10 and an estimated variance of 4. A sample of size 40 is taken from the population.

    (a) Determine $E(X)$, the expected value of the sample mean, and interpret this value.

    (b) Determine the estimated standard error of the sample mean.

    (c) What is the sampling distribution of $\overline{X}$? What assumptions are needed to determine this distribution?

10. Suppose that independent random samples of size $n_X = 9$ and $n_Y = 16$ are taken from normal distributions with common variance. Determine the distribution of:

    (a) S_X^2/S_Y^2

    (b) S_Y^2/S_X^2

11. A sample of size $n = 5$ is taken from a normally distributed random variable with a standard deviation of 12.

    (a) If we took a large number of samples of size 5 from this population, what would you expect the average value of the sample variance to be?

    (b) What is the distribution of $S^2/36$?

    (c) Determine the integral that gives $P(S^2 < 120)$ by converting to an appropriate χ^2 variable. You do not need to evaluate the integral.

12. Consider independent random variables $X \sim N(5, 21)$ and $Y \sim N(7, 4)$, with the same units. Suppose that samples of size 20 and 25 are taken from populations 1 and 2 respectively. Determine the sampling distribution of the following variables:

    (a) $\overline{X}$

    (b) $\overline{Y}$

    (c) $\overline{X} + \overline{Y}$

13. In a certain lab population of mice, the weights at 20 days of age have an approximately normal distribution with a mean weight of $\mu = 9.6$ grams and standard deviation of 2.6 grams. Let X denote the weight of a randomly selected mouse.

    (a) Suppose that a sample of 9 mice is taken. Explain what is meant by the sampling distribution of the sample mean.

    (b) Determine the probability that a randomly selected mouse will have a weight between 9 and 12 grams.

    (c) Suppose a random sample of 9 mice is selected. Determine the probability that the random sample of mice will have a mean weight between 9 and 12 grams.

    (d) Suppose a random sample of 36 mice is selected. Determine the probability that the random sample of mice will have a mean weight of at least 10 grams.

    (e) Suppose that a large number of litters of 10 mice each are to be weighed. If each litter can be regarded as a random sample from the population, what percentage of the litters will have a total weight of 90 grams or more?

    (f) Give two factors that might invalidate the assumption that each litter in (e) can be regarded as a random sample from the population of mice.

14. In a study, rural high school students were asked to complete a comprehensive test. Past studies suggest that the scores on this test by rural students have a standard deviation of 52.

    (a) Use Chebyshev's Theorem to estimate the probability that the sample mean test score from a sample of 36 rural student test scores is within 20 points of the true mean rural student test scores.

    (b) Use the Central Limit Theorem to determine the probability in (a).

15. Explain the difference between a population parameter and a sample statistic.

16. Identify each of the following as being a population parameter or a sample statistic, assuming the notation used throughout the text.

    $$\sigma^2 \quad S \quad \pi \quad \overline{U} \quad \mu \quad S^2 \quad X/n \quad \sigma$$

17. Let X have a binomial $B(n, \pi)$ distribution.

    (a) Show that X/n is an unbiased estimator of π.

    (b) Is X a consistent estimator? Justify your answer.

    (c) Show that $(X + 1)/(n + 1)$ is a biased estimator of π.

    (d) The bias of an estimator is defined to be the difference between the expected value of the estimator and the true value of the parameter. Determine the bias of the estimator $(X + 1)/(n + 1)$.

18. The number of hours W from an initial spill of a toxin onto a flat field until measurable amounts are detected 1 foot below the surface of the field is thought to follow a log normal distribution. After a recent

spill, 20 measurements (from different locations in the field) of the time until the toxin reaches 1 foot below the surface are taken, with the following results:

14.43 10.03 15.86 12.25 12.87 10.95 12.45 11.17 12.07 15.97
16.43 13.76 13.77 10.28 10.19 10.25 11.57 11.33 13.10 9.60

Summary values include: $\sum \ln w_i = 248.33$ and $\sum \ln^2 w_i = 3165.9407$. Determine the maximum likelihood estimates of the parameters μ and σ^2 if $W \sim LN(\mu, \sigma^2)$.

19. Let X be $LN(\mu, \sigma^2)$.

   (a) Establish the formulae for the mean and variance of X in terms of μ and σ^2.

   (b) Derive the estimates of μ and σ^2 if the method of moments is used to estimate the mean and variance of the log normal.

   (c) Use the results in (b) to find the moment estimates of μ and σ^2 for the data in Exercise 18. Compare the results.

CHAPTER 7

DESCRIPTIVE STATISTICS

Synopsis
Descriptive statistics, including summary values of location and scale, and diagrams for preliminary qualitative and quantitative analysis are summarized.

7.1 INTRODUCTION

In Chapter 6, we considered ways to estimate unknown parameters in the distribution of random variable X defined on a sample space with probability P defined on it. However, to do so, a sample has to be taken in a particular way (specifically, to form a simple random sample) and the form of the distribution of probabilities for X has to be known. For example, we typically assume X has a normal distribution with pdf:

$$\frac{1}{\sigma\sqrt{2\pi}} \exp\left(-\frac{(x-\mu)^2}{2\sigma^2}\right), \mu \in \mathbb{R}, \sigma > 0, x \in \mathbb{R}$$

where μ and σ are the parameters, sometimes known, sometimes not known.

There are two problems to consider: How do we know the choice of distribution (normal, Student's t, binomial, Poisson, etc.) is supported by the data? and, How do we know the sample is a simple random sample? Of course, there can be no way to be absolutely certain, but can we at least be confident our assumptions are reasonable?

Example 1: Let the random variable X have a Cauchy distribution, that is, its pdf is given by:

$$\frac{1}{\sigma\pi} \frac{1}{1 + (x-\mu)^2/\sigma^2}, \mu \in \mathbb{R}, \sigma > 0, x \in \mathbb{R}.$$

Let $X_1, \ldots, X_n$ be a random sample from X's distribution, or, in other words, let them be n iid with the same distribution as X. Consider $\overline{X} = \sum_{i=1}^{n} X_i/n$, the sample mean. Is this a good estimator of μ?

The answer is no. In fact, it is easy to show that $E(X)$ does not exist (it isn't $+\infty$ or $-\infty$, it just does not exist). We showed earlier that $\overline{X}$ was always unbiased for the population mean, but here there is no finite mean. It should also be noted that if this function is graphed, it is symmetric about $x = \mu$ (μ is not interpreted as a mean, but as a measure of the distribution) and looks bell shaped. See Figure 7.1 where $\mu = 0$ and $\sigma = 1$ for both the Cauchy and the normal densities for comparison of shapes. There are many similarities, but the normal has more of its distribution "piled up" around 0, with little left for the tails, while the Cauchy has much of its distribution in its tails.

The Cauchy distribution is one where neither the Central Limit Theorem nor Chebyshev's Inequality/Theorem applies because it does not have a finite mean (or variance, or any moment). Obviously, if we assumed X had a normal distribution, but it in fact had a Cauchy distribution, any inferences would not make sense.

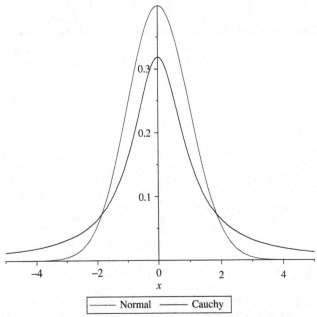

FIGURE 7-1. Cauchy versus Normal Distributions.

This chapter considers certain summary values and diagrams that can provide evidence that our assumptions are reasonable, as well as help organize information in ways that make it easier to understand. We concentrate on methods related to an experiment design called the **completely randomized design**:

Definition 1: Let an experiment have one or more treatments that can be be applied to each member of a population of experimental units. A **completely randomized design** is one where: the sample of experimental units forms a random sample from the population and the treatments are randomly assigned to the different experimental units.

The simplest experiment is one where there is a single treatment. For example, radon is a naturally occurring radioactive element that is odorless and colorless, and can accumulate in basements and attics of houses. To study the level of radon in houses constructed 30 or more years ago, a random sample of 36 such houses is selected from a city's database, and then special detectors are installed in the basement of each and the level of radon recorded over a period of a month. The treatment is the measurement of the level of radon, and the 36 houses form a random sample, so the design is a completely randomized design.

The completely randomized design also applies to problems with more than one treatment. It is important for diabetics to monitor their blood sugar levels on a regular basis. Three testing devices are available, and the variability in the readings by each is under study. Twenty-one diabetics are randomly selected from the records of a support group for diabetics. Twenty-one identical pieces of paper are prepared. The name of treatment 1 is written on seven of the slips, and this is repeated for the other treatments. These are placed in a bowl and thoroughly mixed. A slip is drawn, and one of the diabetics is randomly chosen (both selections without replacement), and this step is repeated until each diabetic has been assigned a treatment. Since the treatments are randomly assigned to the diabetics, this is a completely randomized design.

Even if a design is completely randomized, there can still be questions about the design of the experiment. For the diabetics example, there can be a range of abilities in carrying out the tests; age may play a factor, other health issues may interfere with reading or using the devices, etc. This means other factors should be controlled in running the experiment. Different designs to get the best information in the most efficient manner should be examined to determine how to proceed in any experiment. In this chapter, we will assume the data comes from a completely randomized design and such other factors have been accounted for.

For a given population of experimental units, let Y be the random variable of interest. For the most part, we will assume Y is a measurement and that its values fill out an interval of real numbers. A random experiment is set up to make several observations of Y under fixed conditions. Let the sample size be n and the observations on Y be $y_1, \ldots, y_n$, that is, a realization of the random sample $Y_1, \ldots, Y_n$. Some aspects of Y's nature will be clear before the experiment is run: Y is either continuous or it is discrete, and the possible values of Y must be known. Certain **assumptions** related to Y's distribution are made, such as: Y is normally distributed, or has some other distribution; some of the parameters in the distribution are known, and some are unknown; and Y's distribution may depend on other variables or factors. A major assumption is that $Y_1, \ldots, Y_n$ is a simple random sample, or, in other words, they are n *iid* random variables, each with same distribution as Y. These underlying assumptions drive the statistical analysis that we can perform on a data set, so it is very useful to have ways to qualitatively and quantitatively assess their validity. In this chapter, we consider measures and visual evidence that support or call into question the assumptions made.

Example 2: The range of experiments for which the methods developed in this chapter apply is quite broad. A very brief list includes:

1. Quantifying the ability to remember recent information by a person with mild Alzheimer's disease may lead to early diagnostic tests. A random selection of those with mild Alzheimer's and no other mental issues are given a list of words to study for five minutes, and then an hour later, asked to repeat as many of the words on list as they can in one minute. This request may be repeated several more times to determine the duration of memories.

2. In trying to optimize a chemical process, we may run several batches of chemicals through the process to determine the percentage purity of the yield under given conditions, including different temperatures and pressures.

3. To determine the presence of pollutants and their effect on game fish, a lake downstream from a suspected polluter is selected and several game fish captured. The level of contaminants in each is measured, and aspects of their health are recorded.

4. A new computer processor is under study. Several problems utilizing that type of processor are run to see how long it takes computers with this new processor to solve these problems is recorded, to be compared to the time taken by similar computers, but with an older version of processor as part of the hardware.

5. A new drug for the treatment of an aggressive form of cancer is under study. Several patients with the disease and who have similar characteristics are selected at random and given the new drug. The new drug is ruled a success if a patient survives at least five years from the beginning of treatment, and the success or failure for each patient is recorded.

In any of these cases, we may have some idea of how Y is distributed, but not know one or more of the parameters. In a distribution, any constant that characterizes the distribution is called a **parameter**, and one of the reasons for running the experiment is to estimate the value of unknown parameters. There are many aspects to the design of the experiment that will impact the quality of the estimates: how uniform are the experimental units? how many parameters are to be estimated? how accurate must the estimates be? how much money do we have to run the experiment? Answers to such questions lead to an appropriate **experimental design**, discussed in Chapter 10. At this point, we will assume such matters have already been taken into account and that the protocols for getting the data have been followed.

7.2 DESCRIPTIVE STATISTICS

An experiment has been run, and data has been recorded. Although some experiments may involve small samples, say $n \leq 20$, many have hundreds or thousands or more observations. Looking at page after page of numbers will not be of use in trying to assess the validity of the various assumptions made. That is why we use **descriptive statistics** to help verify assumptions and to summarize important aspects of the data.

Definition 2: **Descriptive statistics** are numerical values and graphical displays used to illustrate basic properties of the observations from a random experiment.

There will be at least one unknown parameter of interest, such as μ, σ, and π. Different letters can be used for population and sample values, but we will use μ for the mean, σ^2 for the variance, and π for the proportion, for simplicity. As per convention, we use X or Y as generic random variables, Z for the standard normal variable, and corresponding lowercase letters for "realizations," that is, observed values of the variable when the experiment is run.

If the random variable is Y, then n *iid* random observations are $Y_1, \ldots, Y_n$. The values $y_1, \ldots, y_n$ are a **realization** of these random variables. The assumption that these variables are **independently and identically distributed**, or simply *iid*, is crucial for any analysis we want to undertake. It is assumed that we are sampling with replacement or from a sufficiently large population so that sampling without replacement has little impact. This is why we can assume Y_i has the same distribution as Y, the "parent" distribution. If the population is small and we sample without replacement, we would have to consider other methods.

When we take our observations, we record them as they occur. Once we have these values, it is often convenient to introduce the **order statistics** of a random sample:

Definition 3: Let $Y_1, \ldots, Y_n$ be n *iid* random variables. We define n new variables, called the **order statistics**, as follows: $Y_{(1)} = min\{Y_1, \ldots, Y_n\}$, $Y_{(2)}$ = second smallest observation, $\ldots$, $Y_{(n)} = max\{Y_1, \ldots, Y_n\}$. We call $Y_{(1)}$ the first order statistic, $Y_{(i)}$ the ith order statistic, and $Y_{(n)}$ is the nth order statistic. In the event two or more Y_i values are equal, we assign consecutive order statistics to each of the ties (see the example immediately following this definition).

Example 3: In theory, if Y is a measurement, there should not be ties among the observations. However, in practice, ties can occur because we can record the Y values to only so many decimals. Suppose our $n = 10$ observations are 3,5,2,4,6,4,2,1,8,5. Then the order statistics are:

$$Y_{(1)} = 1, Y_{(2)} = 2, Y_{(3)} = 2, Y_{(4)} = 3, Y_{(5)} = 4,$$
$$Y_{(6)} = 4, Y_{(7)} = 5, Y_{(8)} = 5, Y_{(9)} = 6, Y_{(10)} = 8.$$

Definition 4: The **range of a random sample** is $Y_{(n)} - Y_{(1)}$, and its value for a given data set is $y_{(n)} - y_{(1)}$. The **range** of a distribution is:

$$max\{Y\} - min\{Y\}.$$

The range of a distribution may be infinite.

Graphical representations of data require a horizontal and a vertical axis. The horizontal axis is related to the possible values of the variable, while the vertical axis typically represents the frequency with which values of the variable occur. We need a scale on each axis, and the range, along with the minimum and maximum, provides the necessary information for the horizontal axis. There are other kinds of plots, including pie charts, but many graphical summaries of the data involve the variable plotted against some measure of frequency.

Depending on the data set, there can be thousands of observations, and a mere list will not be revealing. If the observations are ordered, at least we know the maximum and minimum values (and hence the range). Spreadsheets help organize and sort data by particular criteria, including size. In many problems, we are interested in what might be considered the typical value, or the most commonly occurring value, or how the data is spread out. We may also want some relatively simple pictorial or graphical representations that give support to some of the assumptions made about the distribution of Y.

Chapter 7 — Descriptive Statistics

Example 4: Consider the following 40 observations on the random variable Y:

1.0113	−0.9717	0.01880	0.7246	2.8240	7.9848	0.1501	−2.1609
−1.9832	−0.9570	1.8371	−0.9767	1.5207	−1.0450	4.4865	−0.5115
−1.4129	−0.9965	0.3831	−0.0845	−0.5049	1.7082	0.2756	0.1582
3.7990	−0.7995	1.0498	1.0271	−0.3959	0.2173	−4.1683	−5.9560
0.0975	1.2065	4.8258	−2.3266	0.2206	−0.0969	−29.5215	−0.5636

This list of 40 numbers is unrevealing until it is organized and examined. For example, the mean is −0.4977 and the standard deviation is 5.264. The minimum is −29.5215, and the maximum is 7.9848, so the range is 37.5063. Within one standard deviation, we have 38 of the observations, many more than expected if this were from a normal distribution of the mean, based on the Empirical Rule. We will use this example to illustrate other features as we introduce graphical representations.

The next few sections introduce some important summary values and diagrams that help organize the information in the data.

7.3 MEASURES OF CENTRAL TENDENCY

Definition 5: (Measures of Location: The Mode). The **mode**, whether for a population or for a sample, is the most frequently occurring value. In populations, it is denoted μ_o, and in samples, it is denoted Y_o when considered as a random variable and y_0 if it is a realization. In some cases, two values are tied for the greatest frequency, and then we say the data is **bimodal**. This can be generalized, but for our purposes, if more than two values are tied for the greatest frequency, we will say the data is **multimodal**. This can be calculated for categorical or for continuous data.

If a continuous random variable has a density that has two local maxima, this is said to be a **bimodal distribution**, even if the maximum values are not the same. This can be extended to discrete distributions as well.

Example 5: Monarch butterflies spend the winter in conifers. In a study of their habits, a scientist selects at random 20 Pinyons in a large stand and counts the butterflies on each of the twenty trees, with the results, rounded to the nearest tens:

80, 90, 80, 60, 100, 70, 80, 30, 90, 210, 90, 110, 70, 80, 50, 50, 80, 40, 70, 150

The mode is 80 in this example.

Definition 6: (Measures of Location: The Median). The **median**, whether for a population or a sample of numerical data, is a value that divides the data in two parts, so that at least 50% of the data is less than or equal to this value, and at least 50% of the data is greater than or equal to the value. For populations, the median is denoted $\tilde{\mu}$ and satisfies:

$$P(Y \geq \tilde{\mu}) \geq 0.50 \text{ and } P(Y \leq \tilde{\mu}) \geq 0.50.$$

In samples, the median as a random variable is denoted $\tilde{Y}$ and $\tilde{y}$ denotes realization of the sample median.

For continuous distributions, the median of the population is unique. For discrete distributions or samples, the median may not be unique. If a and b are values of the variable or sample, with $a < b$, and they both satisfy the probability statements to be a median, then so will any number in between them. In such cases, we will take the median to be $(a + b)/2$.

Example 6: For the butterfly data in Example 5, the median is 80. For this data, the first-order statistic value is 30, the 20th-order statistic is 210. The 10th- and 11th-order statistics are both 80. Generally, for even sample sizes, the median is the average of the $n/2$ and the $n/2 + 1$ order statistics.

Example 7: Determine the mode and median of the continuous random variable Y with density:

$$f(y) = \begin{cases} 280y^3(1-y)^4, & 0 \le y \le 1 \\ 0, & \text{otherwise} \end{cases}.$$

Solution: Since this is a continuous random variable, the mode (if there is one) occurs at a critical number. The only critical number is $y = 3/7$, and $f(0) = 0$ and $f(1) = 0$, then the maximum occurs uniquely at $y = 3/7$

The median will also be unique since the density function is continuous. With $\tilde{\mu}$ the median, we have to solve:

$$280 \int_0^{\tilde{\mu}} y^3(1-y)^4 \, dy = 1/2.$$

From a numerical approximation scheme such as the Newton-Raphson method, the median is 0.4401552, approximately.

Definition 7: (Measures of Location: The Mean). The **mean** of a sample is simply the numerical average of the numerical values observed:

$$\bar{y} = \frac{1}{n} \sum_{i=1}^{n} y_i$$

For populations, we use the term **mean** and **expected value** interchangeably. For discrete distributions, where the possible values of Y are $y_1, \ldots, y_k$, we have:

$$E(Y) = \mu = \mu_Y = \sum_{i=1}^{k} y_i P(Y = y_i).$$

Here, k may be infinite. In effect, we weight each possible value of Y with the probability of seeing that value of Y, and then sum these up.

For a continuous random variable Y with probability density function (pdf) $f(y)$, we define:

$$E(Y) = \mu = \mu_Y = \int_{-\infty}^{\infty} y f(y) dy.$$

Thus, the mean is just the first moment of a distribution.

Certain values cannot be computed on all types of data — for example, if the data consists of observing which category (of k categories) each observation falls in, then the mean is not a sensible calculation and neither is the median. Here, the mode is the only reasonable measure of central tendency of those we are discussing. In the case of continuous variables, the mean, mode, and median are all applicable measures, although the mode sometimes conveys no real information about samples (if, for example, the observations are all different).

As has been shown in Chapter 6, under most conditions, the sample mean is a good estimate of the population mean. In general, we take the maximum likelihood estimate of a parameter, since this is often unbiased, of minimum variance or asymptotically of minimum variance, and has a known and tractable distribution.

However, as noted earlier, by itself, the mean does not convey much information without some measure of how the distribution of the variable is spread out about the mean.

7.4 MEASURES OF VARIABILITY, DISPERSION, SPREAD OR SCALE

Measures of variability (dispersion or spread) are very important in summarizing data. These values are often coupled with the mean or median (rarely the mode) to give an idea of how concentrated the data is about the mean, as quantified through Chebyshev's Theorem and the Empirical Rule.

Definition 8: (Measures of Dispersion: The Variance). The **sample variance** of a random sample from a continuous distribution is:

$$S^2 = \frac{1}{n-1}\sum_{i=1}^{n}(Y_i - \bar{Y})^2 = \frac{1}{n-1}\left[\sum_{i=1}^{n}Y_i^2 - \frac{1}{n}\left(\sum_{i=1}^{n}Y_i\right)^2\right].$$

The equality at the end is sometimes called the **computing formula for the sample variance**.

The **population variance** is:

$$\sigma^2 = E((Y - \mu_Y)^2) = E(Y^2) - \mu_Y^2$$

For discrete distributions, the variance is:

$$\sigma^2 = \sum_{i=1}^{k}(y_i - \mu_Y)^2 P(Y = y_i) = \sum_{i=1}^{k}y_i^2 P(Y = y_i) - \mu_Y^2$$

and for continuous random variables:

$$\sigma^2 = \int_{-\infty}^{\infty}(y - \mu_Y)^2 f(y)dy = \int_{-\infty}^{\infty}y^2 f(y)dy - \mu_Y^2.$$

The last equality at the end of each of these last two formulae is the computing formula for variances in populations. The variance is the **second central moment** of the distribution.

The computing formulae for samples or populations are very useful results, and it is good practice to prove these results. One observation for populations is:

$$E(Y^2) = var(Y) + \mu^2 = var(Y) + E^2(Y).$$

A value related to the variance is the standard deviation, namely, $\sqrt{\sigma^2} = \sigma$ for populations and $\sqrt{S^2} = S$ for samples. It is sometimes called a **measure of scale**. The variance has units that are the square of the random variable's, while the standard deviation has the same units as the random variable.

The mean and variance/standard deviation are the most commonly computed values for a data set, although others are used in cases where distributional assumptions are suspect. One reason for the use of the mean and variance arises from:

Chebyshev's Theorem/Inequality: Let X be a random variable with finite mean μ and variance σ^2. Then for each $k > 0$, $P(|X - \mu| \geq k\sigma) \leq 1/k^2$.

As noted earlier, Chebyshev's inequality applies to sample data as well as populations; that is, if we replace the population mean and variance by the sample mean and variance, the inequality remains true.

The effect of making distributional assumptions is illustrated in the comparison of what can be said for general distributions (Chebyshev's Inequality) and for normal or bell-shaped distributions:

Empirical Rule: If a distribution is bell shaped, that is, symmetric about its mean, unimodal, with thin tails and roughly shaped like the normal distribution, then:

1. about 68% of the data will be within one standard deviation of the mean;
2. about 95% of the data will be within two standard deviations of the mean; and
3. virtually all (about 99.7%) of the data falls within three standard deviations of the mean.

One simple application of the Empirical Rule is in determining if a given data set supports the assumption of a normal distribution. For the data in Example 4, normality would be highly suspect.

Example 8: Consider the following 30 random observations of the time X (in minutes) from the application of an icepack to a swelling caused by being kicked in the shin during a soccer match until sufficient reduction in the swelling:

52.2863	50.7812	49.5684	48.8967	54.5934	48.8390	46.5186	48.1714
51.5975	47.1476	53.2417	50.8456	51.5318	46.2927	48.2167	47.9392
48.1951	52.5491	49.5108	51.1132	49.7843	47.8915	49.3946	48.1437
51.1300	49.5702	54.0702	47.1648	47.8930	53.3626		

Summary values include: $\sum x_i = 1496.241$ and $\sum x_i^2 = 74766.036$. From these values, the sample mean is 49.875, the variance is 5.223, and the standard deviation is 2.285.

For this data, we find that 20 observations are within one (sample) standard deviation of the mean, and 29 are within two standard deviations. Thus, two-thirds are within one standard deviation of the mean, and about 97% are within two standard deviations. This data set supports the assumption that the distribution of times is normally distributed.

Example 9: Consider the data in Example 4, and assume the values are the differences Y in liters (L) of milk produced by a random selection of Guernsey milk cows and the average 16.5 liters produced by such cows per day. Summary values include $\sum y_i = -19.9066$ and $\sum y_i^2 = 1090.647$. From these values, $\bar{y} = -0.4977L$ and $s^2 = 27.7113L^2$ [discrepancies in calculations in the last decimal may occur due to round-off procedures]. The standard deviation is $5.2642L$.

One of the drawbacks of a mean as measure of central tendency is that is adversely affected by one or two extreme values. There is a similar problem with the variance or the standard deviation as a measure of dispersion or scale.

Example 10: Again consider the data in Example 4. Suppose the single value -29.5215 is removed, and calculations proceed on the remaining 39 observations. Removing highly suspect values is called **censoring** and should not be done without very careful consideration.

For our remaining data, $\sum_{i=1}^{39} y_i = 9.6149$ and $\sum_{i=1}^{39} y_i^2 = 219.130$. The sample mean is 0.2465 and variance is 5.7042. The standard deviation is 2.3883. Eliminating one extreme value has changed the variance and standard deviation by a large amount.

Chapter 7 — Descriptive Statistics

The last example illustrates why other measures of location and scale are used.

Definition 9: (Quantiles). The term **quantile** refers to any number used to partition samples or distributions into two parts based on frequencies of sizes of values.

Quartiles: Quartiles are numbers that break data into four parts: The **first quartile** Q_1 is a value so that at least 25% of the observations are less or equal this number and at least 75% are greater or equal to it. The **second quartile** Q_2 is a value so that at least 50% of the observations are less or equal this number, and at least 50% are greater or equal to it. The **third quartile** Q_3 is a value so that at least 75% of the observations are less or equal this number, and at least 25% are greater or equal to it.

Deciles: Deciles are numbers that break data into 10 parts: The **first decile** D_1 is a value so that at least 10% of the observations are less or equal this number and at least 90% are greater or equal to it. The **second decile** D_2 is a value so that at least 20% of the observations are less or equal this number and at least 80% are greater or equal to it. The remaining deciles are defined in the obvious way.

Percentiles: Percentiles are numbers that break data into 100 parts: The **first percentile** P_1 is a value so that at least 1% of the observations are less or equal this number and at least 99% are greater or equal to it. The **second percentile** P_2 is a value so that at least 2% of the observations are less or equal this number and at least 98% are greater or equal to it. The remaining percentiles are defined in the obvious way.

Note that the median is just the 50th percentile, the second quartile, or the fifth decile.
Some other measures of dispersion include:

Definition 10: The **variation ratio** is denoted V, and defined by $V = (n-\text{modal frequency})/n$, and is used for categorical data. The **modal frequency** is the frequency of the mode, i.e., how often the mode occurs. The variation ratio is also $V = 1-(\text{modal frequency})/n$, or the relative frequency of all of the non-mode values.

Example 11: For the butterfly data of Example 5, the mode is 80, with frequency 5, so $V = 1 - 5/20 = 3/4$.

Definition 11: The **mean absolute deviation** is denoted mad and is defined by: $mad = \frac{1}{n}\sum_{i=1}^{n} | X_i - \overline{X} |$.

Example 12: Consider the data in Example 8. In this case, $mad = 1.9069$.

Definition 12: The **coefficient of variation** is denoted cv and is defined by: $cv = s/|\bar{x}|$ for samples, and $cv = \sigma/|\mu|$ for populations. The absolute value ensures this measure is positive, and cv is not defined for means of 0. This quantity is meaningful only when the variable is measured on a ratio scale, that is, when values are measured relative to a fixed unit.

Example 13: Consider the data in Example 8. For this data set, the coefficient of variation is 10.58, indicating the data is widely dispersed about the mean. For the icepack data in Example 8, the coefficient of variation is 0.049 indicating the data is compactly distributed about the mean.

Definition 13: The **semi-interquartile range** is denoted SIR, and is defined as $\text{SIR} = (Q_3 - Q_1)/2$. $Q_3 - Q_1$ is the interquartile range.

Example 14: For the data in Example 8, $Q_1 = 48.13$, $Q_2 = 49.54$, and $Q_3 = 51.54$. Therefore, the semi-interquartile range is $(51.54-48.13)/2=1.71$ (approximately).

A useful observation is that virtually all of the time, $Range/4 \geq s \geq Range/6$, with $s \approx Range/4$ for bell-shaped data. This gives us a way to either quickly estimate the variance of a sample or check that our calculations are correct. If asked for an approximate value of s, we will use $Range/4$, which will usually overestimate the value by a small amount.

Example 15: Consider the data in Example 8. The first-order statistic is $x_{(1)} = 46.2927$ and the 30th-order statistic is $x_{(30)} = 54.5934$, so the range is 8.3007. Then Range/4 = 2.0752. Note that the standard deviation is 2.285.

7.5 VISUAL DISPLAYS

Visual displays of data can suggest that it came from a bell-shaped distribution, i.e., one for which the shape of the normal is a good overall approximation. If so, the distributional assumption is justified, and the methods for dealing with normal distributions can be used with confidence. Alternatively, a graphical display may suggest the distribution is skewed, or that the tails are thicker than one would expect from a distribution that is at least close to normal, or that the distribution is bimodal. Other assumptions such as a mean of zero or other assumed parameter values may not be supported by an appropriate visual display, and such displays may help pinpoint specific problems such as outliers in the data. Although there are specific procedures and more formal quantitative approaches to assessing the validity of assumptions, such as **goodness of fit** tests, or tests of normality, etc., in this section, we concentrate on types of visual displays and what we might glean from them.

There are many different displays, some specific to categorical data, and others best used with measured data. We will look at some basic displays for such data.

7.5.1 Visual Displays: Categorical Data

Suppose that we have **categorical data**. This occurs when we have a fixed number of mutually exclusive and exhaustive **categories** (also called **bins**), and there is no order relation between the categories. The simplest version of this is Bernoulli trials where the data consists of 1s and 0s, or successes and failures. Other examples include blood types, ruminants in a forest, and eye color. We might be interested in the distribution of combinations of different alleles or species of trees.

Let k be the number of categories, and typically, k is a relatively small number. A standard display is the **bar chart**. Code the categories from 1 to k, and label the horizontal axis in a bivariate coordinate system the categorical factor, with the numbers 1 to k equally spaced. The vertical axis is the frequency axis, and its scale is determined by the most frequently occurring category. Bars, starting from or centered on each of the numbers 1 to k, of equal width and with small gaps between them, are erected over the horizontal axis. The heights of the bars are the individual **(absolute) frequencies**, that is, the actual number of occurrences, of the categories. This kind of display reflects the proportion of time each category occurs, relative to the other categories. Sometimes, the vertical axis is the **relative frequency**, determined by taking each n_i, the number of times category i occurs, and dividing by $n = \sum_{i=1}^{k} n_i$, the total sample size. The ith bar then has height n_i/n.

Example 16: In the ABO and Rh Blood typing system, there are eight types: A+, A−, B+, B−, AB+, AB−, O+, and O−. The distribution of these types varies between populations and across populations. In a study to determine the medical needs of an isolated community in Canada, a first step was to determine the distribution of these blood types and compare the results to the general Canadian population. One hundred residents of the community were selected at random, and their blood types determined, with the following results:

Chapter 7 — Descriptive Statistics

	ABO and Rh Blood Type							
	O+	O−	A+	A−	B+	B−	AB+	AB−
Canada (%)	39	7	36	6	7.6	1.4	2.5	0.5
Community (%)	35	6	40	8	7	2	2	0

A bar chart presents the information in a more readily understood way:

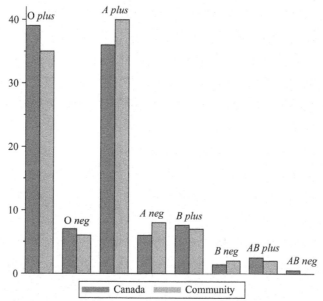

FIGURE 7-2. Comparison of AB and Rh Blood Types.

It is easier to examine the data across all categories at once with such a diagram. There are discrepancies, but they may very well be based on the particular sample.

Another simple display is the **pie chart**. A disc is divided into sectors, one for each category, with the area of the sector for category i equal to $n_i/n \times$ the total area of the circle.

Example 17: A farmer is trying to maximize his profits. One factor to be considered is how much of his feed budget is being used by the different types of animals he raises. He randomly selects a month and determines the total feed cost and the total cost per type of animal being raised to be:

Total Cost of Feed in the Month: $ 7,000

Feed Cost: Cattle: $ 3,000; Sheep: $ 500; Horses: $ 1,000; and Chickens: $ 2,500.

He draws a pie chart of the data. From such diagrams, the farmer can readily compare this particular cost element in an overall effort to determine the mixture of animals being raised to maximize profits.

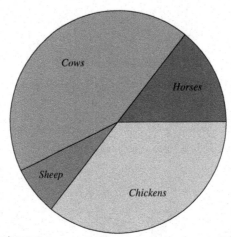

FIGURE 7-3. Pie Chart of Feed Costs per Animal Type.

Other displays are used, depending on what is under study. For example, one may use cut-out figures to represent elements of the workforce, with the relative sizes indicating the percentage of workers in particular industries.

7.5.2 Visual Displays: Continuous Distributions

Stem-and-leaf displays are simple displays, in which we choose as bins or **stems** the main part of the observations, and then the leaf is the difference between the actual value and the stem. Normally, the stem will be some integer.

Example 18: In a study of the production of raspberries, a farmer randomly selected 30 plants at the beginning of July, counted the number of immature berries on the plant, and recorded these values:

43	51	37	39	46	43	50	39	44	47
43	55	61	29	45	50	43	39	49	38
37	40	48	37	38	46	51	47	52	61

A stem and leaf display, using tens as the stem and single digits as the leaves, takes the form:

Stem	Leaf
20	9
30	77788999
40	0333345667789
50	001125
60	11

This display indicates that analyzing the data as if it were normally distributed would be supported. Although, strictly speaking, this is not continuous data (the possible values are whole numbers and so do not fill out an interval); it is scale data, and a continuous approximation to the distribution would be appropriate.

Chapter 7 — Descriptive Statistics

For data from a continuous distribution, the typical kind of plot is the **frequency histogram** or the **relative frequency histogram**. It should be emphasized that we want simple displays, so that our creativity in developing a display comes into play. From our data, we have a largest value and a smallest value, and we choose to create "bins" or classes to put our observations into. The number of bins depends to a great extent on the number of observations: There are several ways to choose this number, but we will use as a rule of thumb that we want an average of at least 5 observations per bin, coupled with the number of bins taken to be approximately the square root of the number of observations, usually rounded up. If we have 20 observations, we may choose 4 or 5 bins, while if we have 100 observations, we might choose about 10 bins. The number of bins is flexible, and we use this flexibility to keep the display as simple as possible.

Once we have the number of bins, then we know the approximate length of each bin. In most cases, we want our bins to be of equal length, and the bins cover all of the data. An exercise is to go to the literature and find examples of bins of unequal length and how this is handled. We set class boundaries, which are values to one-half of the smallest decimal unit recorded. For example, if our data consists of whole numbers, the boundaries will be numbers of the form $m.5$. Normally, if the smallest value observed is 10, then the first-class boundary may be 9.5; but again, we want to choose this value for simplicity, so we may choose a value a bit further to the left. If the largest value is 60, then the upper-class boundary will be 60.5 or something close, but to the right. This value is actually determined once the first lower-class boundary is set and the class length is determined. The class boundaries result from adding the class length to $m.5$ until all the data has been covered.

The class limits are the largest and smallest values that can actually fall into a class, say l_i and u_i for class or bin i. We sometimes write the classes in the form $[l_i, l_{i+1})$. This means l_i is a value that can be in this class, but l_{i+1} is not. The value l_{i+1} is then the lower class limit of the next class. These are then placed along a horizontal axis. The vertical axis is the count of the bins for a frequency histogram, or the relative frequency for a relative frequency histogram, or the percentage frequency for a percentage frequency histogram.

One other point to keep in mind is that the fewer the bins, the less the detail that can be modeled. On the other hand, more bins will show finer details. Increasing the number of bins also increases the complexity of the diagram, and one significant reason for using such displays is to make the information easier to comprehend. Again, it is our creativity in selecting an appropriate display that plays a significant role.

The **dot plot** for numerical data is simply the plot of data along a line parallel to the horizontal axis labeled with the particular variable and scaled so that all observations can be plotted. Repeated values are stacked on top of each other. From this, we can see where values are bunching up, and where there are major gaps between values. A **line chart** is, typically, a plot of the observations against the numbers 1 to n, the order in which they are taken, with consecutive observations joined by a line segment. One use of this type of plot is in determining if there are patterns to the observations and if there are extreme values.

Box plots or box-and-whisker plots are also called five-point plots. A box plot uses the median, first and third quartiles, and extremes [or 5th and 95th percentiles for large data sets]. On a two-dimensional display, the vertical axis is scaled to reflect the range of the observations. A rectangular box is drawn in, with the top of the box at height equal to the third quartile, and the bottom of the box at the first quartile. A horizontal line in the box is drawn at the height of the median. At the midpoint of the top of the box, a vertical line extends to the greatest value, and at the bottom, a line from the midpoint down to the smallest value is drawn.

An **outlier** is deemed to be any value that is more than $3 \times$ SIR (the semi-interquartile range) above the third quartile or any value that is more than $3 \times$ SIR below the first quartile. An asterisk or other special symbol is used to denote the outliers, and if two values are the same and deemed outliers, then a "$\times 2$" is placed beside the asterisk.

Ogives are plots of cumulative probability or absolute frequency, connected with straight lines, through class boundaries. An ogive can be interpreted as an approximation to the cumulative distribution function of the parent distribution.

Example 19: An ornithologist is studying effects of winter weather on the spring weight of adult northern cardinals living in Southern Ontario. These are non-migratory birds and are often seen in urban back yards with hanging bird feeders. With the help of an assistant, she randomly selects 50 such cardinals from randomly chosen locations in a large urban area at the end of March and records their weights in grams. The original data is:

39.93	38.03	42.56	52.41	41.28	39.34	37.69	39.93
42.01	54.45	39.66	37.56	45.91	38.36	46.92	40.32
36.73	39.63	39.10	50.19	46.10	39.66	40.92	40.47
54.95	50.69	38.07	39.95	40.41	48.87	38.77	40.22
41.79	39.83	39.12	37.67	54.04	43.14	37.07	39.80
38.76	40.34	42.69	38.65	37.86	40.43	40.54	39.89
43.46	39.99						

and then ordered:

36.73	37.07	37.56	37.67	37.69	37.86	38.03	38.07
38.36	38.65	38.76	38.77	39.10	39.12	39.34	39.63
39.66	39.66	39.80	39.83	39.89	39.93	39.93	39.95
39.99	40.22	40.32	40.34	40.41	40.43	40.47	40.54
40.92	41.28	41.79	42.01	42.56	42.69	43.14	43.46
45.91	46.10	46.92	48.87	50.19	50.69	52.41	54.04
54.45	54.95						

Some summary values are:
mean = 41.92, var. = 22.86, skew. = 1.48, kurtosis = 4.10, median = 40.11, min = 36.73, max = 54.95, Q_1 = 39.07, Q_2 = 40.11, Q_3 = 42.73, P_{10} = 37.75, P_{90} = 50.51

From these values, we can see that the sample has characteristics that do not support the assumption of a normal distribution — the skewness coefficient should be nearer 0 and the kurtosis coefficient should be closer

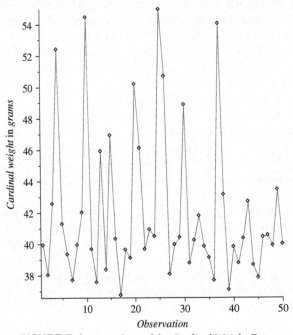

FIGURE 7-4. Line Chart of the Cardinal Weight Data.

to 3. The semi-interquartile range is 1.83, and so any value that is greater than $42.73 + 3 \times 1.83 = 48.22$ or less than $39.07 - 3 \times 1.83 = 33.58$ will be deemed an outlier. In this case, there are seven outliers, all greater than 48.22. A sample of size 50 from a normal distribution may have one outlier as determined by this method, since under the assumption of normality, this calculation determines those values that are more than 2.7 standard deviations away from the mean. Recall that the normal is symmetric, so the mean and median should be close.

A line chart of the data is given in Figure 7-4.

This line chart suggests that the data is not from a normal distribution. Several values are detached from the main body of observations — there seems to be a significant gap between about 10 of the observations and the other 40.

A box plot, also called the five-point plot or the box-and-whiskers plot, uses Q_1, Q_2, Q_3 and the maximum and minimum. The box plot of this data follows.

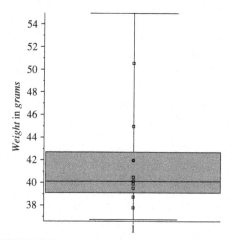

FIGURE 7-5. Box Plot of the Cardinal Weight Data.

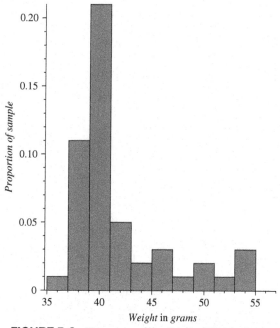

FIGURE 7-6. Histogram of the Cardinal Weight Data.

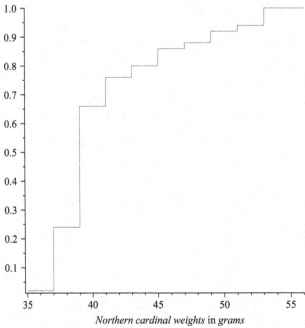

FIGURE 7-7. Ogive of the Cardinal Weight Data.

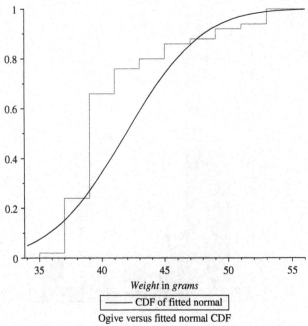

FIGURE 7-8. Ogive with Normal Approximation of the Cardinal Weight Data.

Chapter 7 — Descriptive Statistics

The box plot further solidifies the view that the data is not likely to have come from a normal distribution, since nonzero skewness is very much in evidence.

A histogram of the data can be thought of as a discrete approximation to the parent distribution. The number of bins and the start, end, and length of the bins (categories) are chosen to provide insight into the distribution without being either too complicated or too simple. There are several ways to determine the number of bins, with the simplest being to have about five observations on average in the bins; so here, about 10 bins are appropriate, and nine were chosen. It is also convenient to have the bins centered at simple values, such as whole numbers or integers +1/2. In this case, with bin width set to 1, starting at 35 (as the first lower-class limit, making 34.995 the first lower-class boundary), the resulting histogram is given in Figure 7-6.

An ogive of the northern cardinal weight data is an approximation to the cumulative distribution function. In this case, the ogive is given in Figure 7-7.

The weights of the northern cardinals are not easily analyzed in raw form, or even in sorted form. These diagrams begin to reveal some interesting facets of the data: If our researcher assumes the weights are normally distributed, the box plot and the identification of possible outliers begin to call the normality assumption into question. The histogram also calls this assumption into question, since the histogram is skewed to the right. The ogive supports the possibility of non-normality, since it illustrates again that the distribution is skewed to the right. An ogive compared to a fitted normal cumulative distribution function, meaning we have used the sample mean and the sample variance as the values of the corresponding parameters in a normal distribution, by plotting the two on the same axes suggests strongly that a normal distribution is not a good fit to this data.

The assistant's assigned sampling areas included a sector that had received significantly less snow in the past year, and the area was more affluent, so the birds had more feed and less winter stress. The assistant captured (and released) 10 cardinals from this area, plus 15 from other areas. The researcher sampled 25 birds only from areas where the winter snows had been much heavier. Such situations where two different populations (in this case, differentiated by the winter conditions and food supply) are sampled as if they were one are not uncommon.

This example illustrates the value of retaining all (or at least as many as possible) aspects of how the data was gathered. It may be important to know the geography or weather patterns in the sampling areas, or even pictures of the birds, since there may be two species that look very similar but can be separated by a close examination of characteristics.

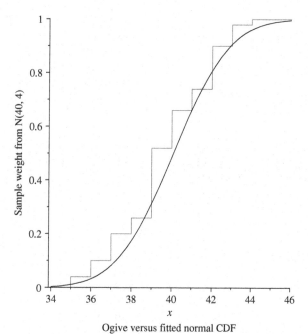

Ogive versus fitted normal CDF

FIGURE 7-9. Ogive with Normal Approximation of the Second Set of Cardinal Weight Data.

A similar study in a separate urban area that had been uniformly blanketed with snow over the winter yielded the following ordered data:

35.23	35.91	36.37	36.43	36.86	37.20	37.25	37.76
37.89	37.92	38.03	38.26	38.75	39.05	39.11	39.21
39.33	39.37	39.45	39.62	39.62	39.67	39.82	39.85
39.93	39.98	40.06	40.17	40.31	40.36	40.47	40.67
40.83	41.24	41.28	41.32	41.60	42.22	42.25	42.26
42.37	42.56	42.59	42.62	42.94	43.03	43.28	43.54
43.74	44.30						

In this case, the ogive and the cdf of the normal approximation given in Figure 7-9 are much better matches, and the assumption of normality would not be seriously in doubt.

It should be noted that, frequently, two or more populations are under study, and we can then compare the sample information from each by constructing on the same axes histograms, or box plots, or ogives (as well as other displays).

7.6 SUMMARY

Summary statistics, including the mean, variance, skewness coefficient, and the kurtosis coefficient, as well as particular quantiles, can help highlight information in data sets. Visual displays help us understand data in ways simple lists of raw data or even ordered data cannot. We can get a useful overview of the entire data set from these summary displays.

It is also important to make use of visual displays in a way that is properly representative of the data. For example, histograms can have bins of unequal width, but if that option is selected, then an adjustment has to be made to the heights of the bars to ensure the data is in proper perspective.

There are formal tests of the fit of an assumption with the data. Often, such tests are conservative in nature, meaning that in order to identify, say, non-normality, the sample size has to be quite large. In many problems in the life sciences, samples are modest in size, so the summary values and the visual displays presented in this chapter may help pinpoint problems formal tests cannot. It is, therefore, very important to analyze the data with descriptive statistics techniques.

Exercises

1. Twenty-five students at a local college were selected at random from all of the students at the college, and their IQs, X, recorded:

111	108	165	97	122
105	104	120	99	114
118	103	117	119	111
108	109	101	111	123
136	105	118	109	115

(a) Determine the mode of this data, if there is one.

(b) Draw a stem and leaf display of the IQs.

(c) Order the data, and then determine the first, second, and third quartiles, as well as the maximum and minimum. Use these values to create a box plot.

(d) Based on the results in (c), determine the semi-interquartile range, and determine if there are suspected outliers.

(e) Given $\sum x_i = 2948$ and $\sum x_i^2 = 328992$, determine the mean and variance of this data set.

(f) With bins starting at 94.5 of width 10, draw a histogram of this data.

(g) Draw a line chart for this data.

2. At a medium-size university, a total of 1,400 students is enrolled in Faculty of Science programs. This Faculty consists of nine departments: Biology, Chemistry, Computer Science, Geography, Health Science, Kinesiology, Mathematics, Physics, and Psychology. The distribution of enrollments are:
Biology: 200; Chemistry: 100; Computer Science: 200; Geography: 200;
Health Science: 150; Kinesiology: 150; Math: 100; Physics: 50; Psychology: 450

(a) Draw a bar chart of this data.

(b) Draw a pie chart of this data.

(c) Determine the modal class, that is, the most frequently occurring class.

(d) Assume this is a population and individual students are enrolled in only one of these departments. Records of students are randomly selected, and each student record is equally likely to be selected. Determine the probability that a randomly chosen record is that of a student in either Psychology or Physics.

3. Eighteen mature earthworms were randomly gathered from damp lawns on a given night and their lengths in inches recorded as:

3.156	3.684	3.933	3.945	4.054	4.077	4.399	4.464
4.555	4.580	4.765	4.823	4.876	5.073	5.163	5.315
5.596	6.252						

Use the methods of this chapter to determine if the assumption of a normal distribution to the lengths of mature earthworms (in the general area sampled, at least) is supported by this data.

4. One hundred and fifty market-ready farmed Atlantic cod were selected at random from a cod farm, and various health measurements, including weight X in kilograms, were recorded to ensure the farming process produced healthy and edible fish. The weights are given in the following chart.

(a) Determine the following quantiles: Q_1, Q_2, Q_3; P_{10}, P_{90}; D_2, D_8.

(b) Determine the range of the data.

(c) Calculate the semi-interquartile range, and determine if there any suspected outliers.

(d) Create a histogram of this data, using bin widths of 0.1, with an appropriately chosen first class boundary.

(e) Does this data reflect the assumption of a normal distribution for the weights of market-ready cod in this fish farm? Use techniques of this section to support your argument either for or against such an assumption.

3.12	3.27	3.36	3.42	3.44	3.46	3.46	3.49	3.51	3.51
3.52	3.52	3.52	3.53	3.53	3.53	3.54	3.55	3.55	3.56
3.56	3.56	3.59	3.60	3.60	3.60	3.61	3.63	3.63	3.63
3.65	3.65	3.65	3.67	3.68	3.68	3.68	3.69	3.69	3.69
3.69	3.70	3.70	3.70	3.71	3.71	3.72	3.72	3.73	3.74
3.74	3.75	3.76	3.76	3.76	3.77	3.77	3.78	3.78	3.78
3.78	3.78	3.79	3.80	3.80	3.81	3.81	3.81	3.82	3.83
3.83	3.84	3.84	3.85	3.85	3.85	3.86	3.86	3.86	3.87
3.87	3.87	3.87	3.87	3.88	3.89	3.90	3.90	3.90	3.90
3.90	3.91	3.91	3.91	3.91	3.92	3.92	3.94	3.95	3.95
3.95	3.95	3.96	3.97	3.97	3.97	3.98	3.98	3.99	4.00
4.00	4.03	4.03	4.03	4.04	4.05	4.07	4.08	4.09	4.09
4.10	4.11	4.12	4.12	4.14	4.14	4.14	4.14	4.14	4.15
4.16	4.16	4.16	4.16	4.17	4.17	4.18	4.18	4.20	4.22
4.23	4.26	4.28	4.29	4.30	4.30	4.30	4.32	4.38	4.39

5. Mature cattails average 120 cm at full growth, with a variance of 81 cm^2. Use the methods of this chapter to determine if the 30 observations below of mature cattails from a remote swampy area are consistent with the overall population, which is assumed normally distributed.

102.26	103.67	109.08	115.27	116.13	116.92	116.96	117.83	118.58	121.29
121.90	121.97	122.30	122.49	122.50	123.58	124.10	124.30	125.55	126.24
126.45	126.45	128.56	128.66	128.94	129.85	133.56	134.18	137.29	143.51

6. Twenty mature white mice are fed in a collective-feeding station. The amount of food available is 10% less than what is normally required for this many mice, to determine effects of competition. This continues for two weeks, and then the difference between the current weight and the weight at the start of the experiment is recorded for each. The ordered differences are:

−2.05	−1.44	−0.75	−0.73	−0.65	−0.55	−0.48	−0.19	−0.11	−0.03
−0.03	0.02	0.05	0.12	0.20	0.32	0.63	0.87	1.55	1.92

Use the methods of this chapter to determine if the change in weight data supports the assumption that the random variable Y = weight change under the specified diet follows a normal distribution.

7. The following 15 values are the number of months Y a patient survives from the point of diagnosis of end-stage cancer (of a particularly virulent type):

0.17	0.79	1.04	1.31	1.71	1.98	2.42	3.19	4.43	5.11
5.90	8.34	13.32	19.66	27.16					

(a) Prepare a line chart of the data, and comment on whether a normal distribution would be a good model for the life expectancy for such patients. What general observations about an ordered normal sample's line chart does not fit with this ordered data's line chart?

(b) Find the 15 percentage points of the exponential distribution with parameter 5 by solving $\int_0^{x_i} \exp(-x/5)/5 \, dx = i/16$, $1 \le i \le 15$. Now plot the pairs (x_i, y_i). Does this $Q-Q$ plot suggest such an exponential distribution is a reasonable fit?

Chapter 7 — Descriptive Statistics

8. Consider the following 20 ordered observations on random variable U:

10.02	10.13	10.65	10.99	11.29	12.31	12.54	12.64	12.87	12.92
13.28	13.53	13.85	14.02	14.24	15.02	15.14	16.90	17.20	18.47

(a) Prepare a line chart of this data, and comment on whether this data is likely from a normal distribution. If a normal distribution is not supported, suggest a distribution that might be a better fit.

(b) Evaluate the mean, median, and variance of this data.

(c) Compare the Empirical Rule results to the proportions of the data between 1, 2, and 3 (sample) standard deviations away from the sample mean. Do your conclusions about the normal distribution assumption differ from that in (a)?

(d) Using methods of this chapter, examine the suggestion that the data is a random sample from a uniform distribution on [10, 20].

9. Consider the following 40 observations of the random variable W, the time in days until a fresh cut of approximately three cm in length and no more than 1 cm deep has healed:

14.84	15.61	18.71	20.14	20.29	21.16	21.17	21.42	21.43	22.06
22.27	22.61	22.74	22.81	22.97	23.85	24.15	24.33	24.72	24.72
24.88	25.02	25.06	25.15	25.36	26.07	26.25	26.34	26.67	27.03
27.63	27.79	28.40	31.81	33.41	35.14	35.17	35.28	36.77	38.02

(a) Make a line chart, histogram, and other appropriate diagrams to explore the assumption that the data comes from a normal distribution.

(b) Determine the mean and variance of this data. Compare the results of the Empirical Rule to similar calculations of proportions within 1, 2, and 3 standard deviations of the mean.

(c) Evaluate the three quartiles, and calculate the semi-interquartile range. Also plot the box plot of this data.

(d) From an appropriate source, find information about the Laplace distribution, and determine if a Laplace distribution with parameters $a = 25$ and $b = 5$ is a suitable fit to this data. The Laplace distribution has density:

$$f(x) = \frac{1}{2b} \exp\left(-\frac{|x - a|}{b}\right), x \in \mathbb{R}.$$

10. The total rainfall (with snowfall converted to equivalent rainfall) in a month has been recorded at an environmental station for many years. A random sample of 20 months from across the years is selected, and the total rainfalls in those months are:

1.89	38.48	28.45	2.65	9.30
32.62	2.01	5.07	7.86	16.48
6.49	8.27	13.88	7.24	3.22
5.68	3.67	52.23	5.70	6.81

(a) Determine the mean and variance of this data.

(b) Plot a histogram, and determine if a normal distribution is a reasonable assumption for the monthly accumulated rainfall at this station.

(c) The natural logarithms of the monthly accumulated rainfall are:

0.64	3.65	3.35	0.97	2.23
3.48	0.70	1.62	2.06	2.80
1.87	2.11	2.63	1.98	1.17
1.74	1.30	3.96	1.74	1.92

(i) Make a histogram of this data.

(ii) Determine whether a normal distribution would be a reasonable assumption for this transformed data.

CHAPTER 8

CONFIDENCE INTERVALS AND HYPOTHESIS TESTING

Synopsis

Procedures for generating confidence interval estimates for a range of parameters are developed and applied. Tests of hypotheses concerning parameters are outlined, along with the general procedure for carrying out such tests of hypotheses. The link between confidence intervals and hypothesis tests is explained.

8.1 INTRODUCTION

Example 1: Let the distribution of the height X of mature golden delicious apple trees in large orchards be normal, with mean μ meters and standard deviation of 0.3 meters. It is possible that we do not know the mean in a particular orchard because of differences in soil, weather patterns, etc., but past history has shown that the variance of the heights of such trees is constant, regardless of these other factors. The problem is to estimate the average height of these trees in a large orchard. Since the trees were planted at the same time, in well-laid-out rows and columns, it is easy to select a random sample of reasonable size from the orchard, and there are too many trees to do a complete census. From what we have discussed in Chapters 6 and 7, a good estimator of the average height will be the sample mean. We set up the experiment to randomly select 25 mature trees from the orchard and then measure the heights of these trees, denoted $x_1, \ldots, x_n$. In terms of previous discussions, these are the respective values of the *iid* random variables $X_1, \ldots, X_{25}$, each with the same distribution as X.

In Example 1, what is the probability that our sample average will be exactly the same as the true mean height of the trees in this large orchard? This probability must be 0. We are taking measurements, so the random variable, the height of a mature golden delicious tree, is a continuous random variable. The theory of probability for continuous variables requires that the probability of any individual value be zero. Second, since we are taking a sample, it is possible more of the taller trees are selected by chance than are representative of the entire population of trees. Thus, the sample mean $\overline{X}$ can be "significantly" larger than μ, the true population mean (we will discuss what we mean by "significantly" shortly). Similarly, we might have a sample that includes a preponderance of trees that are shorter than average.

Under our assumptions in Example 1, $\overline{X} \sim N(\mu, 0.09/25)$. This means $\overline{X}$ satisfies the condition of having a bell-shaped distribution, so about 67% of such means should be within 0.06 meters of the true mean, about 95% within 0.12 meters, and virtually all within 0.18 meters of the true mean. If we take a large number of random samples of size 25, calculate the average height of the trees in each sample, and then average the averages, that final average should be very close to the true mean. Within 0.120 meters of this average of averages, we should find about 95% of all of the sample averages. However, we will in fact be running the experiment once and have a single average or sample mean. We then want to quantify how close it should be to the true average height in the population. Basically, in 100 runs of the entire experiment, we expect 95 of the sample means to be within 0.12 meters of the true height (this applies to the long-run average, not to just 100 observations, but helps illustrate the point being made). If we randomly select one of these averages, it should be in the interval $(\mu - 0.12, \mu + 0.12)$ 95 times out of 100 (roughly speaking). About 5% of the time it will be further than 0.12 meters from the true mean.

Another way to look at the analysis in the previous paragraph is that any one of the randomly chosen sample means $\overline{X}$ satisfies $\overline{X} \in (\mu - 0.12, \mu + 0.12) \Leftrightarrow \mu - 0.12 < \overline{X} < \mu + 0.12$, 95% of the time. We can reverse this: If $\mu - 0.12 < \overline{X} < \mu + 0.12$ then $\overline{X} - 0.12 < \mu < \overline{X} + 0.12$ 95 times out of 100 (on average). So, for a given sample mean $\overline{X}$, the true population mean will be within 0.12 meters of the particular sample mean 95 times out of 100. We call $(\overline{X} - 0.12, \overline{X} + 0.12)$ a 95% confidence interval. When we carry out the experiment, find the actual sample mean $\overline{x}$ from the data, and then compute $(\overline{x} - 0.12, \overline{x} + 0.12)$, this is said to be a 95% symmetric confidence interval for the true mean. This does not mean 95% of all orchard means fall in this interval, or 95% of sample means fall in this interval. What it does mean is that if we were to repeat the entire experiment a large number of times, this method of creating a confidence interval would capture the true mean 95% of the time. We don't know the true mean (or we wouldn't be estimating it), so a specific confidence interval either captures the true mean or it doesn't, and we can't tell. The confidence interval gives us a range of possible values of the true mean supported by the data.

In the next section, the arguments and ideas above are put on a rigorous basis.

8.1.1 Interval Estimates of a Normal Mean, Variance Known

If the point estimate of a parameter based on an estimator with a continuous distribution is always different from the true value, what good is it? One value is that even if it is not exactly the population value, it is likely to be close to it (assuming the estimator is "best," as described in Chapter 6). If the sample size n on which the estimate is based is large, we would expect that the estimate would be very close, with that kind of confidence increasing as the sample size grows. Also, it may not be essential that we know the value of the population parameter exactly, but only within some tolerance. It is important to note that we will not know from taking one sample just how close we are to the true population value—we would need a complete census to be 100% accurate and confident.

Often, what is needed is a way to quantify how frequently our experimental procedure produces an estimate that is further away than some tolerance from the true population value. One way to do this is to determine in advance the probability α of making such a "large" error in using the estimator to generate a point estimate of the parameter. Typically, experimenters want to be conservative in the sense of making big errors rarely, so the value of α is small, often 0.05, 0.025, and 0.01, and similar. Thus, $1 - \alpha$ will be the probability that our estimate is sufficiently close to the value of the parameter being estimated, based on how often we allow ourselves to make a large error. The question then is: How do we incorporate these probabilities into our experimental and estimation procedures?

Example 2: Consider the problem of estimating the mean in a normal distribution $N(\mu, \sigma^2)$. To simplify matters, assume σ^2 is known. Resources are available to take a random sample of size n from this distribution. From earlier considerations, the sample mean $\overline{X}$ is considered the best point estimate of μ, since it is unbiased, has least variance amongst unbiased estimators, and has a tractable distribution, namely $N(\mu, \sigma^2/n)$. Clearly, as the sample size increases, the variance in $\overline{X}$'s distribution gets smaller and tends to 0 as $n \to \infty$, so it is consistent. From the information about the distribution of this estimator, we want to generate an interval of values, based on data to be collected, in which we have a good level of certainty that the true mean μ will lie. We formulate this requirement as the following interval estimate problem:

> For a fixed sample size n, the estimation procedure is to produce an interval estimate that captures the true mean $(1 - \alpha) \times 100\%$ of the time, in large (read this as infinite) repetitions of the experiment.

The only way to be absolutely certain would be to take $(-\infty, \infty)$ as the interval, which means $\alpha = 0$. However, all this interval would tell us is that the mean is some number, and that is not helpful. Alternately, we can take a complete census of the population, and that is rarely practical. Thus, $\alpha > 0$ is assumed, but since we want to have a good level of confidence that the interval does in fact contain μ, we take α small, usually 0.10, 0.05, 0.025, or 0.01.

Chapter 8 — Confidence Intervals and Hypothesis Testing

Other values can be specified as well. There is a trade-off between the size of α and the boundaries, and hence, the length, of confidence intervals.

Before we can proceed, we have to decide whether we want to guard against overestimation only, underestimation only, or the possibility of either an over- or underestimation of the true mean. For the moment, we will assume we want to guard against both under- and overestimation. This means the interval will be of the form (L, U), that is, $L < \mu < U$ with a given level of confidence, where L and U are functions of the data, and hence, are random variables. In advance of carrying out the experiment, we will not know what L and U are, but we should have functions of the data that generate them, just as we need a function of the data to generate a point estimate of the mean.

Let μ be the true mean. If the interval estimate is (L, U), and $\mu \in (-\infty, L)$, we have overestimated the mean. On the other hand, if $\mu \in (U, \infty)$, we have underestimated the mean. The standard procedure is to guard against over- and underestimation equally, which means $P(\mu < L) = P(\mu > U) = \alpha/2$. The interval (L, U) is called a symmetric confidence interval because we guard against over- and underestimation equally, not because of the form of the interval.

So far, we have determined that the interval estimate should take the form of (L, U), with the conditions that $P(L < \mu < U) = 1 - \alpha$ and $P(\mu < L) = P(\mu > u) = \alpha/2$. This is where a pivotal statistic for μ enters. As noted at the end of Chapter 6, the pivot for μ under the given conditions is $\sqrt{n}(\overline{X} - \mu)/\sigma$ and its distribution is standard normal. Essentially, we want to introduce the pivot by algebraic manipulations that do not alter the solution set of the inequalities. So, first, we multiply by -1 and that reverses inequalities, giving us:

$$L < \mu < U \Leftrightarrow -L > -\mu > -U.$$

Now we add $\overline{X}$ to all three sides, and then multiply by $\sqrt{n}/\sigma$ so that:

$$L < \mu < U \Leftrightarrow \overline{X} - L > \overline{X} - \mu > \overline{X} - U$$
$$\Leftrightarrow \sqrt{n}(\overline{X} - L)/\sigma > \sqrt{n}(\overline{X} - \mu)/\sigma > \sqrt{n}(\overline{X} - U)/\sigma.$$

Since we know $\sqrt{n}(\overline{X} - \mu)/\sigma \equiv Z \sim N(0, 1)$, the last inequality can be written as:

$$\sqrt{n}(\overline{X} - U)/\sigma < Z < \sqrt{n}(\overline{X} - L)/\sigma.$$

Since all we have done is added or multiplied by constants, the inequality and its solution set have not changed. Thus:

$$P(L < \mu < U) = P(\sqrt{n}(\overline{X} - U)/\sigma < Z < \sqrt{n}(\overline{X} - L)/\sigma) = 1 - \alpha.$$

Further, by following the appropriate steps,

$$P(Z > \sqrt{n}(\overline{X} - L)/\sigma) = P(Z < \sqrt{n}(\overline{X} - U)/\sigma) = \alpha/2.$$

Because Z is continuous, there is only one value, say $z_{\alpha/2}$, for which $P(Z > z_{\alpha/2}) = \alpha/2$ and so $\sqrt{n}(\overline{X} - L)/\sigma = z_{\alpha/2}$. The standard normal is symmetric about 0, and therefore, $\sqrt{n}(\overline{X} - U)/\sigma = -z_{\alpha/2}$. These equations can be solved for L and U, respectively to yield:

$$L = \overline{X} - \frac{\sigma}{\sqrt{n}} z_{\alpha/2} \text{ and } U = \overline{X} + \frac{\sigma}{\sqrt{n}} z_{\alpha/2}.$$

The interval we want, then, is:

$$\left(\overline{X} - \frac{\sigma}{\sqrt{n}} z_{\alpha/2}, \overline{X} + \frac{\sigma}{\sqrt{n}} z_{\alpha/2}\right).$$

The end result has been written in the given form for a reason: the term $\frac{\sigma}{\sqrt{n}}$ is the standard deviation of $\overline{X}$'s distribution, and we have labeled it the **standard error** of the estimator. The other factor, $z_{\alpha/2}$, represents how many units of standard error from $\overline{X}$ are allowed to reflect the level of confidence we have set in capturing the true mean. In the case of the standard normal, some commonly required values are:

$$z_{0.10} = 1.34, \ z_{0.05} = 1.645, \ z_{0.025} = 1.96, \ z_{0.01} = 2.33, \text{ and } z_{0.005} = 2.575.$$

When we run the experiment, we substitute the sample values $x_1, \ldots, x_n$ into the formula for the estimator, i.e., the sample mean, and label this $\bar{x}$ and call the resulting interval:

$$\left(\bar{x} - \frac{\sigma}{\sqrt{n}} z_{\alpha/2}, \bar{x} + \frac{\sigma}{\sqrt{n}} z_{\alpha/2}\right)$$

the $(1 - \alpha) \times 100\%$ symmetric confidence interval or realization of the symmetric confidence interval. It is not that this particular interval catches the true mean this frequently, but it does mean that intervals generated by this method will. We can also take the following interpretation: If we did run the experiment many times, then $(1 - \alpha) \times 100\%$ of such confidence intervals should capture the true mean. By running the experiment once, the probability is $1 - \alpha$ that this randomly chosen sample is one that generates a confidence interval that captures the true mean.

Summary: Example 2 illustrates all of the essential features of generating and interpreting confidence intervals. Let θ be an unknown parameter in the distribution of the random variable X, and assume we want a symmetric confidence interval estimate of θ, in addition to a point estimate.

1. The sample size n is given.
2. A point estimate (usually from the best point estimator, as determined from maximum likelihood) of θ is required.
3. A pivot or pivotal statistic for θ and the pivot's distribution is available. Let the random variable representing the pivot be Y.
4. A confidence interval estimate for θ of the form (L, U) is required, based on the sample and the distribution of the pivot. L and U are functions of the data, to be determined.
5. The probability that a randomly generated confidence interval contains the true value of θ is specified. This is denoted $1 - \alpha$ with $\alpha \in (0, 1)$. A symmetric confidence interval requires $P(\theta < L) = P(\theta > U) = \alpha/2$.
6. The pivot is introduced by using algebraic steps to change $P(L < \theta < U) = 1 - \alpha$ into $P(A < Y < B) = 1 - \alpha$ with $P(Y < A) = P(B < Y) = \alpha/2$.
7. The functions L and U are determined in terms of A and B and the distribution of the pivotal statistic.
8. A sample of the requisite size and consistent with the experimental design is chosen, and the realizations of the values L and U are computed using the data. Call the actual numerical values l and u, respectively.
9. Report the interval (l, u) as the $(1 - \alpha) \times 100\%$ symmetric confidence interval.

Definition 1: Let $0 < \alpha < 1$ be given, and let θ be a parameter to be estimated in the distribution of the random variable X. Assume a simple random sample of size n (fixed) from the distribution can be selected as often as necessary. We call the random interval (L, U), with L and U functions of the random sample, a $(1 - \alpha) \times 100\%$ **symmetric confidence interval** for θ if $P(L < \theta < U) = 1 - \alpha$ and $P(U < \theta) = P(\theta < L) = \alpha/2$. The value $1 - \alpha$ is the called the **confidence coefficient** or the **confidence level**, and the value α is called the **significance level**.

In some cases, we may be interested only in protecting against overestimation of the parameter, and sometimes, we are only interested in protecting against an underestimate.

Chapter 8 — Confidence Intervals and Hypothesis Testing

Definition 2: Let $0 < \alpha < 1$ be given. Let n be the sample size of a sample to be taken from the distribution of the random variable X with unknown parameter θ. We call the random variable U a $(1 - \alpha) \times 100\%$ **confidence upper bound** and the interval $(-\infty, U)$ a $(1 - \alpha) \times 100\%$ **confidence interval** for θ if $P(\theta > U) = \alpha$. We call the random variable L a $(1 - \alpha) \times 100\%$ **confidence lower bound** and the interval (L, ∞) a $(1 - \alpha) \times 100\%$ **confidence interval** for θ if $P(\theta < L) = \alpha$.

Example 3: Consider the information in Example 1, where we have assumed the heights of golden delicious trees in a large orchard satisfy $X \sim N(\mu, 0.09)$, with μ unknown and $\sigma^2 = 0.09$. We have taken a random sample of size $n = 25$ in order to estimate μ. The best point estimator of μ is $\hat{\mu} = \overline{X}$, and we know $\overline{X} \sim N(\mu, \sigma^2/n)$. From the data, it is determined that $\overline{x} = 7.2$. How do we determine a 95% symmetric confidence interval for μ? Based on our definitions, we have to evaluate l and u, the realization of L and U as in Definition 1. From Example 2, the values A and B are such that $P(Z < A) = P(Z > B) = 0.05/2 = 0.025$, where Z has a standard normal distribution. This means $A = -z_{0.025}$ and $B = z_{0.025} = 1.96$. Thus, $L = \overline{X} - 1.96\sigma/\sqrt{25} = \overline{X} - 1.96(.06) = \overline{X} - 0.1176$ and $U = \overline{X} + 0.1176$. For our particular data set, the 95% symmetric confidence interval for the true mean height of the trees in this orchard is $(7.0824, 7.3176)$.

Suppose, instead, we need a 95% confidence lower bound for the heights of the trees. One reason for such a requirement may be in a study of the effect of a drought early in the growth cycle of the trees, and another may be to determine settings necessary for a piece of machinery to be used in spraying the trees with insect repellents. In this case, we want L so that $\mu \in (L, \infty)$ with probability 95%, or $P(\mu < L) = 0.05$. The procedure is exactly as for two-sided confidence intervals, except all of the probability of making an error is put in the left tail. This means $L = \overline{X} - \frac{\sigma}{\sqrt{n}} z_{0.05} = \overline{X} - 0.06 \times 1.645$. Given the data, the realization of L is $l = 7.2 - 0.09870 = 7.10130$, and the 95% confidence interval is $(7.10130, \infty)$.

8.1.2 Interval Estimates of a Normal Mean, Variance Unknown

Example 4: The time H (in days) from when a fertile chicken egg is laid until it hatches is considered to have a normal distribution. A new incubation method is being considered, so a farmer randomly selects 100 fertile eggs and places them in the incubator. She wants to determine a 98% symmetric confidence interval for the mean time for the chicken eggs to hatch using the new incubator. Fortunately, all of the eggs hatched, and some summary statistics based on the 100 values $h_1, \ldots, h_{100}$ include: $\overline{h} = 20.4$ days, with $s = 0.2$ days, where S^2 is the variance of the sample, and s is the standard deviation of the actual data.

Solution: In this case, neither the mean nor the variance is known. In cases where a parameter is to be estimated, any other unknown parameter is called a **nuisance parameter**. Since the population variance of H is not known, we cannot use the method of 8.1.1. The overall methodology will be the same, but the real problem is that the pivot in 8.1.1 is not a pivot here. However, we did show that under the conditions that $H \sim N(\mu, \sigma^2)$ and that the sampling is random:

$$\sqrt{n}(\overline{H} - \mu)/S \sim t(n - 1)$$

that is, the distribution of the random variable on the left-hand side is Student's t with $n - 1$ degrees of freedom. Since this statistic is free of unknown parameters, it is a pivot for μ. The upper 0.01 percentage point of Student's t with 99 degrees of freedom is (approximately) 2.364. By following appropriate algebraic steps (outlined below), the values (L, U) that correspond to a 98% symmetric confidence interval will be given by $\overline{H} \pm \frac{S}{\sqrt{n}} t_{0.01}$. For the given data, the realization of the confidence interval is $(l, u) = (20.4 - 0.0475, 20.4 + 0.0475) = (20.3525, 20.4475)$, with days as the units.

The development of the functions L and U follow the same lines as in the case of a known variance, except the pivot is the t variable. In particular,

$$L < \mu < U \Leftrightarrow -L > -\mu > -U \Leftrightarrow \overline{H} - U < \overline{H} - \mu < \overline{H} - L$$
$$\Leftrightarrow \frac{\sqrt{n}}{S}(\overline{H} - U) < \frac{\sqrt{n}}{S}(\overline{H} - \mu) < \frac{\sqrt{n}}{S}(\overline{H} - L) \Leftrightarrow \frac{\sqrt{n}}{S}(\overline{H} - U) < t(99) < \frac{\sqrt{n}}{S}(\overline{H} - L)$$

where $t(99)$ is the t variable with 99 degrees of freedom. The t distribution is always symmetric about the origin, and it is continuous. Thus, there is a unique value $t_{0.01}$ such that $P(t(99) > t_{0.01}) = 0.01$ and, from the tables, that number is 2.364. Thus:

$$\frac{\sqrt{n}}{S}(\overline{H} - U) = -2.364 \text{ and } \frac{\sqrt{n}}{S}(\overline{H} - L) = 2.364.$$

We can then isolate L and U.

The exact value of $t_{0.01}$ will be slightly larger than 2.364, and in fact, is 2.3646. Here, 2.364 is the upper 1% point of the t variable with 100 degrees of freedom and is used as an approximation to the required value, since in most tables, percentage points are not tabulated for all required degrees of freedom. We use the values corresponding to the degrees of freedom in the table closest to the degrees of freedom from the problem.

In the case that the assumption of normality of the parent is suspect and the variance is either known or unknown, but the sample size is large ($n \geq 30$ is usually enough), we use the same pivot $\sqrt{n}(\bar{X} - \mu)/S$ as in the case of the unknown variance problem above, or $\sqrt{n}(\bar{X} - \mu)/\sigma$ when the variance is known, and apply the Central Limit Theorem (CLT). From this, we can say that the **approximate confidence interval** with confidence level $1 - \alpha$ is $(\bar{X} - z_{1-\alpha/2}S/\sqrt{n}, \bar{X} + z_{1-\alpha/2}S/\sqrt{n})$. If σ is known, we replace S by σ. Note that we use the percentage points of the standard normal distribution, which are virtually identical to any t distribution when $n \geq 30$. A somewhat conservative value would use $t_{\alpha/2}$ in place of $z_{\alpha/2}$. The same procedure would be used in calculating upper and lower bounds when the Central Limit Theorem applies.

One question that needs to be addressed is, How many decimals should be kept in calculations? This is not always clear. For example, the confidence interval (20.3525,20.4475) includes four figures after the decimal. Is that many decimals justified? Probably not. The sample value for the mean was 20.4, so we cannot be certain of the second decimal. One reason for such uncertainty is that we are not given information about the accuracy of the measuring devices. In the case of time, we can measure it quite accurately, but we also need to make sure of when the egg was laid and what it means for the egg to hatch. To set a rule, we will treat all values as exact in all intermediate calculations, but record no more than two decimals beyond the least accurate value in the given information. Thus, this confidence interval should be written as (20.353,20.448).

8.1.3 Interval Estimate of a Normal Variance

So far, the algebra that is used to determine confidence intervals for means from normal distributions has been the same whether standard normal or Student's t is the pivot. Variances are different because: the values are in $(0, \infty)$; the pivot has a χ^2 distribution, and it is not symmetric; and the algebraic steps producing a confidence interval for variances are different. Otherwise, however, the overall goals of confidence interval estimation remain
the same.

Suppose we want a symmetric confidence interval for the variance σ^2 of the variable X with a normal distribution $N(\mu, \sigma^2)$, and we have the resources to take a random sample of size n from X's distribution. Let $1 - \alpha$ be the confidence coefficient of the random interval estimate of σ^2, and assume we want a $(1 - \alpha) \times 100\%$ symmetric confidence interval. As in the case of finding confidence intervals for μ, this means that the interval, say (L, U), is such that $P(\sigma^2 \leq L) = P(\sigma^2 \geq U) = \alpha/2$. The interpretation of such an interval is the same: in long runs of repeating the experiment (with a fixed sample size) and calculating confidence intervals from each

Chapter 8 — Confidence Intervals and Hypothesis Testing

sample, then the proportion of intervals that capture the true value of σ^2 is $1 - \alpha$, the proportion where the true value is to the left of L is $\alpha/2$ and the proportion where the true value is to the right of U is $\alpha/2$. To find L and U (which again are random variables and are functions of the data), we transform the problem to one involving the pivotal statistic $(n-1)S^2/\sigma^2$. We will discuss how this is modified if μ is known shortly. Under the assumption that the parent distribution is normal, or nearly so, we have:

$$P(\sigma^2 \leq L) = \alpha/2 \Leftrightarrow P\left(\frac{1}{L} \leq \frac{1}{\sigma^2}\right) = \alpha/2$$

$$\Leftrightarrow P\left(\frac{(n-1)S^2}{L} \leq \frac{(n-1)S^2}{\sigma^2}\right) = \alpha/2 \Leftrightarrow P\left(\frac{(n-1)S^2}{L} \leq \chi^2(n-1)\right) = \alpha/2$$

where $\chi^2(n-1)$ represents the χ^2 variable with $n-1$ degrees of freedom. Using tables or direct calculation, we can find $\chi^2_{\alpha/2}$, the $\alpha/2 \times 100\%$ upper percentage point of this distribution. With this number, and recalling that χ^2 variables are continuous, we can then find L:

$$\chi^2_{\alpha/2} = \frac{(n-1)S^2}{L} \Leftrightarrow L = \frac{(n-1)S^2}{\chi^2_{\alpha/2}}.$$

In a completely analogous manner, we have:

$$U = \frac{(n-1)S^2}{\chi^2_{1-\alpha/2}}$$

where $\chi^2_{1-\alpha}$ is the $(1-\alpha) \times 100\%$ upper percentage point or, equivalently, the $\alpha \times 100\%$ lower percentage point, of the $\chi^2(n-1)$ distribution. It is important to remember that the notation χ^2_α means the $\alpha \times 100$ upper percentage point of the $\chi^2(n-1)$ variable, and that $n-1$ is a parameter of the distribution, not a factor.

At first, these results may look strange, because the left endpoint L involves a value from the right tail of the χ^2 distribution, and U involves a value closer to 0, i.e, in the left tail of the distribution. However, since we are dividing, a larger value in the denominator gives a smaller value in a fraction, while a smaller value in the denominator increase the overall size of the fraction. Thus, $L < U$.

It should be noted that the methods outlined above for a variance depend heavily on the requirements that the parent distribution be normal and the sample be random. If either of these conditions is violated to any great extent, these results are not justified. Again, it is important to verify that the assumptions are supported by the actual data and by the experimental design.

Example 5: In some experiments, preliminary studies are done to try to estimate parameters in order to determine the sample size necessary to achieve a given level of precision. Consider Example 4: Before using the new incubator on 100 eggs, the farmer decides to test the machine on 12 viable eggs. In this sample, the average time to hatch was 22.3 days, and the standard deviation of the sample was 0.4 days. A 95% symmetric confidence interval for the population variance σ^2 would be:

$$\left(\frac{(11)S^2}{\chi^2_{\alpha/2}}, \frac{(11)S^2}{\chi^2_{1-\alpha/2}}\right) = \left(\frac{(11)0.16}{\chi^2_{0.025}}, \frac{(11)0.16}{\chi^2_{0.975}}\right) = \left(\frac{(11)0.16}{21.9}, \frac{(11)0.16}{3.82}\right) = (0.0804, 0.4607).$$

This interval appears to be quite wide, but that is primarily a function of the sample size. Suppose we want to estimate σ^2 so that the symmetric confidence interval for estimating the mean hatching time at the level of significance in Example 4 is no longer than 0.04, based on the preliminary information in Example 5. Since the total length is to be 0.04, then $\frac{S}{\sqrt{n}} z_{0.01} = 0.02$. We use the standard normal percentage points because we expect the sample size will be greater than 30. If the sample size needed is less than 30, we can modify this argument by starting with normal values, then once the sample size is determined, use the t values for these degrees of freedom and revise the calculation. Doing this once or twice will give a good estimate of n. In this case, $z_{0.01} = 2.33$, so $0.02 \approx 0.4 \times 2.33/\sqrt{n}$ or $n = 46.6^2 = 1171.56$. We would round this up to the nearest integer, so the sample size would be at least 1172.

The sample size required is very large and is the result of the required precision and the level of confidence specified, along with the estimated variance. It may be more cost effective to take a larger sample to get a better estimate of the variance. We can solve for n in general as $n = S^2 \times z_{\alpha/2}^2/m^2$, where m is the precision as expressed as one-half of the length of the confidence interval. If all else is held fixed, but we decrease m by a factor of one-half, the sample size needed increases by a factor of 4.

Suppose that in Example 4 the population mean μ was in fact known. Clearly, we do not have to estimate it now. In this case, we should use the fact that the mean is known, which means the maximum likelihood estimate of σ^2 is going to be $\hat{\sigma}^2 = \sum_{i=1}^{n}(X_i - \mu)^2/n$. The distribution of $n\hat{\sigma}^2/\sigma^2 \sim \chi^2(n)$, and so we can then use $n\hat{\sigma}^2/\sigma^2$ as a pivot for σ^2. The method for finding confidence intervals or confidence bounds doesn't change, just the form of the pivot and the degrees of freedom in the χ^2 distribution used. Some researchers continue to use the method as outlined in Example 4, even if μ is known, since this provides a somewhat more conservative interval estimate (it will be slightly longer).

8.1.4 Interval Estimate of a Proportion in Bernoulli Trials: Large Samples, Large Populations

Example 6: In a large hospital, records on the individuals who use the emergency room are kept, including insurance status. An administrator wants to study aspects of the issues surrounding admitting patients without insurance, and as a first step, she randomly selects 200 files of emergency room admissions from the past two years. In this selection, 73 did not have insurance. Find a 95% confidence interval of the proportion of all emergency admissions without insurance.

A symmetric confidence interval for π, the proportion in Bernoulli trials, satisfies $P(L < \pi < U) = \alpha$ with $P(\pi < L) = \alpha/2$ and $P(\pi > U) = \alpha/2$. One immediate problem is that we do not have an exact pivot for π. However, if our population and sample size are large enough to ensure the Central Limit Theorem applies, we can use the normal approximation to the binomial distribution. Let $\hat{\pi} = a/n$, where a is the number of successes or observed 1s in n trials. This has an approximately normal distribution, by the Central Limit Theorem, if:

$$\hat{\pi} \pm 3\sqrt{\hat{\pi}(1 - \hat{\pi})/n} \in (0, 1).$$

Another rule of thumb is to require the number of 1s and the number of 0s observed in the n Bernoulli trials each be at least 10. If either of these conditions holds, then $\hat{\pi} \sim N(\pi, \pi(1 - \pi)/n)$ approximately, and hence, a confidence interval for π can be determined as follows:

First, $\sqrt{n}(\hat{\pi} - \pi)/\sqrt{\pi(1 - \pi)} \sim N(0, 1)$, at least approximately (why?). The problem is that the variance contains the parameter that we are trying to estimate. We get around this by using the Central Limit Theorem when the variance is not known, so that at least approximately,

$$\sqrt{n}(\hat{\pi} - \pi)/\sqrt{\hat{\pi}(1 - \hat{\pi})} \sim N(0, 1).$$

This is justified by the fact that the sample variance of a Bernoulli random sample of size n is:

$$\frac{n}{n-1}\hat{\pi}(1 - \hat{\pi}) \approx \widehat{\pi(1 - \pi)}$$

and the variance of a Bernoulli random variable is $\pi(1 - \pi)$. With this, we can then write our approximate $(1 - \alpha)$ confidence interval as:

$$(\hat{\pi} - z_{\alpha/2}\sqrt{\hat{\pi}(1 - \hat{\pi})/n},\, \hat{\pi} + z_{\alpha/2}\sqrt{\hat{\pi}(1 - \hat{\pi})/n}).$$

Example 6 Cont'd: Since $\hat{\pi} = 73/200$ then $\hat{\pi} \pm 3\sqrt{\hat{\pi}(1-\hat{\pi})/n} = 0.365 \pm 3(0.0340) \in (0,1)$. Thus, the Central Limit Theorem applies, which means we can use the normal approximation to the binomial distribution as derived above. Since $Z_{0.025} = 1.96$, the required confidence interval is $(0.365 - 1.96(0.0340), 0.365 + 1.96(0.0340)) = (0.298, 0.432)$.

8.1.4.1 Interval Estimate of a Proportion in Bernoulli Trials: Exact Calculations including Small Samples

Example 7: A study of the impact of the introduction of the eastern gray squirrel into Ireland on the numbers of native red squirrels is being undertaken. In a large urban area, 15 squirrels were caught and their species determined (assume only these two types of squirrels were living in the area under study). Four of the sample are eastern gray squirrels. Determine an 80% symmetric confidence interval for π, the proportion of squirrels that are eastern gray.

In Example 7, the sample size is too small to make use of the Central Limit Theorem, and so a different approach is needed than in solving Example 6. In this case, we revert to the original definition of a confidence interval. A $(1 - \alpha) \times 100\%$ symmetric confidence interval is of the form (p_L, p_U) such that $P(\pi < p_L) = P(\pi > p_U) = \alpha/2$ (with appropriate modifications should just an upper bound or just a lower bound be required). Basically, p_L is the smallest value of π that could reasonably generate a sample with the observed number X of 1s, and p_U is the largest value of π that could reasonably generate the given sample. The **Clopper-Pearson** exact method (see C. Clopper and E. S. Pearson, 1934) finds p_L and p_U that solve the following two equations:

$$\sum_{i=0}^{X} \binom{n}{i} p_U^i (1-p_U)^{n-i} = \alpha/2 \text{ and } \sum_{i=X}^{n} \binom{n}{i} p_L^i (1-p_L)^{n-i} = \alpha/2.$$

The value of p_U that solves the first equation gives the upper bound, and the value of p_L that solves the second gives the lower bound. Note that:

$$\sum_{i=X}^{n} \binom{n}{i} p_L^i (1-p_L)^{n-i} = \alpha/2 \Leftrightarrow \sum_{i=0}^{X-1} \binom{n}{i} p_L^i (1-p_L)^{n-i} = 1 - \alpha/2$$

which, when X is less than $1/2$ of n, has fewer terms and hence is easier to solve.

Example 7 Cont'd: The two equations to be solved are:

$$\sum_{i=0}^{4} \binom{15}{i} p_U^i (1-p_U)^{15-i} = 0.1 \text{ and } \sum_{i=0}^{3} \binom{15}{i} p_L^i (1-p_L)^{15-i} = 0.9$$

with the results: $p_L = 0.122$ and $p_U = 0.464$, so our confidence interval is $(0.122, 0.464)$.

One observation to make about this kind of confidence interval is that it is not symmetric about the estimate $\hat{\pi}$. However, since the binomial distribution is not symmetric except when $\pi = 1/2$, or when n and π are sufficiently large so that the Central Limit Theorem is applicable, this lack of symmetry is not a significant issue.

Another observation is that the confidence interval in Example 7 required $\alpha = 0.20$, leaving significant room for error. However, for small samples, having a very small α value means a very wide confidence interval, generally giving only marginal restrictions on what π can be, based on the sample.

8.2 HYPOTHESIS TESTING

In the last section, we concentrated on finding reasonable bounds on what the value of the true population parameter was, based on information gathered. In this section, we consider the problem of finding evidence in support or against an assumed value of the parameter. Mechanically, the calculations will be similar, but the interpretation will be different.

8.2.1 Introduction to Hypothesis Testing

Let X be a random variable with a distribution containing the parameter θ. The parameter θ is also called the **true population value** (i.e. mean, variance, etc.). Suppose that we have in mind a possible value for θ, say $\theta = \theta_0$, and we want to use the data from an experiment to determine if it is a statistically reasonable value. Of course, we can never know for certain whether this hypothesized value is correct, but we can have good evidence from the data whether it is supported or not.

Definition 3: Let the parameter θ have possible values in Θ, and let $\Theta = \Theta_0 \cup \Theta_1$, where Θ_0 and Θ_1 are mutually exclusive and exhaustive of Θ. The **null hypothesis**, denoted H_0 is the hypothesis that $\theta \in \Theta_0$. The **alternate hypothesis** or **alternative hypothesis** or **research hypothesis**, denoted H_1 or H_a, is the hypothesis that $\theta \in \Theta_1$.

Note: How do we tell H_0 and H_1 apart or, equivalently, decide which set of values constitute Θ_0 and which constitute Θ_1? The typical situation for θ is that its possible values fill out an interval (finite or infinite in length), so $\Theta = (a, b)$ say, or some other form of an interval. Typically, Θ_0 and Θ_1 are either single intervals or at most the union of two intervals. For our purposes, the possible cases are: $\Theta_0 = \{\theta_0\}$ and $\Theta_1 = \Theta \backslash \{\theta_0\}$; $\Theta_0 = \{\theta \leq \theta_0\}$ and $\Theta_1 = \{\theta > \theta_0\}$; or $\Theta_0 = \{\theta \geq \theta_0\}$ and $\Theta_1 = \{\theta < \theta_0\}$. In each case, H_0 is the hypothesis $\theta \in \Theta_0$, and in each case, we write $H_0 : \theta = \theta_0$, regardless of the form of Θ_0. It would not be wrong to write $H_0 : \theta \leq \theta_0$ if $\Theta_0 = \{\theta \leq \theta_0\}$. If $\Theta_1 = \Theta \backslash \{\theta_0\}$, the alternate hypothesis is written $H_1 : \theta \neq \theta_0$; if $\Theta_1 = \{\theta > \theta_0\}$ the alternate hypothesis is written $H_1 : \theta > \theta_0$ while if $\Theta_1 = \{\theta < \theta_0\}$, we write $H_1 : \theta < \theta_0$. Once θ_0 is specified, then the null hypothesis is always associated with the possibility of taking θ equal to θ_0.

Definition 4: We call $H_1 : \theta \neq \theta_0$ a **two-sided alternate hypothesis**, while $H_1 : \theta > \theta_0$ and $H_1 : \theta < \theta_0$ are called **one-sided hypotheses**, or **directional hypotheses**.

Definition 5: An hypothesis that completely specifies the distribution of the random variable is called a **simple hypothesis**. An hypothesis that does not completely specify the distribution of the random variable is called a **composite hypothesis**.

Example 8:

(a) Suppose that from past history, the mean of a particular normal population has been $\mu = 10$. It is believed that this has changed, but it is not known in which direction. In this case, H_0 is taken to be $H_0 : \mu = 10$, while the alternative hypothesis is $H_1 : \mu \neq 10$. Also, we take $\mu_0 = 10$. It is not always clear from the definition why H_0 and H_1 are chosen as they are in the above definition. However, the null hypothesis is often described as the **no change** hypothesis, so H_0 reflects that here.

(b) A device to fill petri dishes with an average amount of nutrients has a variability of fill of $\sigma^2 = 0.5$ cc^2. A new device may be purchased because it is claimed by the manufacturer to be more accurate, that is, the variability of fill is less than 0.5 cc^2. Before buying the new machine, the head of the lab wants to test the appropriate hypotheses. In this case, for the new machine $H_0 : \sigma^2 = 0.5$ cc^2 while $H_1 : \sigma^2 < 0.5$ cc^2.

(c) It is believed that a new teaching method will increase the proportion in classes of eighth grade students who score higher than the national average on a standardized reading test. It will cost a significant amount of money to implement the new method, so a random selection of eighth grade classrooms is selected and the new method used. If π is the proportion of an eighth grade class that score higher than the national average under the new teaching method, and if π_0 is the average proportion in classes of eighth graders who achieve the national average or higher under current teaching methods, then the null hypothesis is $H_0 : \pi = \pi_0$ and $H_1 : \pi > \pi_0$.

In (a), the null hypothesis is simple if the variance is known, and is composite if the variance is unknown. The alternate hypothesis is composite. In (b) and (c), the hypotheses are all composite.

We need a method for deciding which hypothesis is supported by the information in a sample. The term for this kind of procedure is a **test of hypothesis**:

Definition 6: A **test of hypothesis** or **statistical test of hypothesis** is a procedure based on sample data used to decide whether to **accept** or **reject** the null hypothesis. If the null hypothesis is rejected, the alternate hypothesis is accepted.

The procedure constituting a test of hypothesis is based on the information in a sample, typically a simple random sample. From that information, an estimator of the parameter is selected, usually the best point estimator from maximum likelihood. We will assume our estimator is unbiased. As in confidence intervals, the estimator is a function of the sample and provides a point estimate of the parameter. Further, the estimate is not going to be the population value, so even if the null hypothesis is true, our estimate of θ will not be θ_0. In fact, assuming Θ is an interval:

if we are testing $H_0 : \theta = \theta_0$ vs $H_1 : \theta \neq \theta_0$, the point estimate $\hat{\theta}$ will not equal θ_0 even if H_0 is true;

if we are testing $H_0 : \theta = \theta_0$ vs $H_1 : \theta > \theta_0$, the point estimate could satisfy $\hat{\theta} > \theta_0$ and H_0 may be true;

if we are testing $H_0 : \theta = \theta_0$ vs $H_1 : \theta < \theta_0$, the point estimate could satisfy $\hat{\theta} < \theta_0$ and H_0 may be true.

The reason is that, under the null hypothesis, the estimator has a distribution over Θ, so we expect that the realization of the estimator could fall in Θ_1 at least sometimes. The test of hypothesis procedure must reflect these observations.

A test of hypothesis is based on a sample rather than a complete census, so it cannot prove or disprove an hypothesis: The data provides evidence in support of the null or the alternative. It is then possible that the data gathered leads to an incorrect inference. Tests of hypothesis should be structured to keep instances of incorrect inferences to a small, reasonable number. There are two types of incorrect inferences, illustrated in Figure 1:

		Reality	
		H_0 True	H_1 True
	H_0 Accepted	No error	Type II error
Decision			
	H_1 Accepted	Type I error	No error

FIGURE 8-1. Possible Decisions and Errors.

Definition 7: A **type I error** is the error of rejecting a true null hypothesis, or accepting a false alternate hypothesis. A **type II error** is accepting a false null hypothesis (i.e., rejecting a true alternate).

We quantify how frequently we expect to commit either a type I or a type II error by setting some limits on what is good evidence in support of each hypothesis. In particular:

Definition 8: The **significance level** of a test of hypothesis is the probability of making a type I error and is denoted α. The probability of a type II error is denoted β. The **power** of a test is $1 - \beta$, and measures how frequently true alternatives are detected by the test. These probabilities are illustrated in Figure 8-2.

		Reality	
		H_0 True	H_1 True
Decision	H_0 Accepted	No error	β = P(Type II error)
	H_1 Accepted	α = P (Type I error)	No error

FIGURE 8-2. Probabilities of Errors.

As defined above, the power of a test is its ability to detect true alternatives. If there are choices among tests, the one with the highest power over all values of θ under the alternate hypothesis is often the one to use. Detecting changes in the value of the parameter under investigation is essential in determining if conditions have caused an increase or decrease in the response to a given treatment.

It is usually not possible to set both α and β simultaneously, and so we have to decide which one. The practice is to set α, with typical values of 0.10 (probably not significant), 0.05 (probably significant) and 0.01 (highly significant) as the standard choices. In a problem, if α is not specified, it is assumed to be 0.05. Often this is accompanied by a **simulation study** to determine how well the test detects true alternatives. These studies make use of computer methods involving many repetitions of the experiment under different assumed values of the parameter, and then examining the variability and other factors in the resulting observations.

So now we set α. What does this mean? It means P(reject H_0 | H_0 is true) = α, i.e., the probability of drawing a sample that yields a value of our estimator of θ that does not support the null, even though it is true, is α. This gives us a way to set in advance the values of the estimator that will be assumed to be evidence in support of the null hypothesis, and those that will be assumed to be evidence against the null.

To illustrate the choice of critical value, consider testing the mean in a normal distribution with variance 1. In particular, suppose $H_0 : \mu \leq 1$ vs $H_1 : \mu > 1$. We will take a random sample of size $n = 1$, call it X, and we now need a rejection region. Consider the following diagram:

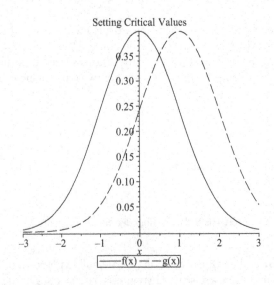

The solid line graph is of $N(0, 1)$, with density $f(x) = \exp(-x^2/2)\sqrt{2\pi}$ while the dash line graph is of $N(1, 1)$, with density $g(x) = \exp(-(x-1)^2/2)/\sqrt{2\pi}$. If in fact the value of μ is 0, and we want to set the rejection region so that the area under $N(0, 1)$ to the right of the critical value is 0.05, we would use $X_{crit} = 1.645$. However, if we did use that value, but the true mean was 1, which is a possible value under the null and may represent historical values (i.e. no change), then the area under $N(1, 1)$ from 1.645 to infinity is 0.24 (approx.). The reader should locate 1.645 on the horizontal axis, and then note the difference in tail areas for the two graphs. That means we would reject a true null too often. If instead we take as the critical value 2.645 (locate this on

Chapter 8 — Confidence Intervals and Hypothesis Testing

the x-axis), then the area from that value to infinity, under $N(1, 0)$ is 0.05. Note that the area under $N(0, 1)$ from 2.645 to infinity is 0.004, so if the true distribution had been $N(0, 1)$, we would still have rejected it for any observed value greater than 2.645. By choosing the largest value (in this case) under the null, we have protected ourselves from rejecting the null too often and ensured we would have rejected the null for sufficiently large values of X if the true mean was less than 1. This explains why we choose to express the null hypothesis in general as $H : \theta = \theta_0$, and use $\theta = \theta_0$ in generating critical values.

Our testing procedure sets $P(\text{reject } H_0 \mid H_0 \text{ is true}) = \alpha$. We can now restate this as $P(\text{reject } H_0 \mid \theta = \theta_0) = \alpha$. In the example in the previous paragraph, we can write the condition to reject the null hypothesis as: If $\bar{x} > \bar{X}_{crit}$, where $\bar{X}_{crit}$ is the upper $\alpha \times 100\%$ point of $\bar{X}$'s distribution then reject the null hypothesis. $\bar{X}_{crit}$ is the **critical score** of the test. This provides one version of the decision rule for this example: If $\bar{x} > \bar{X}_{crit}$ we reject the null hypothesis, and if $\bar{x} < \bar{X}_{crit}$, we accept the null. The notation $\bar{X}_{calc}$, read the **calculated $\bar{X}$** score, is often used in place of $\bar{x}$.

Definition 9: The **critical region** in a test of hypothesis is the set of values of the estimator of the parameter being tested for which the null hypothesis is rejected, even if the null is true.

Thus, we have an acceptance region and a rejection region for the null, based on the observed value of the estimator. One small point from the example of testing a mean: What is the conclusion if $\bar{x} = \bar{X}_{crit}$? Theoretically, this should not happen when Θ is an interval. However, in practical terms, our values will be rounded to a certain number of decimals, and so it becomes a possibility. One approach is to go back to the data and the calculations to determine if greater accuracy resolves the problem. The other approach is to randomly select another observation and add it into the sample, and revise the estimates.

8.2.2 Hypothesis Testing and Pivotal Statistics

Let the random variable X have a given distribution with parameter $\theta \in \Theta$, where Θ is assumed an interval, and suppose we want to test $H_0 : \theta = \theta_0$ versus one of the alternative hypotheses, $H_1 : \theta \neq \theta_0$, $H_1 : \theta > \theta_0$, or $H_1 : \theta < \theta_0$. Let $\hat{\theta}$ be the point estimate of θ, and assume its distribution is known. Further, assume there is a pivotal statistic, or pivot, for θ based on the point estimate's distribution and for which percentage points are available. The **critical region** for the test is the set of values that lead to rejection of the null hypothesis. This can be expressed in terms of the original estimator's distribution, or in terms of the pivotal statistic's distribution.

The Critical Region: Original Estimator's Distribution:

(a) Let the alternate hypothesis be $H_1 : \theta \neq \theta_0$, and let the significance level of the test be α. Just as in confidence intervals, we split α in two and determine A and B, under the assumption of the null hypothesis, so that $P(\hat{\theta} < A \mid H_0) = \alpha/2$ and $P(\hat{\theta} > B \mid H_0) = \alpha/2$. The critical region is:

$$C = \{\hat{\theta} \mid \hat{\theta} < A\} \cup \{\hat{\theta} \mid \hat{\theta} > B\}.$$

If $\hat{\theta}_{calc} \in C$, reject the null.

(b) Let the alternate hypothesis be $H_1 : \theta > \theta_0$, and let the significance level of the test be α. Determine B, under the assumption of the null hypothesis, so that $P(\hat{\theta} > B \mid H_0) = \alpha$. The critical region is:

$$C = \{\hat{\theta} \mid \hat{\theta} > B\}.$$

If $\hat{\theta}_{calc} \in C$, reject the null.

(c) Let the alternate hypothesis be $H_1 : \theta < \theta_0$, and let the significance level of the test be α. Determine A, under the assumption of the null hypothesis, so that $P(\hat{\theta} < A \mid H_0) = \alpha$. The critical region is:

$$C = \{\hat{\theta} \mid \hat{\theta} < A\}.$$

If $\hat{\theta}_{calc} \in C$, reject the null.

The Critical Region: The Pivot's Distribution: Let the variable for the pivot be denoted Y.

(a) Let the alternate hypothesis be $H_1 : \theta \neq \theta_0$, and let the significance level of the test be α. Determine Y_L and Y_U, such that $P(Y < Y_L) = \alpha/2$ and $P(Y > Y_U) = \alpha/2$. The critical region is:

$$C = \{Y \mid Y < Y_L\} \cup \{Y \mid Y > Y_U\}.$$

Let $\hat{\theta}_{calc}$ be the value of $\hat{\theta}$ calculated from the sample. The pivot Y is a function w of $\hat{\theta}$. Determine $w(\hat{\theta}_{calc})$ and call this Y_{calc}. If $Y_{calc} \in C$, reject the null hypothesis.

(b) Let the alternate hypothesis be $H_1 : \theta > \theta_0$, and let the significance level of the test be α. Assume Y is an increasing function of $\hat{\theta}$, and determine Y_U, such that $P(Y > Y_U) = \alpha$. The critical region is:

$$\{Y \mid Y > Y_U\}.$$

Let $\hat{\theta}_{calc}$ be the value of $\hat{\theta}$ calculated from the sample. Determine $w(\hat{\theta}_{calc})$ and call this Y_{calc}. If $Y_{calc} \in C$ reject the null hypothesis.

(c) Let the alternate hypothesis be $H_1 : \theta < \theta_0$, and let the significance level of the test be α. Assume Y is an increasing function of $\hat{\theta}$, and determine Y_L, such that $P(Y < Y_L) = \alpha$. The critical region is:

$$C = \{Y \mid Y < Y_L\}.$$

Let $\hat{\theta}_{calc}$ be the value of $\hat{\theta}$ calculated from the sample. Determine $w(\hat{\theta}_{calc})$ and call this Y_{calc}. If $Y_{calc} \in C$ reject the null hypothesis.

Example 9: Consider our problem of the apple trees in Examples 1 and 3. It is believed that the true average height is $\mu_0 = 7.5$ meters. We want to test that this hypothesis versus the average height is not 7.5 m. This is summarized as: test $H_0 : \mu = 7.5$ versus $H_1 : \mu \neq 7.5$. In general, if a significance level is not given, take $\alpha = 0.05$, which we do here. Our best point estimate is the sample mean, $\bar{x} = 7.2$ based on a sample of size $n = 25$. If $\bar{X}$ is the estimator of μ, then under the null hypothesis, $\bar{X} \sim N(7.5, 0.9/25) = N(7.5, 0.036)$. Given the form of the alternative hypothesis, we would have evidence against the null if our calculated value is too far to the left of 7.5 or too far to the right. Specifically, the testing procedure requires that the probability of rejecting a true null hypothesis for values to the left of 7.5 to be the same as to the right. Thus, the critical values A and B using the estimator's distribution satisfy $P(\bar{X} < A \mid H_0 \text{ is true}) = \alpha/2 = 0.025$ and $P(\bar{X} > B \mid H_0 \text{ is true}) = \alpha/2 = 0.025$. This means:

$$A = 7.5 - z_{0.025} \times 0.036 \text{ and } B = 7.5 + z_{0.025} \times 0.036$$

so that our critical values are:

$$A = 7.5 - 1.96 \times 0.036 = 7.43 \text{ and } B = 7.5 + 1.96 \times 0.036 = 7.57.$$

The critical region would be the set of values of $\bar{X} \in (-\infty, A) \cup (B, \infty)$. Since $\bar{X}_{calc} = \bar{x} = 7.2 < A$, we reject the null hypothesis.

Let's repeat this test, but using the pivotal statistic $Z = (\bar{X} - \mu_0)/(\sigma/\sqrt{n}) = \sqrt{n}(\bar{X} - \mu_0)/\sigma$. Because the variance is assumed known, $Z \sim N(0, 1)$, our critical numbers for the pivot are $-z_{0.025}$ and $z_{0.025}$, where $z_{0.025} = 1.96$ is the upper 2.5% point of the standard normal. The calculated z-score under the null is
$z_{calc} = \sqrt{n}(\bar{X}_{calc} - \mu_0)/\sigma = 5(7.2 - 7.5)/0.3 = -5 < -z_{0.025}$, and so again, we reject the null hypothesis.

If the test procedures using the distribution of the point estimator and the pivot yield the same results, why do we introduce, say, the hypothesis testing procedure based on the pivot? When we carry out a test of hypothesis, we either accept or reject the null. However, the test does not indicate just how strong the evidence against the null is. In the above exercise, we would have rejected the null hypothesis if $z_{calc} = -1.98$, but in fact $z_{calc} = -5$. The probability that the standard normal variable is less or equal to -5 is 2.87×10^{-7}. Clearly, the evidence against the null is very strong. For this reason, the **p-value** is used to quantify just how strong the evidence against the null is, and it is more readily computed by using the distribution of the pivot.

Chapter 8 — Confidence Intervals and Hypothesis Testing

Definition 10: The *p*-value of a test of the hypothesis $H_0 : \theta = \theta_0$ is defined as follows: Let $\hat{\theta}_{calc}$ be the calculated value of the estimator $\hat{\theta}$ of parameter θ.

1. if the rejection region is $\hat{\theta}_{calc} > \theta_{crit}$, then the *p*-value of the test is $P(\hat{\theta} \geq \hat{\theta}_{calc})$;
2. if the rejection region is $\hat{\theta}_{calc} < \theta_{crit}$, then the *p*-value of the test is $P(\hat{\theta} \leq \hat{\theta}_{calc})$; and
3. if the rejection region is $\hat{\theta}_{calc} < \theta_{crit_1}$ or $\hat{\theta}_{calc} > \theta_{crit_2}$, then the *p*-value of the test is $2 \times P(\hat{\theta} \geq \hat{\theta}_{calc})$ or $2 \times P(\hat{\theta} \leq \hat{\theta}_{calc})$, depending on whether the calculated score is to the left or to the right of θ_0.

Calculating *p*-values: To calculate the *p*-value of a test when a pivotal statistic is available, and if Y is the pivot, we determine the value of Y corresponding to $\hat{\theta}_{calc}$ and call it Y_{calc}. Since the function linking the pivot and the estimator is increasing, we have:

1. if the rejection region is $Y_{calc} > Y_U$, then the *p*-value of the test is $P(Y \geq Y_{calc})$;
2. if the rejection region is $Y_{calc} < Y_L$, then the *p*-value of the test is $P(Y \leq Y_{calc})$; and
3. if the rejection region is $Y_{calc} < Y_L$ or $Y_{calc} > Y_U$, then the *p*-value of the test is $2 \times P(Y \geq Y_{calc})$ or $2 \times P(Y \leq Y_{calc})$, depending on whether the calculated score is to the left or to the right of $Y_0 = w(\theta_0)$, where w is the function linking the estimator and the pivot.

It is now evident why the pivotal statistic is useful: Usually, it has a well-defined distribution and easily computed percentage points, or at least such values have been tabulated. This would not be the case for the distribution of the estimator of θ.

It should be noted that the *p*-value is between 0 and 1. The closer it is to 0, the stronger the evidence against the null. We use the "rule-of-thumb" values: If *p* is near 0.10 or greater, then the strength of evidence against the null is weak. If the *p*-value is near 0.05 or less, the strength of evidence against the null is good. If the *p*-value is near 0.01 or less, the strength of evidence against the null is strong. In some instances, the *p*-value may be very small and may be reported in the form $p << 0.001$ or similar. The symbol $<<$ means "very much smaller than," and if this is the case, the strength of evidence against the null is very strong. This symbol is used when the *p*-value is beyond the precision of the usual table for the pivotal statistic's percentage points.

Should Y_{calc} fall between two percentage points of the pivot, simply state that the *p*-value is between these percentage points. For example, if the pivot is Student's t with 14 degrees of freedom, and the calculated t value is 2.45, this value is between the upper 2.5% and 1.0% points. We would report that the *p*-value as $0.01 < p < 0.025$. **Caution**: Because of the way percentage points are organized in t tables (and others), the 2.5% point is listed to the left of the 1.0% point. It is a frequent error that p values are reported as $0.025 < p < 0.01$, and of course, this is wrong.

Example 9 Cont'd: In the case of the apple tree heights problem, the *p*-value is 2.87×10^{-7}. This would be very strong evidence against the null hypothesis. This is an instance where the *p*-value would be reported as $p << 0.001$.

8.2.3 Structure of a Test of Hypothesis

There are several standard steps to follow in a properly formulated test of hypothesis:

1. State the null and the alternative hypotheses.
2. State the significance level of the test. If a significance level is not given, use $\alpha = 0.05$.
3. State the assumptions made, such as: independent random values, X is normal, CLT applies, etc.
4. State the pivotal statistic and its distribution generally and under the null hypothesis.
5. Determine the critical region or the critical value(s) for comparison.
6. Compute the value of the pivot, based on the data observed, under the null hypothesis.

7. Draw an inference about the hypotheses based on the pivot and the critical region. Include a statement about the p value and what it implies.

8. Comment on the test and any problems that may have arisen, such as whether model assumptions seem to be supported by the data.

Example 10: In a study of decibel levels perceptible one-quarter mile away from a "wind farm," 20 readings of D, the decibel level, were recorded at a station one-quarter mile away from the wind farm at randomly chosen times of a given day. Summary statistics include: $\overline{D} = 75$, and $\sum_{i=1}^{20} D_i^2 = 121200$. Test the appropriate hypotheses if the claim by local farmers is the average decibel reading at the station exceeds 70 decibels.

Solution: Following the steps above, we have:

1. $H_0 : \mu_D = 70$ vs $H_1 : \mu_D > 70$.
2. Since the significance level is unspecified, take $\alpha = 0.05$.
3. We assume the observations are independently and identically distributed, and further, D is $N(\mu_D, \sigma^2)$, at least approximately. In this case, the variance is unknown.
4. Because the variance is unknown, the pivot is $\sqrt{n}(\overline{D} - \mu_D)/S$. Under the null hypothesis, the pivot is $\sqrt{n}(\overline{D} - 70)/S \sim t(19)$.
5. Using the pivot, the critical or rejection region will be $t_{calc} > t_{0.05} = 1.729$. The critical value is 1.729.
6. By calculation from the summary statistics, $s^2 = 457.89$ (approximately). Thus, $t_{calc} = \sqrt{20}(75 - 70)/21.40 = 1.045$.
7. We accept the null hypothesis. It is also acceptable to say we do not reject the null hypothesis. From the tables available, we can say the p-value is > 0.1, meaning the sample provides little evidence against the null.
8. Since the data wasn't provided, it is not readily apparent if the data supports the normality assumption. However, we could comment that the weather conditions may have played a factor, since sound travels better in denser air and prevailing winds may impact the way sound travels. Other factors can include the time of day, since it is likely fewer extraneous sounds may be present at night time than during the day.

8.3 CONFIDENCE INTERVALS AND TESTS OF HYPOTHESIS

In sections 8.1 and 8.2, we introduced the basic ideas and procedures related to confidence intervals and tests of hypotheses. It is important to recognize that the mechanics are the same, just applied differently. In fact, we can use confidence intervals in carrying out tests of hypothesis: if we use the same value of α in calculating a symmetric confidence interval with confidence coefficient $1 - \alpha$, then the critical region for a two-sided test of hypothesis is the set of values outside this confidence interval. To illustrate this, suppose we are interested in the mean of a normal population, and suppose the population variance is not known. If $\overline{x}$ and s^2 are the realizations of the sample mean and variance, respectively, then a symmetric confidence interval for μ, the population mean, would be based on the pivotal statistic, $t(n-1)$, and would take the form $(\overline{x} - t_{\alpha/2}s/\sqrt{n}, \overline{x} + t_{\alpha/2}s/\sqrt{n})$. Let μ_0 be the hypothesized value of μ. For a two-sided test of hypothesis with significance level α, we would accept the null if $\overline{x} \in (\mu_0 - t_{\alpha/2}s/\sqrt{n}, \mu_0 + t_{\alpha/2}s/\sqrt{n})$. This is the equivalent of accepting the null if $\mu_0 - t_{\alpha/2}s/\sqrt{n} < \overline{x} < \mu_0 + t_{\alpha/2}s/\sqrt{n} \Leftrightarrow \overline{x} - t_{\alpha/2}s/\sqrt{n} < \mu_0 < \overline{x} + t_{\alpha/2}s/\sqrt{n}$. Comparing this to the form of the confidence interval, we see that we would accept the null if μ_0 falls in the confidence interval, and otherwise reject it (with the ambiguous cases $\mu_0 = \overline{x} - t_{\alpha/2}s/\sqrt{n}$ or $\mu_0 = \overline{x} + t_{\alpha/2}s/\sqrt{n}$ decided by further study).

A similar observation links one-sided tests of hypotheses with confidence upper and lower bounds.

In the next section, we look at a number of different problems involving either tests of hypotheses or confidence intervals. The procedures outlined in sections 8.1 and 8.2 for generating confidence intervals and carrying out tests of hypotheses are the same: we have a parameter of interest and a best-point estimator. Related to this estimator is a pivot, or pivotal statistic, and the pivot's distribution is assumed known (essentially, this is either normal, Student's t, χ^2 or F), and it is important to become familiar with the setup of tabulated values for these distributions. Since the form of confidence intervals or of the critical values in some cases are somewhat different from for a single mean from a normal population, we will look at how these values are generated.

8.4 ONE- AND TWO-PARAMETER INFERENCE

We have looked at the special case of a single normal population and the estimation of the mean when the variance is known. The general case is summarized as follows:

8.4.1 A Single Mean, Normal Population

To test $H_0 : \mu = \mu_0$ v.s. $H_1 : \mu \neq \mu_0$, use either the pivot $(\bar{X} - \mu)/(\sigma/\sqrt{n}) \sim N(0, 1)$ when σ is known, or $(\bar{X} - \mu)/(S/\sqrt{n}) \sim t(n-1)$ when the variance is not known. In the first case, the critical region is $(-\infty, -z_{\alpha/2}) \cup (z_{\alpha/2}, \infty)$ and the critical values are $\pm z_{\alpha/2}$. If we must use the t variable, replace $z_{\alpha/2}$ by $t_{\alpha/2}$. A symmetric confidence interval for μ would take the form:

$$(\bar{X} - t_{\alpha/2}\sigma/\sqrt{n}, \bar{X} + t_{\alpha/2}\sigma/\sqrt{n})$$

where t_α is the upper $\alpha \times 100$ percentage point of the Student's t distribution with $n-1$ degrees of freedom. If we are to test $H_0 : \mu = \mu_0$ v.s. $H_1 : \mu > \mu_0$, we have as rejection region or critical region (z_α, ∞) when the pivot has a standard normal distribution. A similar result holds for the other one-sided alternative hypothesis. Again, if Student's t is the appropriate pivot's distribution, we use percentage points of the t variable with $n-1$ degrees of freedom.

We have used the fact that the ratio of a standard normal to the square root of an independent $\chi^2(n-1)$ variable divided by its degrees of freedom has a Student's t distribution, and that in the case of a normal distribution, the sample mean and the sample variance are independent. It has been shown that $(n-1)S^2/\sigma^2 \sim \chi^2(n-1)$, and this is in fact the pivot used for estimating the variance of a normal population. We then have:

$$\sqrt{n}\frac{(\bar{X} - \mu)}{\sigma} \sim N(0, 1),$$

$$(n-1)\frac{S^2}{\sigma^2} \sim \chi^2(n-1)$$

and these are independent. That is why:

$$\sqrt{n}\frac{(\bar{X} - \mu)}{\sigma} \bigg/ \sqrt{(n-1)\frac{S^2}{\sigma^2}\bigg/(n-1)} = \sqrt{n}(\bar{X} - \mu)/S \sim t(n-1).$$

The form of the pivot when Student's t distribution is used is the same as in the case that the variance is known; we just replace σ by S. The derivation of critical values for a test of hypothesis and the construction of confidence intervals, following the procedure outlined earlier, when the pivot's distribution is Student's t rather than standard normal, is left as an exercise.

In the case that the assumption of normality of the parent is suspect and the variance is unknown, but the sample size is large ($n \geq 30$ is usually enough), we use the same pivot $\sqrt{n}(\bar{X} - \mu)/S$ as in the case of the unknown variance problem above and apply the Central Limit Theorem (CLT) to say that the **approximate confidence interval** with confidence level $1 - \alpha$ is $(\bar{X} - z_{\alpha/2}S/\sqrt{n}, \bar{X} + z_{\alpha/2}S/\sqrt{n})$. Note that we use the upper percentage points of the standard normal distribution, which are virtually identical to any t distribution when $n \geq 30$. We would follow this procedure in a test of hypothesis and call the p-value, calculated from the standard normal distribution, the approximate p-value.

This last observation can be summarized as: if you know the value of the parameter, use that value; do not use an estimate. This theme is repeated in certain special cases later in this chapter.

8.4.2 A Single Proportion

The Clopper-Pearson method (Clopper-Pearson (1934)) is used for small samples and finding confidence intervals in estimating a proportion. We will, however, concentrate on large sample problems, and the reader is encouraged to look up material related to this technique.

In order to apply the procedure for proportions and large samples, we must be able to use the normal approximation to the binomial. Recall that this is justified if:

$$(\hat{\pi} - 3\sqrt{\hat{\pi}(1-\hat{\pi})}/\sqrt{n}, \hat{\pi} + 3\sqrt{\hat{\pi}(1-\hat{\pi})}/\sqrt{n}) \subset (0, 1).$$

In a confidence-interval problem for proportions, we have the result that:

$$(\hat{\pi} - z_{\alpha/2}\sqrt{\hat{\pi}(1-\hat{\pi})}/\sqrt{n}, \hat{\pi} + z_{\alpha/2}\sqrt{\hat{\pi}(1-\hat{\pi})}/\sqrt{n})$$

is a $(1-\alpha)100\%$ symmetric confidence interval. This requires that the normal approximation to the binomial can be applied. When we consider hypothesis testing, the results are somewhat different. The reason is that we assume the null is true when we calculate the pivot. Under $H_0 : \pi = \pi_0$, we assume π is known, and in particular, is equal to π_0. Thus, the distribution of $\hat{\pi}$ under H_0 is at least approximately $N(\pi_0, \pi_0(1-\pi_0)/n)$. Thus, $(\hat{\pi} - \pi_0)/\sqrt{\pi_0(1-\pi_0)/n} \sim N(0, 1)$. The main observation is that since we are assuming $\pi = \pi_0$, we use that value. Everything else about testing one- or two-sided hypotheses for a proportion (at least when the normal approximation is valid) is the same as for testing a mean.

Example 11: Suppose that we want to determine if the proportion of male and female hedgehogs in the wild is the same. Let π be the proportion of female hedgehogs. A sample of 50 hedgehogs is taken (assume this is a simple random sample) to test the hypothesis that $H_0 : \pi = \pi_0 = 1/2$ vs $H_1 : \pi > 1/2$. In the sample, 32 are female. Is this good evidence that the proportion of female hedgehogs is in fact greater than the proportion of males?

Solution: Assuming all of the steps in a proper test of hypothesis have been followed, we need the computed score and the critical scores. We will base this on the pivotal statistic, namely, $Z = (\hat{\pi} - \pi_0)/\sqrt{\pi_0(1-\pi_0)/n} \sim N(0, 1)$, at least approximately. It is left as an exercise to verify that the given data supports the assumption a normal distribution can be used as an approximation to the binomial distribution. Here $\hat{\pi}_{calc} = 32/50$, and $z_{calc} = 1.98$. Since this is a one-sided test, $z_{crit} = 1.645$, we reject the null. The p value is approximately 0.025.

We can summarize the results for a single mean and for a single proportion as: the endpoints of the confidence intervals are a certain number of standard deviations above or below the sample mean or the sample proportion. Recall that we have termed the standard deviation of the point estimator the standard error, so we can also say the endpoints are a certain number (based on the confidence coefficient) of standard errors above or below the sample mean or the sample proportion. In a test of hypothesis, the critical value or values are again just a certain number of standard errors above or below the mean, based on the significance level and whether it is a one- or two-sided test.

8.4.3 A Single Variance

Up to now, the algebra that is used to determine confidence intervals or critical values for tests of hypotheses has been the same when dealing with means or proportions, or if we are using the normal distribution or Student's t. Variances are different, because the $\chi^2(n-1)$ distribution is not symmetric, and in fact, has domain $(0, \infty)$. We have worked out how to generate a symmetric confidence interval for the variance, or upper or lower confidence bounds. Clearly, the method of calculation of endpoints of a confidence interval is the same as for generating critical numbers for a one- or two-sided test of hypothesis.

Under the null $H_0 : \sigma^2 = \sigma_0^2$, the pivotal statistic is $(n-1)S^2/\sigma_0^2 \sim \chi^2(n-1)$. The critical score(s) are: χ_α^2, $\chi_{1-\alpha}^2$, or both $\chi_{\alpha/2}^2$ and $\chi_{1-\alpha/2}^2$ (where χ_α^2 is the notation for upper percentage point of the $\chi^2(n-1)$ distribution) depending on whether it is a right, left, or two-sided test of hypothesis. The calculated score is $(n-1)S^2/\sigma_0^2$ and the rest of the test of hypothesis is the same as others already described.

Example 12: A supplement to the standard diet for lab rats is supposed to produce more reliable results in weight gain, meaning the variability in weight gain is smaller than without the supplement. The variance in weight gain without the supplement (over a given time period, and for rats of the same age) is 5 gm^2. It is claimed that the variance is less than this when the supplement is used. If 25 rats are selected and fed the diet with the supplement, and the variance in weight gain is 2.2, is this good evidence the supplement does produce more reliable results?

Solution: Again assuming all other steps in a full test of hypothesis have been carried out, then the null hypothesis will be $H_0 : \sigma^2 = 5$ versus $H_1 : \sigma^2 < 5$. Under the assumption that the weight gain is normally distributed, or nearly so, we can assume under the null hypothesis $24S^2/5 \sim \chi^2(24)$, so the critical score is 13.85 from tables for the χ^2 distribution. The calculated score is $24(2.2)/5 = 10.56$, so we reject the null hypothesis, meaning we accept that the supplement produces more reliable results.

It should be noted that the methods outlined above for a single variance depend heavily on the requirements that the parent distribution be normal and that the sample be random. If either of these conditions is violated to any great extent, these results are not justified. Again, it is important to verify that the assumptions are supported by the actual data and by the experimental design.

Suppose instead we want to test an hypothesis about σ, or find a confidence interval for σ, the population standard deviation. How should one proceed? This is left as an exercise.

8.4.4 Two Means

In some experiments, a factor has two levels, and what we are interested in is the comparison of the means from the two populations generated by the two levels of the factor. There are two cases to consider: (1) the populations are independent, and (2) the populations are not independent.

8.4.4.1 Two (Independent) Normal Populations

The basic problem is that we have two independent populations, and we are interested in the same value in each—the height of asters with either two dominant alleles of a gene controlling height, or the combination of the dominant and recessive genes controlling height; the proportion of men who survive a first heart attack versus the proportion of women who do so; or the rate of absorption of a drug through two different types of skin patches. The notation usually uses subscripts of 1 and 2 to denote the two populations and then X_1 and X_2 are the two random variables, with X_1 measured on Population 1 and X_2 measured on Population 2. This notation allows us to easily generalize to more populations. However, sometimes we use X and Y to denote these values as well. Other notation includes $E(X_1) = \mu_1$, $var(X_1) = \sigma_1^2$, $E(X_2) = \mu_2$, $var(X_2) = \sigma_2^2$, and, if we assume these variables are normally distributed, then $X_i \sim N(\mu_i, \sigma_i^2)$, $i = 1, 2$.

The most common question is whether the two means are equal, versus one of the standard alternatives. Thus, we may want to test $H_0 : \mu_1 = \mu_2$, or, equivalently, $H_0 : \mu_1 - \mu_2 = 0$, versus $H_1 : \mu_1 - \mu_2 \neq 0$ or $H_1 : \mu_1 - \mu_2 > 0$ or $H_1 : \mu_1 - \mu_2 < 0$. In such problems, we introduce the notation $\Delta = \mu_1 - \mu_2$, and Δ becomes the parameter of interest. It should be noted that instead of determining if the two means are equal, we may want to determine if the means are some particular distance apart. We can express the appropriate null hypothesis as $H_0 : \Delta = \delta_0$, where δ_0 is some number given in the problem. The method for carrying out tests of hypotheses or generating confidence intervals when $\delta_0 = 0$ is the same for $\delta_0 \neq 0$, and the general case is left as an exercise.

Our first problem is to determine how to estimate $\Delta = \mu_1 - \mu_2$. We can estimate μ_1 and μ_2 separately by the respective sample means. Since $E(\bar{X}_1 - \bar{X}_2) = \mu_1 - \mu_2 = \Delta$, and, when $X_1 \sim N(\mu_1, \sigma_1^2)$, $X_2 \sim N(\mu_2, \sigma_2^2)$, it is also true that $\bar{X}_1 - \bar{X}_2 \sim N(\mu_1 - \mu_2, \sigma_1^2/n_1 + \sigma_2^2/n_2)$. Here, n_i represents the size of the sample from population i. Thus, $\bar{X}_1 - \bar{X}_2$ is unbiased for Δ, and as the sample sizes grow, the variance tends to 0. We would also have:

$$(\bar{X}_1 - \bar{X}_2 - \Delta) \Big/ \sqrt{\sigma_1^2/n_1 + \sigma_2^2/n_2} \sim N(0, 1)$$

so there is a pivotal statistic with a relatively simple distribution.

The above argument shows that if σ_1^2 and σ_2^2 are known, then we can use the standard normal to find critical values in a test of hypothesis or find a confidence interval for Δ. Further, if n_1 and n_2 are both large, i.e., ≥ 30, and σ_1^2 and σ_2^2 are unknown, then we can use the Central Limit Theorem and replace σ_1^2 and σ_2^2 by S_1^2 and S_2^2, respectively, and continue to use the standard normal, only report that the results are approximate. In this case, the assumption of normality is not needed, just that the Central Limit Theorem applies, and the sample sizes are adequate.

Small sample problems rely more heavily upon the assumption of normality of the parent distributions. However, we can still proceed as above, with some modifications that reflect the small sample issue. Recall that it is standard practice to try to ensure samples from different populations are of the same size, that is, the design is balanced. Consider then the case of estimating the difference of means, or of carrying out a test of hypothesis when the sample sizes are small (typically, under 30). We shall now assume our variables are very nearly normal, if not exactly normal. It should be noted there are methods for dealing with non-normality, including non-parametric methods and methods based on families of distributions that are better fits than the normal. As indicated earlier, it is important to check if the normality assumption is at least reasonable, and we are assuming this has been done. There are two cases:

(i) **the population variances are equal** (or very nearly so). If this is the case, then each sample variance is an estimator of this common variance, call it $\sigma^2 = \sigma_1^2 = \sigma_2^2$. However, a better estimator would be one that uses both samples. This is because the combined sample size is $n_1 + n_2$ and the larger the sample, the more precise the estimate. Basically, we take a weighted average of the variances, and call the result the **pooled estimate of the variance**, S_p^2. It is defined as follow:

$$S_p^2 = \frac{(n_1 - 1)S_1^2 + (n_2 - 1)S_2^2}{n_1 + n_2 - 2}.$$

We will then have:

$$(n_1 + n_2 - 2)\frac{S_p^2}{\sigma^2} \sim \chi^2(n_1 + n_2 - 2).$$

This comes from the fact that the two populations are independent, and so the sample variances are independent. Also the sum of independent χ^2 variables is again χ^2 with degrees of freedom the sum of the degrees of freedom of the summands. Here,

$$(n_1 + n_2 - 2)\frac{S_p^2}{\sigma^2} = (n_1 - 1)\frac{S_1^2}{\sigma^2} + (n_2 - 1)\frac{S_2^2}{\sigma^2} \sim \chi^2(n_1 - 1) + \chi^2(n_2 - 1) = \chi^2(n_1 + n_2 - 2).$$

Now our pivot for Δ is:

$$(\bar{X}_1 - \bar{X}_2 - \Delta) \Big/ \sqrt{S_p^2\left(\frac{1}{n_1} + \frac{1}{n_2}\right)} \sim t(n_1 + n_2 - 2).$$

Specifically, with the assumption of equal variances,

$$\bar{X}_1 - \bar{X}_2 \sim N\left(\Delta, \sigma^2(\frac{1}{n_1} + \frac{1}{n_2})\right)$$

we standardize this normal variable and take its ratio to the square root of an independent $\chi^2(n_1 + n_2 - 2)$ variable divided by its degrees of freedom.

(ii) **the population variances are not equal**. In this case, we cannot pool the estimates of the individual population variances. However, it has been shown that, as long as both n_1 and n_2 are at least five, and the populations are normal, or nearly so, then:

$$(\bar{X}_1 - \bar{X}_2 - \Delta) \Big/ \sqrt{\frac{S_1^2}{n_1} + \frac{S_2^2}{n_2}} \sim t(r)$$

approximately, where the value r of the degrees of freedom is given by:

$$r = \left(\frac{S_1^2}{n_1} + \frac{S_2^2}{n_2}\right)^2 \Big/ \left[\frac{1}{n_1 - 1}\left(\frac{S_1^2}{n_1}\right)^2 + \frac{1}{n_2 - 1}\left(\frac{S_2^2}{n_2}\right)^2\right].$$

With this pivotal statistic, we can then carry out tests of hypotheses or find confidence intervals as before. Note that r will rarely be a positive integer. Student's t is actually defined for non-integer values, and most statistical software will evaluate percentage points for non-integer r. However, little precision is lost if we use the values associated with the t variable with integer degrees of freedom given by the greatest integer less than or equal to the computed value of r.

8.4.4.2 Two Dependent Normal Populations: Paired Comparisons

Consider the problem of comparing a person's health before the administration of a drug and that person's health some time later. Similar problems involve comparisons between pairs of experimental units that are fundamentally similar. For our example, we would have some measure X of health to be measured twice on the same individuals. In agricultural experiments, often, a set of n plots of land are randomly selected, then each is split in two, and one treatment is applied to one half, another to the second half. In this way, virtually all of the characteristics of the land are the same in comparing the two half-plot results. The application of the two different treatments yields two distributions, each with a mean. The standard question is whether the two means are the same, versus one of the typical alternatives, so we have $H_0 : \mu_1 - \mu_2 = 0$ just as in the independent population model. However, we don't have independent populations, so the previous analysis isn't valid here.

To keep the notation simple, let the random variables X and Y be measured on each experimental unit [or on pairs of experimental units that have been matched]. Our sample size will be n, that is, we have chosen n experimental units, and we have recorded the ordered pairs (X_i, Y_i), $1 \leq i \leq n$, with each (X_i, Y_i) measured on unit i. Now form $D_i = X_i - Y_i$, $1 \leq i \leq n$, the set of n differences (note that it is imperative these differences be computed systematically as $D = X - Y$). Since $E(X - Y) = E(X) - E(Y)$ and $D = X - Y$ is a random variable, then $E(D) = E(X) - E(Y) = \mu_X - \mu_Y$. Thus, $H_0 : \mu_X - \mu_Y = 0$ is equivalent to $H_0 : \mu_D = 0$. If we think of D as our random variable, we would estimate μ_D by $\hat{\mu}_D = \sum_{i=1}^{n} D_i/n$, that is, the mean of the observed sample differences. Note that $\hat{\mu}_D = \sum_{i=1}^{n} D_i/n = \overline{D} = \sum_{i=1}^{n}(X_i - Y_i)/n = \bar{X} - \bar{Y}$, the difference of the X and Y sample means.

We do not need to assume much about the distributions of X and Y, but we will assume when n is small that $D \sim N(\mu, \sigma_D^2)$ and the same for n large or at least that the Central Limit Theorem applies to $\overline{D}$. Since we do not typically know σ_D^2 (even if both σ_X^2 and σ_Y^2 are known), we estimate it with the sample variance $S_D^2 = \sum_{i=1}^{n}(D_i - \overline{D})^2/(n-1)$ and use the pivotal statistic:

$$\sqrt{n}(\overline{D} - \mu_D)/\sqrt{S_D^2} \sim t(n-1)$$

in carrying out tests of hypotheses on μ_D or finding confidence intervals for it, just as in the case of a single mean above. As has been stated several times before, for small samples, departures from normality can invalidate any inferences drawn, so it is important to check that this is a reasonable assumption. Also, the procedure outlined readily generalizes to the case that $H_0 : \mu_D = (\mu_D)_0$, that is, a value that is not necessarily 0. This is left as an exercise.

8.4.5 Correlation in Dependent Populations

We will consider the problem of estimation and tests of hypotheses related to the correlation coefficient when we introduce linear regression in Chapter 11.

8.4.6 Two Proportions from Independent Populations

If we want to compare two proportions from two independent populations, we proceed as follows: Take random samples of n_1 and n_2 from the respective populations X and Y, then count in each sample of the number with the requisite characteristic. Call the realizations of X and Y a_1 and a_2, respectively, and then $\widehat{\pi_1 - \pi_2} = a_1/n_1 - a_2/n_2$. This estimator, for large n_1, n_2 and for π_1, π_2 such that the normal approximation can be used in each sample respectively, is approximately normal with mean $\pi_1 - \pi_2$ and variance $\pi_1(1-\pi_1)/n_1 + \pi_2(1-\pi_2)/n_2$. Thus, confidence intervals and tests of hypotheses follow along the same lines as for a difference of means, and details are left to the reader. A difference from confidence intervals does occur in the testing of $H : \pi_1 = \pi_2$. When computing the value of the pivot, we use a pooled estimate of the assumed common proportion $\pi = \pi_1 = \pi_2$, namely $\widehat{\pi}_p = (a_1 + a_2)/(n_1 + n_2)$. Note that we do not know the common proportion, so it has to be estimated. Our pivot now takes the form, under the null hypothesis,

$$(\widehat{\pi}_1 - \widehat{\pi}_2) \bigg/ \sqrt{\widehat{\pi}_p(1-\widehat{\pi}_p)\left(\frac{1}{n_1} + \frac{1}{n_2}\right)} \sim N(0,1)$$

approximately.

What if the null hypothesis had been $H_0 : \pi_1 - \pi_2 = d$, where d is a nonzero difference? In this case, we cannot pool the proportions, and so our pivot would take the form:

$$(\widehat{\pi}_1 - \widehat{\pi}_2 - d) \bigg/ \sqrt{\frac{\widehat{\pi}_1(1-\widehat{\pi}_1)}{n_1} + \frac{\widehat{\pi}_2(1-\widehat{\pi}_2)}{n_2}} \sim N(0,1).$$

8.4.7 Two Variances

We test the ratio of variances or generate confidence intervals for ratios of variances, often using the notation $\lambda = \sigma_1^2/\sigma_2^2$, by referring to Snedecor's F-distribution. This involves two parameters, the numerator degrees of freedom and the denominator degrees of freedom. This would require a very extensive set of tables, so often texts give only very specific percentage points—i.e., for a very limited set of values of α. Thus, the tests and confidence intervals are somewhat restricted. However, many computer programs give other values so that this problem is not a severe one.

The general procedure follows from the following argument: Let X and Y be independent populations, with $X \sim N(\mu_1, \sigma_1^2)$ and $Y \sim N(\mu_2, \sigma_2^2)$, at least approximately. We want to find a confidence interval for $\lambda = \sigma_1^2/\sigma_2^2$ or test the hypothesis $H_0 : \sigma_1^2 = \sigma_2^2$, or, equivalently, $H_0 : \sigma_1^2/\sigma_2^2 = 1$. We have unbiased estimates of σ_1^2 and σ_2^2, namely the sample variances S_1^2 and S_2^2, respectively. We also know that:

$$(n_1 - 1)\frac{S_1^2}{\sigma_1^2} \sim \chi^2(n_1 - 1) \text{ and } (n_2 - 1)\frac{S_2^2}{\sigma_2^2} \sim \chi^2(n_2 - 1).$$

Further, these are independent χ^2 variables. If we take the ratio of two independent χ^2 variables, each divided by its degrees of freedom, the result is a random variable with (Snedecor's) F-distribution. There is an analytic form of this variable, but for our purposes, it is enough to know that percentage points have been tabulated. This becomes our pivotal statistic. Note that the means of the distributions play no direct role in this analysis, and as stated before, are nuisance parameters for this problem. The F-distribution has two parameters: numerator degrees of freedom, determined from the numerator in the ratio of the χ^2s, and denominator degrees of freedom, determined from the denominator in the ratio of the χ^2s.

Under our assumptions, then, we have:

$$\left[(n_1 - 1)\frac{S_1^2}{\sigma_1^2(n_1 - 1)}\right] \bigg/ \left[(n_2 - 1)\frac{S_2^2}{\sigma_2^2(n_2 - 1)}\right] = \frac{\sigma_2^2}{\sigma_1^2}\frac{S_1^2}{S_2^2} \sim F(n_1 - 1, n_2 - 1).$$

It is useful to note that:

$$\left[(n_2 - 1)\frac{S_2^2}{\sigma_2^2(n_2 - 1)}\right] \bigg/ \left[(n_1 - 1)\frac{S_1^2}{\sigma_1^2(n_1 - 1)}\right] = \frac{\sigma_1^2}{\sigma_2^2}\frac{S_2^2}{S_1^2} \sim F(n_2 - 1, n_1 - 1)$$

because tables of values are not extensive. It may be easier to use one version rather than the other. If we want to find a symmetric confidence interval for $\lambda = \sigma_1^2/\sigma_2^2$, we find the values a and b so that $P(F > b) = \alpha/2 = P(F < a)$ for the appropriate F-distribution, and then solve for λ from $a < \lambda(S_2^2/S_1^2) < b$. We can denote $a = F_{1-\alpha/2}(n_2 - 1, n_1 - 1)$ and $b = F_{\alpha/2}(n_2 - 1, n_1 - 1)$ to emphasize that these are percentage points of a particular F-distribution. A bit of care is needed in finding confidence upper or lower bounds, since we are dealing with fractions.

Note that if we assume a common variance, and then are asked to find a confidence interval for this common value, say σ^2, we use the pooled estimate of the variance and the related χ^2 distribution.

8.5 NON-PARAMETRIC TESTS—AN INTRODUCTION

In many cases, the assumption of normality is not justified, and so many of the procedures introduced cannot be used in statistical inference. However, there are tests that are based on relatively few assumptions about the parent distributions, and these are collectively called non-parametric tests. It is possible to use these methods to generate confidence intervals for certain quantities [a median, for example], but we will not go into this here.

Sign Test: This is a test of a particular value of the median, $H_0 : m = m_0$. Let a random sample be taken: $X_1, \ldots, X_n$, and then form $X_1 - m_0, X_2 - m_0, \ldots, X_n - m_0$. If m_0 is the true median, we would expect about the same number of + and − signs in this collection of differences. Thus, if Y counts the number of − signs (this works for + signs as well), then $Y \sim B(n, 1/2)$. We can do one, tail or two, tail tests based on this: This binomial is symmetric, so if we want a two-tail test with $\alpha = 0.05$, then we determine the largest value y so that $P(Y \leq y) \leq \alpha/2$. If our observed value of Y is less than this, or greater than $n - y$, we reject the null. If the alternative is that m is greater than m_0, then we would expect many more plus signs than minus signs. For our Y variable, we would find the largest value y so that $P(Y \leq y) \leq \alpha$. If the alternative is that $m < m_0$, we find the smallest value y so that $P(Y \geq y) = \alpha$.

Wilcoxon Rank Test: This test makes use of the differences in signs, as well as the relative size of values. We find $X_1 - m_0, X_2 - m_0, \ldots, X_n - m_0$ as above, then take the absolute values of these numbers. These are ranked from 1 to n, with tied values assigned the average of all similarly tied ranks. We then list the ranks, with the numerical sign of the corresponding difference. For the positive ones, add them together to get W_+. For the negative signed ranks, add up their absolute values to get W_-. Define W_s to be the maximum of these two values. One- and two-tailed tests can be run, using percentage points readily found on the internet.

Wilcoxon-Mann-Whitney: This is a test about the equivalence of two independent populations. Typically, there is a factor with two levels, and the null is that the distributions involved are the same, that is, the variable measured has a distribution that is independent of this factor. One- and two-sided alternatives can be addressed. The procedure is: Let Y_1 and Y_2 be the random variables being measured under the two levels of the factor. Order each sample separately: $Y_{(11)}, \ldots, Y_{(1n_1)}$ and $Y_{(21)}, \ldots, Y_{(2n_2)}$. For each value in each group, count the number of values in the other group that are less or equal to that value. If there are ties, assign 1/2 to each tied value. Add the counts together for each group to get K_1 and K_2. As a check, $K_1 + K_2 = n_1 n_2$. Let $U_s = max\{K_1, K_2\}$. Critical values for this test are readily found on the internet.

8.6 SUMMARY OF PIVOTAL STATISTICS

This summary includes pivots for the generation of confidence intervals and running tests of hypotheses studied in this chapter and in Chapter 11 on regression methods.

1. *Single Normal Population:* $X \sim N(\mu, \sigma^2)$:

    (a) Large Sample Procedures

    $$(\sigma^2 \text{ known or } \sigma^2 \approx S^2) \text{ for } \mu, \quad \frac{\bar{X} - \mu}{\sigma/\sqrt{n}} \sim N(0, 1).$$

    (b) Small Sample Procedures for μ:

    $$\frac{\bar{X} - \mu}{S/\sqrt{n}} \sim t(n-1); \text{ for } \sigma^2, \quad \frac{(n-1)S^2}{\sigma^2} \sim \chi^2(n-1).$$

2. *Two (Independent) Normal Populations:* $X_1 \sim N(\mu_1, \sigma_1^2)$, $X_2 \sim N(\mu_2, \sigma_2^2)$:

    $$\Delta = \mu_1 - \mu_2, \quad \lambda = \sigma_1^2/\sigma_2^2$$

    (a) Large Sample Procedures (σ_i^2 known or $\sigma_i^2 \approx S_i^2$), $i = 1$ and 2):

    $$\text{for } \Delta, \quad \frac{(\bar{X}_1 - \bar{X}_2) - \Delta}{\sqrt{\sigma_1^2/n_1 + \sigma_2^2/n_2}} \sim N(0, 1).$$

    (b) Small Sample Procedures: If $\sigma_1^2 = \sigma_2^2 = \sigma^2$:

    $$\text{for } \Delta, \quad \frac{(\bar{X}_1 - \bar{X}_2) - \Delta}{S\sqrt{1/n_1 + 1/n_2}} \sim t(n_1 + n_2 - 2).$$

    $$\text{For } \sigma^2, \quad \frac{(n_1 + n_2 - 2)S^2}{\sigma^2} \sim \chi^2(n_1 + n_2 - 2)$$

    (c) If $\sigma_1^2 \neq \sigma_2^2$:

    $$(\bar{X}_1 - \bar{X}_2 - \Delta) / \sqrt{\frac{S_1^2}{n_1} + \frac{S_2^2}{n_2}} \sim t(r)$$

    approximately, where the value r of the degrees of freedom is given by:

    $$r = \left(\frac{S_1^2}{n_1} + \frac{S_2^2}{n_2}\right)^2 \bigg/ \left[\frac{1}{n_1 - 1}\left(\frac{S_1^2}{n_1}\right)^2 + \frac{1}{n_2 - 1}\left(\frac{S_2^2}{n_2}\right)^2\right].$$

    For λ, $\dfrac{S_1^2/S_2^2}{\lambda} \sim F(n_1 - 1, n_2 - 1)$.

    Note: if $\sigma_1^2 = \sigma_2^2 = \sigma^2$, the pooled estimator of σ^2 is:

    $$S^2 = \frac{(n-1)S_1^2 + (n_2 - 1)S_2^2}{n_1 + n_2 - 2}.$$

3. *Two (Dependent) Populations*

    $$\Delta = \mu_1 - \mu_2, \quad \rho = \text{Cov}(X_1, X_2)/(\sigma_1 \sigma_2), \quad D = X_1 - X_2.$$

    (a) Large Sample Procedures for Δ, $\dfrac{\bar{D} - \Delta}{\sigma_D/\sqrt{n}} \sim N(0, 1)$

    (b) Small Sample Procedures for Δ, $\dfrac{\bar{D} - \Delta}{S_D/\sqrt{n}} \sim t(n-1)$

Estimator for Δ is:

$$\bar{D} = \bar{X}_1 - \bar{X}_2 = \frac{1}{n}\sum_{i=1}^{n} D_i, \quad S_D^2 = \frac{1}{n-1}\left(\sum_{i=1}^{n} D_i^2 - \frac{1}{n}\left(\sum_{i=1}^{n} D_i\right)^2\right)$$

Estimator for ρ is:

$$r = \frac{\sum X_{1j}X_{2j} - n\bar{X}_1\bar{X}_2}{(n-1)S_1 S_2} = \frac{\sum X_{1j}X_{2j} - n\bar{X}_1\bar{X}_2}{\sqrt{\left(\sum X_{1j}^2 - n\bar{X}_1^2\right)\left(\sum X_{2j}^2 - n\bar{X}_2^2\right)}} = \frac{S_{12}}{\sqrt{S_{11}S_{22}}}$$

(c) Correlation Procedures: X_1, X_2 Bivariate Normal

Test $H_0: \rho = 0$, use $r\sqrt{\frac{n-2}{1-r^2}} \sim t(n-2)$.

Test $H_0: \rho = \rho_0 (\neq 0)$, use $W \sim N\left(\mu_w, \frac{1}{n-3}\right)$ where:

$$W = w(r) = \frac{1}{2}\ln\left(\frac{1+r}{1-r}\right), \quad \mu_w = w(\rho).$$

(d) Estimated Least Squares Regression Line: $\hat{X}_2 = \hat{\beta}_0 + \hat{\beta}_1 X_1$, where:

$$\hat{\beta}_1 = S_{12}/S_{11}, \quad \hat{\beta}_0 = \bar{X}_2 - \hat{\beta}_1 \bar{X}_1$$

4. *Test of Proportions*

   (a) One Binomial Population, Large Sample $\dfrac{\hat{p} - p_0}{\sqrt{p_0 q_0/n_1}} \sim N(0, 1)$, approximately, $q_0 = 1 - p_0$.

   (b) Two Binomial Populations, Large Sample $\dfrac{\hat{p}_1 - \hat{p}_2}{\sqrt{\hat{p}\hat{q}(1/n_1 + 1/n_2)}} \sim N(0, 1)$, approximately where
   $\hat{p} = \dfrac{n_1 \hat{p}_1 + n_2 \hat{p}_2}{n_1 + n_2}$, $H_0: p_1 = p_2$, and $\hat{q} = 1 - \hat{p}$.

Exercises

Throughout these exercises, standard notation will be adhered to: μ for the population mean; σ^2 for the population variance; if the random variable is X, then $\bar{X}$ is the sample mean and $\bar{x}$ is the realization of $\bar{X}$; n is the sample size; S^2 is the sample variance, and s^2 its realization; π is the population proportion. When used, $\hat{\mu}$, and the "hat" notation in general, means the best point estimator as determined earlier.

1. Let $Z \sim N(0, 1)$. Determine the value z for which:

   (a) $P(Z \leq z) = 0.05$
   (b) $P(Z > z) = 0.70$
   (c) $P(|Z| < z) = 0.1$
   (d) $P(|Z| > z) = 0.98$

2. Determine the value of the following upper percentage points z_α of the standard normal distribution. In each case, sketch the normal distribution and the area that corresponds to the associated probability.

   (a) $z_{0.05}$
   (b) $z_{0.95}$
   (c) $z_{0.75}$
   (d) $z_{0.025}$
   (e) $z_{0.975}$

3. Use the table in the text to determine the value of the following upper percentage points t_α of Student's t distribution, where the degrees of freedom parameter is $\nu = 10$.

   (a) $t_{0.05}$
   (b) $t_{0.95}$
   (c) $t_{0.005}$
   (d) $t_{0.975}$

4. For each set of hypotheses (a)–(d) below:

   (i) identify the population parameter(s) of interest (ie. population mean, proportion, etc), based on standard usage of notation;

   (ii) restate each hypothesis in words;

   (iii) state whether the null hypothesis is simple or composite;

   (iv) state whether the non-rejection region would correspond to a symmetric confidence interval, upper bound confidence interval, or lower bound confidence interval.

   (a) $H_0 : \mu = 5$ vs $H_1 : \mu \neq 5$
   (b) $H_0 : \pi = 0.7$ vs $H_1 : \pi < 0.7$
   (c) $H_0 : \sigma^2 = 8$ vs $H_1 : \sigma^2 > 8$
   (d) $H_0 : \mu_1 = \mu_2$ vs $H_1 : \mu_1 < \mu_2$

Inferences for One Population Mean

5. Let the random variable $X \sim N(\mu, \sigma^2)$. You are conducting a test of hypothesis where the null hypothesis is $H_0 : \mu = 10$. Assume that the usual pivotal statistic has a standard normal distribution (i.e., either the sample is large or the variance is known). Use the given alternate hypothesis, significance level, and calculated pivotal value z_{calc} to determine the following in each of (a), (b) and (c):

   (i) the critical region;

   (ii) a decision to reject or "accept" the null hypothesis;

   (iii) the p-value and an interpretation of the value.

   (a) $H_1 : \mu \neq 10$, $\alpha = 0.10$, $z_{calc} = 1.55$
   (b) $H_1 : \mu > 10$, $\alpha = 0.05$, $z_{calc} = 2.15$
   (c) $H_1 : \mu < 10$, $\alpha = 0.01$, $z_{calc} = -1.255$

Chapter 8 — Confidence Intervals and Hypothesis Testing

6. Consider a sample taken from a normal population with sample mean $\bar{x} = 26$ and sample variance $s^2 = 121$.

   (a) Construct a 95% confidence interval for the mean of the population assuming that the sample size is $n = 42$.

   (b) Determine the estimated standard error of the estimate of the mean.

   (c) Suppose, rather unrealistically, that we are able to determine the true mean for the population and find $\mu = 30$. Would the sample be considered unusual? Explain your answer.

   (d) What percentage of samples of size 42 taken would we expect to have sample means falling outside the confidence interval?

   (e) Repeat (a), assuming that $n = 21$.

   (f) Suppose that the variance of the population is known to be $\sigma^2 = 121$ and the sample size is $n = 21$. Construct a 95% confidence interval for the mean of the population. Compare this with your result in (e).

7. Consider the 90% confidence interval (0.55, 0.75) for the true mean of a population obtained from a small sample from a normal distribution with unknown mean and variance. Determine whether the width of the interval would increase, decrease, or stay the same for the following individual changes. (Assume that all else remains the same.)

   (a) The standard error of the mean is increased.

   (b) The number of sampled units is increased.

   (c) The best point estimate of the variance increased.

   (d) The level of confidence was adjusted to 99%.

8. The mean weight of mature lake trout in northern Ontario was estimated to be approximately 3.0 lb. After a number of environmental changes, it was hypothesized that the true mean weight is no longer 3 lbs. A sample of 36 fish was obtained and their weights W recorded. The sample mean weight and sample standard deviation were found to be $\bar{w} = 2.70$ lb and $s_W = 0.42$ lb, respectively.

   (a) For what parameter is $\bar{W}$ the best point estimate?

   (b) Determine the estimated standard error of $\bar{W}$.

   (c) Construct a 95% confidence interval for the true mean weight of the fish.

   (d) Use the data to perform a test of hypothesis at a significance level of 5% of the claim that the average weight of the fish is not 3 lbs by completing the following steps:

   (i) State the null and alternative hypotheses.

   (ii) State the level of significance.

   (iii) State the appropriate pivot and its distribution.

   (iv) Determine the critical region.

   (v) Calculate the value of the pivot using the sample data.

   (vi) Draw a conclusion.

   (vii) Determine the corresponding p-value and comment on the strength of evidence against the null hypothesis.

   (e) For what levels of significance, α, would the data have provided evidence for the null hypothesis? Be specific.

(f) Explain the meaning of a Type I Error, in the context of this study, and give, with appropriate justification, what is known about the probability of making such a mistake in this case.

(g) Explain how your confidence interval in (c) justifies the result of your hypothesis test.

9. (a) Use the data in 8(d) to perform a hypothesis test at a significance level of 1% to determine if the mean weight of the fish is less than 3 lbs.

(b) Calculate the *p*-value for the test in (a), and comment on the strength of evidence against the null hypothesis.

10. To study the growth rate of ragweed, 12 plants of approximately the same age were randomly selected in a large field and measured at the beginning and end of April. The increase in height, measured in millimeters, is given below.

170.5	190.3	189.6	189.2	188.4	195.1
90.9	192.3	191.6	192.7	191.1	189

(a) Use the sample data to determine a point estimate of the mean increase in height of the plants.

(b) Use the sample information to construct a symmetric 95% confidence interval for average increase in height.

(c) What is the maximum error of your point estimate of the true mean average increased growth in April, at the 95% confidence level?

(d) How large a sample of plants would be required to ensure that the sample data yields a point estimate that is accurate to within 0.5mm of the population's average increase in height at the 95% confidence level? Assume the sample standard deviation does not change.

11. Consider the growth rate experiment in Exercise 10. Suppose that the expected average increase in height was approximately 190 mm. A claim is made that, on average, the increase in height over April actually exceeds 190 mm.

(a) Perform a hypothesis test at the 10% significance level to test the claim.

(b) What would a Type I Error mean in this context? What is the probability of such an error?

(c) What would a Type II Error mean in this context?

(d) Without additional calculations, indicate which of the following changes would make rejection of the null hypothesis in (a) more likely. In each case, assume all other relevant quantities involved would remain at their original values.

   (i) The mean of the sample increases.
   (ii) The mean of the sample decreases.
   (iii) The standard error of the mean increases.
   (iv) The standard error of the mean decreases.
   (v) The sample size increases.
   (vi) The sample size decreases.
   (vii) The significance level increases.
   (viii) The significance level decreases.

Chapter 8 — Confidence Intervals and Hypothesis Testing

12. Consider the data in Exercise 18, Chapter 6, concerning a log normal distribution.

    (a) Determine a 95% symmetric confidence interval for the parameter μ.

    (b) Determine a 95% confidence upper bound to the average number of hours until the toxin reaches 1 foot below the surface of the field.

Inferences for One Population Variance

13. Use the table in your text to determine the following chi-square upper percentage points for the given degrees of freedom:

    (a) $\chi^2_{0.05}$, $\nu = 23$
    (b) $\chi^2_{0.02}$, $\nu = 14$
    (c) $\chi^2_{0.001}$, $\nu = 17$
    (d) $\chi^2_{0.95}$, $\nu = 16$
    (e) $\chi^2_{0.98}$, $\nu = 22$
    (f) $\chi^2_{0.999}$, $\nu = 5$

14. Let X be a random variable with a chi-square distribution with ν degrees of freedom. Use the tables in the text to determine the value of a such that:

    (a) $P(X > a) = 0.01$; $\nu = 12$
    (b) $P(X < a) = 0.90$; $\nu = 12$
    (c) $P(X > a) = 0.025$; $\nu = 7$
    (d) $P(X < a) = 0.025$; $\nu = 7$
    (e) $P(X > a) = 0.05$; $\nu = 24$
    (f) $P(X < a) = 0.05$; $\nu = 24$

15. A sample of measurements is taken from a population. The results are:

    $$10, 12, 19, 14, 15, 18, 11, 13.$$

    (a) What are the standard assumptions about the original population needed to determine confidence intervals and to perform tests of hypotheses concerning the population variance? Why are these necessary?

    (b) Determine s^2. What population parameter is estimated with this value?

    (c) Use the sample data and the $\chi^2(7)$ upper percentage points $\chi^2_{0.025} = 16.013$; $\chi^2_{0.975} = 1.690$; $\chi^2_{0.005} = 20.278$ and $\chi^2_{0.995} = 0.989$ to determine an interval estimate for the population variance with a confidence level of:

        (i) 95%
        (ii) 99%

    (d) It is claimed that the population variance exceeds 9. Use a hypothesis test to determine if the claim can be supported at a 5% significance level.

16. Refer to Exercise 10. Past data suggests that the amount of growth in the type of plant under consideration over the one-month period has a variance around 30 mm². Use the data from question 10 and a significance level of 10% to determine whether the current data provides evidence that the variance exceeds 30.

17. For the data in Exercise 18 in Chapter 6, determine a 95% symmetric confidence interval for the parameter σ^2.

Inferences for One Population Proportion

18. It has been estimated that in a certain forest, 60% of the elm trees have dutch elm disease. A biologist examines 150 randomly chosen elm trees and discovers that 103 have the disease.

   (a) If X is the number of elm trees with dutch elm disease, verify that the normal approximation to the binomial distribution of X can be used, assuming there is a large population of elm trees in this forest.

   (b) Construct a 90% confidence interval for the proportion of elm trees in the population infected with the disease.

   (c) Determine the maximum error of the point estimate from (a).

19. **(a)** The point estimator X/n of a proportion π has variance $var(X/n) = \pi(1-\pi)/n = (\pi - \pi^2)/n$. Show that the maximum value of the variance of X/n is $1/(4n)$.

   (b) Use (a) to show that the minimum sample size n required to ensure that the estimated proportion is within E of the true proportion with $(1-\alpha)$ 100% confidence is given by $n \geq (z_{\alpha/2}/(2E))^2$.

   (c) Refer to Exercise 18. Determine the minimum sample that the biologist must use to estimate of the proportion of infected trees to ensure that the estimate is within 5% of the true proportion with 95% confidence.

20. Refer to Exercise 18. Perform a hypothesis test to determine if the biologists data provides evidence that the proportion of infected elm trees in the population exceeds the estimate of 60%. Use a significance level of 5%.

21. Human malaria can be transmitted only by female mosquitoes of the genus Anopheles. To control the spread of the disease, insecticides that contain DDT are used in conjunction with protective netting. See Curtis and Townson (1998). Recently concerns have been raised as mosquitoes are showing an increased tolerance to DDT. Suppose that a researcher is testing the effectiveness of a new, DDT-free, insecticide. The insecticide is provided to 400 individuals living in a region considered to have a high rate of malaria cases. In one year of use, 88 of the 400 individuals tested positive for malaria.

   (a) Let π denote the proportion of the population using the new insecticide for year that test positive for malaria. Use the sample data given to estimate this value.

   (b) Construct a 95% confidence interval for π. First, verify that the normal approximation to the sample proportion can be applied.

   (c) Construct a 99% confidence interval estimate for π.

   (d) If historical estimates suggest that the incidence rate of malaria is approximately 24% in the area under consideration, is there sufficient evidence to suggest that the new insecticide is effective in reducing the rate of malaria incidence? Conduct a hypothesis test at the 95% significance level to test the claim.

Inferences for Two Population Means

22. In a study of the effectiveness of a new drug in weight reduction, a group of 8 individuals participated in a program combining physical exercise with the use of the drug for one month. The results, in pounds, are as follows, in the same order of the individuals for the before and after observations:

   Weight Before (B) 209 178 175 183 197 170 164 190

   Weight After (A) 196 175 161 165 152 161 147 189

   (a) Explain why the sample values are dependent.

   (b) Determine the decrease in weight for each individual, $D_i = B_i - A_i$.

   (c) Using the sample data, determine the sample mean $\overline{D}$ of the differences $D_i = B_i - A_i$. For what parameter is $\overline{D}$ the best point estimate?

Chapter 8 — Confidence Intervals and Hypothesis Testing

(d) Perform a significance test at the 5% level to determine whether the use of the new drug resulted in a mean decrease in weight.

(e) Determine a 95% confidence interval for the true mean decrease in weight.

23. To study the effect of oxygen on mental performance, 100 students from a large first-year psychology class were randomly selected and administered a multiple choice test on course material up to the time of the test. Fifty students from this group were randomly selected and placed in one of two separate rooms, identical except one had a 10% increase in oxygen for the duration of the test. The results for each group on the test (scored out of 120) included:

	Control Group (C)	Treated Group (T)
Number of students	$n_C = 50$	$n_T = 50$
Group average on test	$\bar{c} = 93.6$	$\bar{t} = 95.9$
Standard deviation in test scores	$s_C = 5.3$	$s_T = 6.1$

(a) Evaluate the realization of $\bar{T} - \bar{C}$. For what population parameter is this value the best point estimate? What does the parameter represent?

(b) Determine the estimated standard error of your point estimate in (a).

(c) Determine a symmetric 95% confidence interval for the amount by which the true mean score on the test in the treated group exceeds that of the control group.

(d) Does the data provide evidence to suggest that the mean test scores of the two groups are different at the 5% significance level? Conduct a hypothesis test at the 5% significance level to answer this question.

(e) Compute the strength of evidence against the null hypothesis, and interpret the value.

(f) Perform a significance test to show that the data provide evidence at the 5% level that the true mean score in the treated group exceeds that of the control group.

(g) How strong is the evidence against the null hypothesis? Compute the p-value, and interpret the value.

24. A survey of math students between the ages of 18 and 22 was conducted to determine how many hours of sleep they get in a typical night. In total, 13 male students and 16 female students completed the survey. The average number of hours for males and females, as well as the sample standard deviations from each group, are as follows:

$$n_m = 13, \bar{x}_m = 6.69 \text{ hrs}, s_m = 4.51 \text{ hrs}, n_f = 16, \bar{x}_f = 8.35 \text{ hrs}, \text{ and } s_f = 5.32 \text{ hrs}.$$

(a) Assume that the variances for the two populations are equal. Calculate the best point estimate for the common variance. That is, calculate the pooled sample variance s_p^2 to estimate the common variance in the numbers of hours of sleep obtained by the students.

(b) Compute the best point estimate of $\mu_f - \mu_m$. Explain what this point estimate represents in the context of the problem.

(c) Use the data to test the claim that the average amount of sleep that 18–22-year-old males differs from the average amount of sleep that females aged 18–22 obtain on a typical night.

(d) What assumptions did you make when completing the hypothesis test in (c)?

(e) There are some problems with the assumptions made to carry out the test. What are they? Explain.

25. A research team has suggested that the sex of the individual being treated may be linked to effectiveness of a particular drug treatment for high blood pressure. A group of 200 women and a group of 250 men were

treated with the drug. Out of the group of 200 women, 128 showed a decrease in blood pressure, while 220 men showed a decrease in blood pressure.

   (a) Calculate the best point estimate for the true value by which the proportion of men experiencing positive effects exceeds that of women experiencing positive effects.

   (b) Calculate the estimated standard error of the difference in proportions of females and males experiencing positive effects after taking the drugs.

   (c) Determine a 95% confidence interval for the true difference in proportions.

26. In a study on the relationship between tumor incidence in mice under stress, one group of mice was maintained in a large cage with adequate food and normal lighting, while another group was placed in a small cage, with constant noise, adequate food, and random patterns of light and darkness. After six months, the mice were examined for tumors. Of the 100 mice in the stress-less environment, 11 had tumors, and of the 50 mice in the stressful environment, 12 had tumors. Let π_1 and π_2 denote the proportions of mice with tumors in the stress-less environment and in the stressful environment, respectively.

   (a) Determine the best point estimates of π_1 and π_2.

   (b) Perform a hypothesis test at the 5% significance level to test the hypothesis that the incidence of tumors in mice is greater in mice exposed to stress. (Hint: use the pooled estimate of the common proportion $\hat{\pi}$.)

   (c) Suppose that the claim was that the incidence of tumors in mice in the stressful environment exceeds that of the incidence of tumors in mice in a stress-less environment by more than 2%. How would your hypotheses and calculations change? (You do not need to complete a hypothesis test.)

Two Population Variances

27. It can be shown that $F_{1-\alpha}(\nu_1, \nu_2) = 1/F_\alpha(\nu_2, \nu_1)$. Evaluate the following F upper percentage points.

   (a) $F_{0.05}(15, 12)$

   (b) $F_{0.95}(15, 12)$

28. In Exercise 24, it was assumed that the two populations have equal variances and the pooled sample variance was used.

   (a) Use your answers to Exercise 25 to conduct a hypothesis test to determine if the sample data provides evidence that the two population variances are not the same. Use a significance level of 10%.

   (b) Construct a 90% confidence interval for the ratio of variances in Exercise 24. Explain how your confidence interval supports your decision in part (a).

CHAPTER 9

GOODNESS OF FIT TESTS AND CONTINGENCY TABLES

Synopsis

The χ^2 and Kolmogorov-Smirnov tests of the fit of an hypothesized distribution with sample information are developed. Contingency table methods for testing the independence of two factors in an experiment are examined.

9.1 GOODNESS OF FIT TESTS AND CATEGORICAL VARIABLES

In Chapter 7, descriptive statistics, including summary values and diagrams, were introduced to get evidence in favor of or against an assumed form of a distribution. These are very useful in getting an overview of how a data set fits with an assumed distributional form, such as normal, exponential, etc. However, it is often of value to have formal evidence of the fit of the data with the assumed distribution, and that is given by **goodness of fit** tests.

The basic test structure is as follows: There are k mutually exclusive and exhaustive categories, classes, or cells. These may correspond to k levels of a particular treatment, for example. A sample of size n is taken from the population, and then the number falling into each category is determined. Label these values $n_1, \ldots, n_k$, and note that $\sum_{i=1}^{k} n_i = n$. What we want to test is whether these frequencies are reasonable matches for theoretical frequencies. The theoretical frequencies are based on the hypothesis that the proportion of the population that falls in category i is p_{0i}. Thus, the null hypothesis under consideration is $H_0 : p_1 = p_{01}, p_2 = p_{02}, \ldots, p_k = p_{0k}$. Here, p_i is the true population proportion of category i. The alternative hypothesis is $H_1 : p_i \neq p_{0i}$ for at least one i. This means that, in fact, at least two categories differ in theoretical and observed distributions.

From this basic setup, we can determine a measure of how far the observed frequencies differ from (average) theoretical differences. We expect on average $p_{0i}n = n_{0i}$ of the sample should fall in category i, so $n_i - n_{0i}$ is a measure of the discrepancy between the theoretical model of frequencies and the observed frequencies. Since it does not matter if this is positive or negative, we use $(n_i - n_{0i})^2/n_{0i}$ as the measure. The main reason for using squares is that these terms will be related to tractable distributions.

The reason for dividing by n_{0i} in $(n_i - n_{0i})^2/n_{0i}$ is not obvious. However, it comes from considering the counts as having a multinomial distribution, and that $(n_i - np_i)/\sqrt{np_i(1-p_i)}$ is, by the Central Limit Theorem, approximately $N(0, 1)$. It can then be shown that

$$Q \equiv \sum_{i=1}^{k} \frac{(n_i - n_{0i})^2}{n_{0i}}$$

is approximately $\chi^2(k-1)$, for n sufficiently large. Q is the sum of k squared (approximately) standard normal variables and, if they were independent, then the distribution would be approximately $\chi^2(k)$. However, the summands are not independent, because once $k-1$ of the categories are filled, providing the estimates of p_i, $1 \leq i \leq k-1$, say, then the last category's entries are known and hence the estimate of p_k is determined by the other $k-1$ values, and so one degree of freedom is lost. A full analysis of this important result can be found in Kendall, Stuart and Ord (1987), Chapter 15, Exercise 15.3.

One rule of thumb to ensure Q is approximately $\chi^2(k-1)$ is that there should be at least 5 observations in each category. If that is not the case, smaller categories are combined until it is. Also, the notations $n_i = O_i$, that is, the observed number in cell i, and $n_{0i} = E_i$, the expected number in cell i, are standard.

Example 1: 1,500 residents in a remote community had their blood typed, with the following results:

Blood Type	O+	O−	A+	A−	B+	B−	AB+	AB−
Community (Sample)	576	165	473	111	92	36	22	25

Based on Example 16 in Chapter 7, the expected category counts if the local community reflected the distribution of blood types in Canada would be:

Blood Type	O+	O−	A+	A−	B+	B−	AB+	AB−
Community (Expected)	585	105	540	90	114	21	37.5	7.5

The researcher is interested in testing whether the distribution of blood types is the same in the local community as it is in Canada as a whole. She computes the differences, $D_i = O_i - E_i$:

Blood Type	O+	O−	A+	A−	B+	B−	AB+	AB−
$D_i = O_i - E_i$	−9	60	−67	21	−22	15	−15.5	17.5

The test statistic value is:

$$Q = \sum_{i=1}^{k}(O_i - E_i)^2/E_i = 109.84$$

Asymptotically, this will be $\chi^2(7)$. The upper $0.0001 \times 100\%$ point is 29.88, so the null hypothesis would be rejected, and the p-value would be $<< .0001$, meaning we would have very strong evidence against the null that the community and the general Canadian populace have the same blood type distributions.

One drawback with this test is that it does not identify which classes are different, although a check of the data suggests where the differences lie. An alternate approach is to test each category's proportion versus the population proportion of that category. However, the more tests run, the greater the chance of a type I error.

Example 2: Let X be a non-negative random variable, measuring the length of time (in minutes) processing new arrivals takes at an emergency room at a large local hospital. Long-term studies have established that X has an exponential distribution, with average time processing arrivals of five minutes. A new procedure is put in place, which is meant to speed up processing new arrivals. After a month under the new procedure, 30 processing times were randomly selected:

0.216	0.333	0.391	0.620	1.315	1.476	1.525	1.525	1.617	2.087
2.488	2.736	3.192	3.206	3.434	3.513	3.711	4.130	4.158	4.435
4.506	4.509	4.766	6.005	7.000	7.532	9.008	14.114	14.902	17.961

If this data is from an exponential distribution with mean 5, then the density would be:

$$f(x) = \frac{1}{5}\exp(-x/5), \quad x \geq 0$$

and $f(x) = 0$ otherwise. We will break $[0, \infty)$ into five subintervals, based on the quantiles u_i, where $\int_0^{u_i} f(x)\,dx = .20i$, $i = 1, 2, 3, 4$. Thus, $u_1 = 1.116$, $u_2 = 2.554$, $u_3 = 4.581$, and $u_4 = 8.047$. We can compute these values to more decimals depending on the precision of the original data. If the null hypothesis is that the distribution of processing times is exponential with mean 5, then there would be six observations out of 30, on average, in each of the cells. The numbers of observations in these cells are 4, 7, 11, 4 and 4, respectively. The value of the test statistic Q is approximately 6.3, which generates a p-value of about 0.176. This indicates there is not much support against the null hypothesis.

Another test of fit with a specific distribution is the **Kolmogorov-Smirnov test**. This test compares the theoretical cumulative distribution function, that is, the cdf of the hypothesized distribution, with the empirical (cumulative) distribution function.

Let X be a continuous random variable, with density f and cdf F. Let $X_1 \ldots, X_n$ be a random sample from X's distribution, with realization $x_1, \ldots, x_n$. We order these values, and then the order statistics are $x_{(1)}, \ldots, x_{(n)}$. The empirical distribution function is an estimate of the true cdf of X, and is denoted $\hat{F}_n(t)$ with:

$$\hat{F}_n(t) = \begin{cases} 0 & t < x_{(1)} \\ \frac{1}{n} & x_{(1)} \leq t < x_{(2)} \\ \vdots & \\ \frac{i}{n} & x_{(i)} \leq t < x_{(i+1)} \\ \vdots & \\ 1 & t \geq x_{(n)} \end{cases}$$

We subscript $\hat{F}$ with n to indicate its dependence on the sample, and hence, the sample size. The empirical distribution function is the number of observations up to t, and then this number is divided by n. At each $x_{(i)}$, this cdf jumps by $1/n$. Thus, $\hat{F}_n(t) = \frac{1}{n}$(number of observations $\leq t$).

The empirical distribution is a reasonable estimate of the true distribution of X. First, if A is any set, define $I_A(x) = 1$ if $x \in A$ and $= 0$ otherwise. We call $I_A(x)$ the **indicator function** of A. For a fixed t, the random variable $Y = I_{\{y \leq t\}}(X)$ has a Bernoulli distribution and the probability X falls in $(-\infty, t]$ is $F(t)$, so $P(Y = 1) = F(t)$. When we take a random sample of size n from X's distribution, this generates a simple random sample $Y_1, \ldots, Y_n$ of independent Bernoulli variables with probability of success $F(t)$. We can also recognize $n\hat{F}_n(t) = \sum_{j=1}^n Y_j$. This new variable is then $B(n, F(t))$. The expected value is $nF(t)$ and the variance is $nF(t)(1 - F(t))$. This means $E(n\hat{F}_n(t)) = nF(t)$, or that $E(\hat{F}_n(t)) = F(t)$. Also, $var(n\hat{F}_n(t)) = nF(t)(1 - F(t))$ so $var(\hat{F}_n(t)) = F(t)(1 - F(t))/n$, which goes to 0 as $n \to \infty$. This means $\hat{F}_n(t)$ is a good estimator of the true cdf, since it is pointwise unbiased and consistent.

We can now compare the empirical distribution to the theoretical distribution by examining just how far apart they are. We evaluate F_0, the theoretical cdf, at each of the values $x_{(i)}$. Next, we determine the maximum distance between $F_0(t)$ and $\hat{F}_n(t)$, and this is our measure of the discrepancy between the theoretical and empirical distributions; we denote this value D. Thus, $D = \max_t |\hat{F}(t) - F_0(t)|$.

The calculation of D is straightforward. Because F_0 is monotonically increasing, and $\hat{F}$ is constant on $[x_{(i-1)}, x_{(i)})$, the maximum absolute difference must occur at one of the endpoints $x_{(i)}$. This follows from optimizing continuous functions on closed intervals, and the critical numbers are the values that make the derivative 0 or undefined. The derivative is not defined at the jumps in the empirical distribution function. We call this maximum difference D_i. If $F(x_{(i)}) \geq i/n$, then the maximum difference at this endpoint is $D_i = F(x_{(i)}) - (i - 1)/n$. If $(i - 1)/n \leq F(x_i) < i/n$, the maximum difference is $D_i = \max\{i/n - F(x_{(i)}), F(x_{(i)}) - (i - 1)/n\}$. Finally, if $F(x_{(i)}) < (i - 1)/n$, then the maximum difference will be $D_i = i/n - F(x_{(i)})$. We can summarize this as the maximum difference at this endpoint is:

$$D_i = \max\{i/n - F(x_{(i)}), F(x_{(i)}) - (i - 1)/n\}.$$

Thus,
$$D = \max_{1 \leq i \leq n} D_i.$$

The exact distribution of D is not known, but, accurate percentage points of it have been tabulated and are readily available on the web or in many texts. It is also important to note that the distribution of D does not depend on the specific distribution of the random variable under consideration. Also, the null hypothesis that the distribution is represented by the cdf F_0 is rejected for large values of D; that is, our calculated D score must exceed an appropriate upper percentage point of the distribution of D.

Example 3: Consider the data in Example 2, and again, we want to test whether the data supports the null hypothesis that the distribution of processing times is exponential with mean 5. We need the values of the cdf at each of the (ordered) observations, which, in this case, would be $1 - \exp(-x_{(i)}/5)$. These are:

0.042	0.065	0.075	0.117	0.231	0.256	0.263	0.263	0.276	0.341
0.392	0.421	0.472	0.473	0.497	0.505	0.524	0.562	0.565	0.588
0.594	0.594	0.614	0.699	0.753	0.778	0.835	0.941	0.949	0.972

The value of D for this data is 0.152. For $n = 30$, the critical value of D with significance level $\alpha = 0.05$ is 0.242, so we cannot reject the null hypothesis that this data comes from an exponential distribution with mean 5. Again, we would conclude the new procedure has not sped up processing time.

The data was actually generated from an exponential distribution with mean 4.5. The ability of a test to detect differences from the null is called its **power**. In this case, the true mean was not detected, and one reason for that is the size of the sample: It will require larger samples to generate sufficiently precise estimates of the mean to distinguish between a mean of 5 and 4.5.

These goodness-of-fit tests require that the distribution under the null has all parameters known, i.e., the distribution is completely specified. It is not appropriate for testing whether data supports an assumption of an exponential distribution with unknown mean, or of a normal distribution with unknown mean and variance, say. Such methods that do allow testing of, say, normality or exponentiality will not be considered in this text.

9.2 CONTINGENCY TABLES

Contingency tables are used in the analysis of relationships between different factors or characteristics. For a medicine under investigation, any differences in the survival rates of men and women is important in determining treatment options. Although there are generalizations, we will consider the problem of two factors, say A and B, where A has I levels and B has J levels. This means the different combinations of levels of A and B are mutually exclusive and exhaustive. We want to examine whether what we observe about A is influenced by the different levels of B; that is, we want to examine if the outcomes for A are contingent on the specific levels of B; The approach is to take samples and then count the number of individuals that have the combination of levels (i,j), where $1 \leq i \leq I$ and $1 \leq j \leq J$. The major problem is to determine if the two factors A and B are independent, or if in fact knowing which level of one impacts the probabilities of the different levels of the other.

There are actually three different sampling schemes, based on whether just the total number of observations n is fixed, or the number of observations on the different levels of A is fixed, or all marginal totals are fixed. Fortunately there is just one method of analysis that applies to all three. We will outline these schemes but not develop the reasons why the one method is applied. The details can be found in Kendall and Stuart (1979).

9.2.1 2×2 Contingency Tables

To start, let A and B have two levels each. Regardless of the sampling plan, we record the number n_{ij} in the sample that falls into each cell (i,j), and let $n_{.j} = \sum_{i=1}^{2} n_{ij}$, $n_{i.} = \sum_{j=1}^{2} n_{ij}$ and $n_{..} = n = \sum_{j=1}^{2} \sum_{i=1}^{2} n_{ij}$. Thus, n

Chapter 9 — Goodness of Fit Tests and Contingency Tables

is the total sample size and $n_{.j}$ and $n_{i.}$ are the marginal totals. The "dot" notation is a standard notation to indicate the quantity has been summed over a row or column. We could use $n_{..}$ to indicate the total sample size as well.

The standard visual display:

		B		
		1	2	
A	1	n_{11}	n_{12}	$n_{1.}$
	2	n_{21}	n_{22}	$n_{2.}$
		$n_{.1}$	$n_{.2}$	n

is called a **contingency table**. In general, a contingency table is a matrix of values of the counts of experimental units falling into the different combinations of levels of the factors. Included are the factors and levels, as well as row totals, column totals, and total number of observations.

Example 4: In examining the hardiness of two strains of red roses, one measure of health was the presence or absence of parasites. In a large field of roses containing the two strains, samples of each were selected and examined for parasites. The information is summarized in a contingency table as follows:

	Rose		
	Type 1	Type 2	Total
With Parasites	n_{11}	n_{12}	$n_{1.}$
No Parasites	n_{21}	n_{22}	$n_{2.}$
Total	$n_{.1}$	$n_{.2}$	$n = n_{..}$

This is called a 2×2 contingency table. It is of interest to know if one strain is more resistant to parasites than the other.

This general example illustrates the fact that there can be different sampling procedures to get our information. Do we select $n_{.1}$ and $n_{.2}$ of strains 1 and 2, respectively (that is, choose the sample sizes of both in advance), or do we simply randomly select n roses, and then categorize them by strain and presence or absence of parasites? In the first case, we have one set of marginal values fixed in advance, while in the second case, we do not fix any of the marginal values. There is in fact a third possibility: namely, we fix both sets of marginal values in advance. There are experiments where this third form of the sampling method applies, but we will not go into further details since it is a rare type of experiment.

The problem in either fixing one set of marginal values, or not fixing any of the marginals, just the total, is that there is a different number of random variables. This should mean that each one requires a different testing procedure. Fortunately, the method and distributions used for each coincide. The reasons are beyond the scope of this text (see Kendall and Stuart (1979), Chapter 33 for details), but it is important to understand that there are different sampling schemes.

Remember that our goal is to test the hypothesis that the two factors are independent, versus whether they are dependent. Suppose the factors are independent. What would that mean? Let p_{ij} be the probability that strain j has parasites when $i = 1$, and does not have parasites when $i = 2$. We can also let these values be the proportions of the different strains with or without parasites. Assume that the total sample size n is determined in advance, and then a random selection of n roses is selected from the field, so that none of the marginal values are fixed. Note that $p_{.j}$ is the proportion of the sample that is strain j, while $p_{i.}$ represents the proportions of the two parasite levels, i.e., the existence of parasites or no parasites. We can estimate each of these values by using binomial distribution theory, so $\hat{p}_{.j} = n_{.j}/n$ and $\hat{p}_{i.} = n_{i.}/n$.

Consider the general case of two factors A and B, with A_i representing the ith level of A and B_j the jth level of B. We call each entry (i, j) in the contingency table the (i, j) **cell**. If the two factors are independent, we know that $P(X$ falls in cell $(i,j))$, the probability a randomly selected rose of strain j has parasite condition i, must satisfy:

$$P(X \text{ falls in cell } (i,j)) = P(X \text{ has condition } A_i \cap \text{condition } B_j)$$
$$= P(X \text{ has } A_i)P(X \text{ has } B_j) = p_{i.}p_{.j}.$$

If the strain and parasite condition in Example 3 are independent, then our estimate of p_{ij} is just the product of $\hat{p}_{i.}$ with $\hat{p}_{.j}$, that is:

$$\hat{p}_{ij} = \hat{p}_{i.}\hat{p}_{.j} = n_{i.}n_{.j}/n^2.$$

If we have a sample of size n from the rose population, the expected number that fall into cell (i,j) is np_{ij}. Thus, in the case of independence, a good estimate of the average number that should fall in this cell is:

$$E_{ij} = n\hat{p}_{ij} = n\hat{p}_{i.}\hat{p}_{.j} = n(n_{i.}n_{.j})/n^2 = n_{i.}n_{.j}/n$$

where E_{ij} denotes the estimated expected number of the sample in cell (i,j) under the assumption of independence.

The last equation gives the (approximate) number of roses that should fall in cell (i,j) when independence holds. The notation O_{ij} = the observed number that falls in cell (i,j) is often used. The test statistic used is defined by:

$$\chi^2 = \sum_{i=1}^{2}\sum_{j=1}^{2} \frac{(O_{ij} - E_{ij})^2}{E_{ij}}.$$

The reason for the specific form is not obvious, but this statistic has been shown to have, for sufficiently large values of the n_{ij}, an approximately $\chi^2(1)$ distribution. See Kendall and Stuart (1979), for example, for further details. The numerators do represent measures of deviations of the observed cell frequencies from the (approximate) cell frequencies. We also note that the form of this statistic is similar to that of Q in the goodness-of-fit test in Section 9.1. For small, expected cell frequencies (if at least one cell frequency is less than 5 is a standard rule of thumb), we need a somewhat more complicated procedure. However, if all of the expected cell frequencies are ≥ 5, then it has been shown that this statistic is distributed at least approximately as a $\chi^2(1)$ random variable.

There is only one degree of freedom in this χ^2 distribution. The reason is this: There are four parameters that represent the probabilities of the cells. However, all the probabilities must add up to one, so there are only three free parameters. If we estimate one of the two row factor probabilities, i.e., $p_{1.}$, and one column probability, say $p_{.2}$, then all the probabilities are known, so we need to specify two more parameters, meaning we only have one left over. A formula that generalizes for more complicated contingency tables is that the degrees of freedom = (number of cells $-$ 1) $-$ (number of rows $-$ 1) $-$ (number of columns $-$ 1) = $(2 \times 2 - 1) - (2 - 1) - (2 - 1)$. This is the total number of cells, minus one because the four probabilities add to 1, and then minus however many probabilities we need from the rows and columns in order to fill in the cells in the table. We fill in the expected frequencies, which can be shown to be, for cell (i,j), the ith row total times the jth column total, divided by the grand total. To finish, we find the calculated χ^2 value from the formula and data, and then use the $\chi^2(1)$ distribution to determine the critical value for comparison. Large values of the test statistic tend to justify rejection of the hypothesis of independence, and so this is a right-tail test.

The above argument for the number of degrees of freedom outlines a general principle that is used in other situations, namely, a degree of freedom is lost for every parameter estimated in contingency tables. The result that the χ^2 statistic is $\chi^2(1)$ is an asymptotic result, but is generally a good approximation as long as $E_{n_{ij}} \geq 5$

Chapter 9 — Goodness of Fit Tests and Contingency Tables

for each cell. The use of this statistic and the asymptotic distribution applies to the two main cases of either fixed marginals for one of the factors or no fixed marginals. The difference is in interpretation:

Experiment Type 1: the marginal totals are fixed for one factor only. In this experimental design, with $n_{.1}$ and $n_{.2}$ fixed for factor B, we can interpret the problem of independence as: are the two populations determined by the two levels of B **homogeneous** with respect to the factor A? In terms of Example 3, are the two types of roses equally susceptible to parasites? We determine in advance how many from each of the populations determined by the two levels of B we will sample, and then count in each sample how many have level 1 of factor A and how many have level 2. When we compute the χ^2 value, we are measuring how far from reflecting equal proportions of the different factor levels in each population the data is. We could write the null as H_0 : the two populations are homogeneous with respect to A, versus H_1 : there is a difference in proportions with Level 1 of factor A in the two populations. [We could rephrase this with respect to level 2 of A, since these are equivalent.]

In this situation, we can state the problem as follows: If π_{ij} represents the probability in each cell, then we can write $H_0 : \pi_{11} = \pi_{12}$. The alternative can be any one of: $H_1 : \pi_{11} \neq \pi_{12}, H_1 : \pi_{11} < \pi_{12}$ or $H_1 : \pi_{11} > \pi_{12}$. In this form, we can have **directional** alternatives, as well as a two-sided alternative (which basically says there is a difference, but not what kind of difference, between the proportions). Testing here would be the usual testing related to the difference of proportions.

Example 5: In Example 3, suppose that from the field of roses containing the two strains, 35 of each type were selected and examined for parasites, in testing whether the proportions of the two strains with parasites are the same, versus are they different (i.e., a non-directional alternative). The information is summarized in the following contingency table:

	Rose Type 1	Rose Type 2	Total
With Parasites	14	20	34
No Parasites	21	15	36
Total	35	35	70

If A and B are independent or that the strains of roses are homogeneous with respect to parasite infections, then the estimated cell proportions are $\hat{p}_{11} = 34 \times 35/70^2$, $\hat{p}_{12} = 34 \times 35/70^2$, $\hat{p}_{21} = 36 \times 35/70^2$, $\hat{p}_{22} = 36 \times 35/70^2$. The (estimated) expected cell frequencies are $\hat{n}_{11} = 34 \times 35/70 = 17$, $\hat{n}_{12} = 34 \times 35/70 = 17$, $\hat{n}_{21} = 36 \times 35/70 = 18$, $\hat{n}_{22} = 36 \times 35/70 = 18$. This information is put into the contingency table by putting the expected cell frequency of the cell in brackets as:

	Rose Type 1	Rose Type 2	Total
With Parasites	14 (17)	20 (17)	34
No Parasites	21 (18)	15 (18)	36
Total	35	35	70

The calculated χ^2 value is:

$$\chi^2_{calc} = \frac{3^2}{17} + \frac{3^2}{17} + \frac{3^2}{18} + \frac{3^2}{18} = 2.059.$$

The *p* value of this test statistic is between 0.10 and 0.25, so we have insufficient evidence to reject the null hypothesis; therefore, we would assume the plants are equally susceptible to the parasites.

Directional alternatives can be handled through the χ^2 test. First, the direction of difference between the corresponding proportions if the null is rejected must be specified in advance. A very common error is to wait until the data comes in, then formulate a directional alternative. For all tests of hypotheses, we must specify the null and alternative in advance of carrying out the experiment. Let the significance level be α. The first thing to check if the alternate hypothesis is $H_1 : \pi_{11} > \pi_{12}$ is whether the estimated proportions satisfy this same inequality. If they don't, we will definitely have insufficient evidence to reject the null. Assume the inequality is satisfied for the sample proportions. We calculate the χ^2 value as above, and then determine the p-value of this from the $\chi^2(1)$ table and split it in half. That is because the calculation of the χ^2 value does not take into account direction: half of the values that produce the $\chi^2(1)$ values come from $\pi_{11} > \pi_{12}$ and half would come from $\pi_{11} < \pi_{12}$. The value $p/2$ becomes the p-value for our directional alternative test, and if this is $< \alpha$, we reject the null.

Experiment Type 2: none of the marginal totals are fixed. In this design, we determine only n, the total sample size, in advance, and then classify each individual by both factors A and B. Thus, the row totals and the column totals are random variables. This case becomes a question about the independence of two random variables A and B, and because each is dichotomous, some refer to this as the **double dichotomous or cross-sectional** case. The alternative is that the two variables are dependent. The test statistic remains the same, as does the actual testing procedure, using the $\chi^2(1)$ table. The proof of this is beyond the scope of the text. Example 4 is an experiment of type 2.

Example 6: Should men and women be treated differently after their first heart attack? In a study, 100 volunteers were selected from a large number of volunteers who had a first heart attack, and they were given the same regimen, including exercise and medications. Their health was followed for the next six years, or up to the point of death. The information is summarized in the following contingency table:

	Patient Type		
	Male	Female	Total
Survived 5+ years	30	15	45
Survived < 5 years	25	30	55
Total	55	45	100

If men and women react the same to the treatment, then the estimated cell proportions are $\hat{p}_{11} = 55 \times 45/100^2$, $\hat{p}_{12} = 45 \times 45/100^2$, $\hat{p}_{21} = 55 \times 55/100^2$, $\hat{p}_{22} = 55 \times 45/100^2$. The (estimated) expected cell frequencies are $\hat{n}_{11} = 55 \times 45/100 = 24.75$, $\hat{n}_{12} = 45 \times 45/100 = 20.25$, $\hat{n}_{21} = 55 \times 55/100 = 30.25$, $\hat{n}_{22} = 55 \times 45/100 = 24.75$. This information is put into the contingency table by putting the expected cell frequency of the cell in brackets as:

	Sex		
	Male	Female	Total
Survived 5+ years	30 (24.75)	15 (20.25)	45
Survived < 5 years	25 (30.25)	30 (24.75)	55
Total	55	45	100

The calculated χ^2 value is 4.50, which has a p value between 0.025 and 0.05. If the significance level of the test had been 0.05, we would be able to reject the null hypothesis.

It is notable that even though the setups for the two designs are different, the testing procedure (for the two-sided alternative or the independence alternative) is the same.

Experiment Type 3 : As mentioned, there is a third design in which all the marginal totals are set in advance, and if that is the case, the test statistic and the procedure remains the same. An exercise is to try to set up an experiment that in fact fixes both sets of marginal frequencies in advance.

9.2.2 $r \times c$ Contingency Tables

The testing procedure outlined for 2×2 contingency tables generalizes to more than two factors, and to two factors with two or more levels. We will consider the case of $r \times c$ contingency tables, in which we are trying to determine the independence of two factors A and B over all the levels of each. One difference is that in such analyses, the χ^2 variable has $(r-1) \times (c-1)$ degrees of freedom. Another is the form of the null hypothesis. In 2×2 contingency tables, we had to specify that the proportions for level 1 of A were the same across the two populations or levels of B. Thinking of B as a way to separate the total population into c subpopulations, if there are r levels of A then for A and B to be independent, we would have to specify $r-1$ conditions of the form $p_{i1} = p_{i2} = \cdots = p_{ic}$ for $1 \leq i \leq r-1$. If all of these hold, this will force $p_{r1} = p_{r2} = \cdots = p_{rc}$ to be true as well, so the null hypothesis is stated as $H_0: p_{i1} = p_{i2} = \cdots = p_{ic}$ for all $1 \leq i \leq r$. The alternative hypothesis is H_1: The factors A and B are not independent.

The standard array will take the form:

		B			
		1	$\cdots$	c	
	1	n_{11}	$\cdots$	n_{1c}	$n_{1.}$
A	$\vdots$	$\vdots$	$\vdots$	$\vdots$	$\vdots$
	r	n_{r1}	$\cdots$	n_{rc}	$n_{r.}$
		$n_{.1}$	$\cdots$	$n_{.c}$	n

Under the assumption that the two factors are independent, then in cell (i,j) we would expect to see $n\pi_{i.}\pi_{.j}$ observations in a sample of size n. In this problem of independence, all we set in advance is the overall sample size, and none of the marginal sample sizes. If factor A is listed along the left side of the contingency table, and Factor B along the top, then $\pi_{i.}$ is the probability of any observation falling into the ith level of Factor A, while $\pi_{.j}$ is the probability of falling in the jth level of B. Thus, the probability of falling into the (i,j) cell is the product of these two values when the factors (or variables) are independent, and that leads to the expression for the expected number in the given cell being the same as in the 2×2 contingency table problem. We do not know these marginal probabilities, so we have to estimate them. The best way to estimate $\pi_{i.}$ is to think of this as part of a binomial experiment [falling in the ith level of Factor A, over all levels of B or without considering B], and hence, we would use $\hat{\pi}_{i.} = n_{i.}/n$. We would do this for each of the values of i and then repeat for each value of j. This means the expected number in cell (i,j) would be approximately $E_{ij} = n\hat{\pi}_{i.}\hat{\pi}_{.j} = n(n_{i.}/n)(n_{.j}/n) = n_{i.}n_{.j}/n$. Thus, we take the product of the marginal frequencies and then divide by n. Our statistic takes the form:

$$\chi^2 = \text{the sum over all cells}(i,j) \text{of } (n_{i,j} - n_{i.}n_{.j}/n)^2/(n_{i.}n_{.j}/n)$$
$$= \sum_{i,j}(O_{i,j} - E_{ij})^2/E_{ij}.$$

The degrees of freedom come from a generalization of the calculation in the 2×2 case: The number of cells is rc. The number of free cells is $rc - 1$, since the probabilities in all the cells must add up to 1. We have to estimate $r-1$ and $c-1$ marginal probabilities (since the row probabilities and the column probabilities must each add to one). Thus, the degrees of freedom are $rc - 1 - (r-1) - (c-1) = (r-1)(c-1)$, using the rule that every time we need to estimate a parameter, we lose a degree of freedom. The test is whether the two factors are independent, versus whether they are not. In effect, this is a one-tailed test, using (at least for large samples) the statistic χ^2 defined above, and the fact that it has a $\chi^2((r-1)(c-1))$ distribution (approximately). Because only large values of the calculated χ^2 value will lead to rejection of the null, this is a right-tail test.

The argument about the calculation of degrees of freedom is not rigorous, but provides a convenient way to reach the correct value. The distributional results are again asymptotic, but are reasonably good if the expected number in any cell is at least 5. We also note this procedure is valid if only the total sample size n is fixed and none of the marginals, or if n is fixed and so are the marginal totals for one of the factors.

Example 7: A group studying the impact of social conditions on the educational achievements of children of two-parent families with different employment histories during their children's high school years randomly selected the records of 10,000 students who started high school at least four years before the start of the study. At the end of 10 years from the time starting high school for each individual student, their educational achievements were determined. This included: Did not finish high school; finished high school but no further academic achievements; earned a college diploma; completed a bachelor's degree. It is assumed these categories are mutually exclusive and exhaustive. These are coded 1 to 4, respectively. The employment status in the first four years from when their child started high school of the parents was categorized by: both parents were unemployed the majority of the time; at least one parent had part-time work for the majority of the time, but neither had full-time work for the majority of the time; one parent had full-time work for the majority of the time; and both parents had full-time work for the majority of the time. These are coded 1 to 4, respectively, and are assumed mutually exclusive and exhaustive for the purposes of the study. The employment status of the parents and the educational achievements at the end of the study were:

		\multicolumn{4}{c}{Educational Achievement}	
		1	2	3	4	
	1	925	325	125	390	1765
Employment	2	945	375	150	450	1920
Status	3	100	2600	950	250	3800
	4	130	1500	675	110	2415
		2100	4800	1900	1200	10000

We calculate (approximate) expected cell frequencies for all 16 cells. For example, $E_{23} = 10,000 \times (1,920)(1,900)/10,000^2 = 394.8$. Since the degrees of freedom are $(4-1)(4-1) = 9$, if we set the significance level of the test at $\alpha = 0.05$, then we would reject the null hypothesis that parental employment status (as defined here) and an offspring's educational achievements are independent when the calculated χ^2 value exceeds 16.919. The details of this test are left to the reader.

9.2.3 $2 \times c$ Contingency Tables

In the last section, we considered the general $r \times c$ contingency table analysis. The special case of $2 \times c$ contingency tables arise frequently; we will write out the null hypothesis and other features explicitly. The testing procedure uses the calculated score:

$$\chi^2 = \sum_{i=1}^{2}\sum_{j=1}^{c}(n_{i,j} - n_{i.}n_{.j}/n)^2/(n_{i.}n_{.j}/n) = \sum_{i=1}^{2}\sum_{j=1}^{c}(O_{i,j} - E_{ij})^2/E_{ij}$$

which, asymptotically, has a $\chi^2(c-1)$ distribution. Note that the calculation of the degrees of freedom follows the pattern from the other cases: There are $2c-1$ free cells, and we have to compute 2-1 row probabilities and $c-1$ column probabilities to fill in the table. Thus the degrees of freedom are $2c - 1 - (2-1) - (c-1) = 2c - 1 - 1 - c + 1 = c - 1$. Our null hypothesis takes the form:
$H_0 : \pi_{11} = \pi_{12} = \cdots = \pi_{1c}$, with the alternative expressed as H_1 : at least two of these probabilities are different or that A and B are not independent. Since large values of our calculated score will indicate a lack of fit between the assumption of independence and the data, the percentage points of the right tail of the $\chi^2(c-1)$

Chapter 9 — Goodness of Fit Tests and Contingency Tables

table are used to decide to accept or reject H_0. Note that the standard form of the contingency table is:

		B			
		1	$\cdots$	c	
A	1	n_{11}	$\cdots$	n_{1c}	$n_{1.}$
	2	n_{21}	$\cdots$	n_{2c}	$n_{2.}$
		$n_{.1}$	$\cdots$	$n_{.c}$	n

Example 8: In a study to determine if there is a link between tobacco use and the development of oral cancers, a random selection of 100 professional baseball players who used chewing tobacco, 100 who smoked cigarettes, and 100 who did not chew or smoke tobacco products were chosen since their health records and tobacco use are easily documented. At the time of selection, they were all cancer free. Their health was monitored for the next 10 years, with the following results at the end of that period:

		Ball Player			
		Chewer	Smoker	Non – User	
Status	Oral Cancer	20 (38/3)	12 (38/3)	6 (38/3)	38
	No Oral Cancer	80 (262/3)	88 (262/3)	94 (262/3)	262
		100	100	100	300

The null hypothesis would be that the incidence of cancer is the same across the types (chewers, smokers, non-users) of ball players. The calculated χ^2 score is 8.919, and the p value from the $\chi^2(3-1) = \chi^2(2)$ distribution is between 0.01 and 0.025. If we had set the significance level of the test as $\alpha = 0.05$, we could reject the null. If we had set it at 0.01, we would not be able to reject the null of independence of incidence of oral cancer and tobacco use for professional baseball players.

Exercises

1. There are four phenotypes of mature milkweed plants under investigation: tall plant, large leaves; tall plant, small leaves; short plant, large leaves; and short plant and small leaves. We will code these 1 to 4, respectively. It is believed that these different phenotypes occur in ratio 9:3:3:1. From 320 randomly selected milkweeds, the results were: Type 1 — 182; Type 2 — 77; Type 3 — 51; Type 4 — 10. Use an appropriate goodness of fit test to determine if the theoretical ratio is supported by these observations.

2. The following are the ordered results of the parts per million of impurities in 30 samples of lake water after treatment with a new filter: Using appropriate methods of this chapter, determine if the assumption the

45.81	47.09	47.36	47.95	48.03	48.05	48.40	48.53	48.56	49.08
49.24	49.50	49.63	49.91	50.37	50.52	50.76	50.77	50.79	50.96
50.98	51.30	51.67	51.79	52.49	52.52	52.54	52.65	53.01	53.28

   distribution of the measure of impurities after filtering is $N(50, 4)$ is supported by the data.

3. The following 25 observations are the times, in hours, that it took for a random sample of 25 racehorses to return to their normal body temperature after a race. Determine if this data supports an assumption that the time to return to normal body temperature is exponentially distributed. Use this data to test explicitly if the distribution is exponential with mean 3 hours.

0.025	0.254	0.472	0.791	0.921	1.054	1.307	1.659	1.715	1.781
1.923	2.203	2.698	2.787	3.099	3.167	3.501	4.072	4.192	4.310
5.130	5.834	6.056	11.708	19.730

4. The differences $D = (W - 450)$ of the weight W in grams of a lentic rainbow trout 33 cc in length and the average weight, 450 grams, of such trout are assumed to follow a standard normal distribution. Twenty trout of length 33 cm are selected at random, and the resulting differences D are computed and recorded as:

−3.035	−2.000	−1.570	−1.430	−1.196	−1.169	−0.891	−0.696	−0.600	−0.310
−0.159	0.012	0.088	0.336	0.487	0.749	0.897	0.982	2.507	2.530

Use appropriate methods of this chapter to test the hypothesis that the differences do follow a standard normal distribution (versus they do not).

5. The following observations come from a log normal distribution. Test whether they come from a log normal with parameters $\mu = 2$ and $\sigma = 2$ by converting to the appropriate normal distribution.

0.43	0.53	2.69	2.92	3.79	5.30	5.80	6.28	9.97	10.30
11.96	14.66	19.19	30.53	32.16	37.75	38.52	51.20	56.81	58.48
61.23	276.30	346.05	417.03	1473.71					

6. Two hundred cats and 100 dogs with fleas are selected at random and treated with a flea-killing solution that is supposed to keep fleas from returning for at least a month. After receiving the treatment, which did kill the fleas at the start of the experiment, the animals were examined a month later, and the number with fleas and without were recorded as follows:

	Cats	Dogs	
Fleas	56	40	96
No Fleas	144	60	204
	200	100	300

Test whether the effectiveness of the treatment is the same for the two types of animals.

7. Twenty-five 18-year-old male university students and 25 18-year-old female university students were selected at random and asked whether they planned to have children within the next 10 years. The numbers of each gender indicating they did or did not plan to have children are given in the following contingency table:

	M	F	
Have Children	7	9	16
No Children	18	16	34
	25	25	50

Test whether the proportion of males and females planning children is the same.

8. A standard measure of the state of the economy is the response to the question, Do you intend to spend more, the same, or less during the Christmas season this year as compared to last year? Fifty men and 50 women working in the automobile construction industry were selected at random and asked this question, with the following results:

		Spending			
		More	The Same	Less	
Gender	M	17	19	14	50
	F	21	13	16	50
		38	32	30	100

Determine if this data provides evidence that men and women have the same distributions of planned Christmas spending.

9. A new growing medium for starting plants is to be tested by determining if seedlings of different flowers developed root systems in a similar way. Two weeks after germination, 150 seedlings were randomly selected from a large greenhouse producing Sweet William, Petunias, and Alyssum plants for sale. The root system of a seedling was determined by measuring the length of the longest root and categorized as Small (less than 3cm), Medium (3cm or more, but less than 5 cm), and Large (5cm or longer) with these coded as 1, 2, and 3, respectively. The results are:

		Sweet William	Petunia	Alyssum	
	1	12	13	20	45
Root	2	16	25	17	58
System	3	18	20	9	47
		46	58	46	150

Test the hypothesis that the development of root systems in seedlings with the new growing medium is independent of the type of seedling.

10. A new vaccine for cattle to combat hoof-and-mouth disease is also known to cause allergic reactions in most cattle. It is thought that the form of administering the drug may impact the occurrence and severity of the reaction. To test the hypothesis that the level of allergic reaction is independent of the form of administration of the drug, an experiment was set up as follows: 200 cows were selected at random from herds in a large region of cattle country. Of these, 100 were injected with the vaccine in the hip, 50 were injected in the shoulder, and 50 were given the drug in their food, and these methods were coded 1, 2, and 3, respectively. The reactions were classified as severe, moderate, minimal, or none, and these were coded 1, 2, 3, and 4, respectively, and results recorded as:

		Vaccine Delivery			
		1	2	3	
	1	17	11	15	43
Reaction	2	22	14	13	49
Type	3	23	11	10	44
	4	38	14	12	64
		100	50	50	200

Determine if in fact the distribution of reaction types is the same for the three types of delivery of the vaccine.

CHAPTER 10

ANALYSIS OF VARIANCE

Synopsis

Analysis of Variance is introduced in the context of completely randomized designs and complete block designs. The advantage of balanced designs is discussed. Testing procedures for homogeneity and of linear contrasts are introduced. A brief introduction to experimental design and concepts used in developing experiments is provided at the end of this chapter.

10.1 INTRODUCTION

ANOVA, from **AN**alysis **O**f **VA**riance, is a general method for analyzing how samples deviate from specific assumptions about the populations involved. It is used in connection with both simple designs, such as the single factor with multiple levels, and more complicated experimental designs involving multiple factors and levels. The primary assumptions made throughout this chapter are that the errors (or residuals) are independent and are distributed as $N(0, \sigma^2)$. The assumption that the variances of the errors/residuals are the same (i.e., σ^2) is called homoskedasticity (also spelled homoscedasticity). ANOVA generalizes the t test for equality of two means from independent populations with common variance.

10.2 THE GLOBAL F TEST: A SIMPLE ANOVA

Let the random variable X be measured on a population of experimental units after one of several levels, say k, of a factor has been applied. A typical question is whether the average value of X observed on those receiving level i of the treatment is the same across all levels of the treatment. That is, if μ_i is the mean response of X for those receiving treatment level i, then we want to test the hypothesis $H_0 : \mu_1 = \mu_2 = \cdots = \mu_k$. The alternative hypothesis would be that at least two of these means differ.

We will assume in this section that the experimental design is a **completely randomized design**, often abbreviated **CRD**. This means that the assignment of experimental units to the treatment levels is completely at random. In other designs, there can be restrictions on the assignment of experimental units to treatment levels. A brief overview of experimental design is given at the end of this chapter.

There are two models: **Model 1** is called the **fixed effects** model. The assumption is that the number of levels of the treatment is finite (and usually small) and all are observed. Generally, the researcher sets these levels and determines how many observations will be made of the experimental units under each level of the treatment. It is also assumed that values associated with a given level are exact. By this we mean that, for example, if the levels are different temperatures in a chemical experiment, then the temperatures are known without error. This keeps from having some of the variability seen in the response variable due to fluctuations in individual treatment levels.

Model 2 is called the **random effects** model and involves a large number of treatment levels, not all of which are observed. In this second model, we generally randomly select a certain number of these levels and then apply them to experimental units. Fortunately, the methodologies for analyzing these two models is the same; it is the interpretation of results that changes. We develop the ANOVA method for Model 1 first.

10.2.1 ANOVA for Model 1: Fixed Effects

Consider the case $k = 2$ so that we want to test $H_0 : \mu_1 = \mu_2$ versus $H_1 : \mu_1 \neq \mu_2$. We take independent random samples of size n_1 and n_2 from the experimental units available, apply level 1 and 2 of the treatment to the respective samples, and denote the observations $X_{11}, X_{12}, \ldots, X_{1n_1}$ and $X_{21}, X_{22}, \ldots, X_{2n_2}$. According to our assumptions, the first set of observations is *iid* $N(\mu_1, \sigma^2)$ and the second set is *iid* $N(\mu_2, \sigma^2)$. Earlier, we considered how to test the equality of two means in independent populations where the variances are equal and where they are not equal. We will be assuming the variances are equal throughout.

Under the assumptions in the last paragraph, we applied a t test: the pivotal statistic under the null hypothesis was:

$$T = (\overline{X}_1 - \overline{X}_2) / \left(S_p \sqrt{\frac{1}{n_1} + \frac{1}{n_2}} \right)$$

where $\overline{X}_i$ is the sample mean of the ith sample and S_p^2 is the pooled estimate of σ^2. T has a Student's t distribution with $n_1 + n_2 - 2$ degrees of freedom. Recall how this comes about: it is the simplification of the ratio of a standard normal variable to the square root of an independent $\chi^2(n_1 + n_2 - 2)$ variable divided by its degrees of freedom. Also recall that the square of a standard normal variable is $\chi^2(1)$. So now if we square T, T^2 is the ratio of an independent $\chi^2(1)$ variable, divided by its degrees of freedom (1, in this case) to an independent $\chi^2(n_1 + n_2 - 2)$ variable, divided by its degrees of freedom. Earlier, this was defined to be an $F(1, n_1 + n_2 - 2)$ random variable. Thus, the t test of the given hypotheses is equivalent to an F test.

In general, we have k independent populations (distinguished by the treatment levels) and we want to test whether all of these populations have the same mean. One approach would be to test each pair of means, but if k is, say, 5, then we would have to make as many as 10 comparisons. If each test has significance level 0.05, then the overall probability of making a type I error is about 0.40. To avoid these complications, we use the **global F test**. This test examines deviations between means for all different levels of a single factor simultaneously. Its drawback is that it will not identify where the differences occur; but usually, that will be clear from the data. We can then run a t test on one or two differences to pinpoint which treatment levels produce statistically different means. The test is named for the distribution of the pivotal statistic, which will be F.

Treatment levels are coded from 1 to k. The levels may start off with names, such as: a placebo, dose of drug at one value, dose of drug at a higher level, and then they are coded from 1 to 3. There is no implied order of importance by the size of number used in the code.

One particular description of the problem is as follows: We assume a random variable Y depends on a **design variable** X with a finite number of levels, say k. A design variable is any variable or factor controlled by the designer of the experiment. The dependence of Y on X is assumed a linear one with the model expressed as $Y_{ij} = \mu + \tau_i + \epsilon_{ij}$. Y_{ij} is the jth value of Y observed under level i of X. For level i of the design variable, we observe the response variable Y n_i times, so for Y_{ij}, we have $1 \leq j \leq n_i$. In a balanced design, all n_i values would be equal but reasons do arise calling for or leading to unequal sample sizes.

Let N represent the total number of observations taken on Y, that is, $N = \sum_{i=1}^{k} n_i$. If we randomly assign treatments and carry out the experiment to observe Y, we expect a global mean response, μ. If we concentrate on a particular treatment level i, we assume the mean response is $\mu + \tau_i$, so τ_i is the effect of treatment i. This then requires us to assume $\sum_{i=1}^{k} \tau_i = 0$, since they represent deviations from the global mean. We also assume there are other uncontrolled factors, so we expect an error, random or residual term in each of our observations, regardless of the level; in the linear model, it is an additive error term, typically denoted ϵ_{ij}, where the i corresponds to the treatment level and the j corresponds to the particular observation on Y under the ith treatment level.

Thus, our linear model is $Y_{ij} = \mu + \tau_i + \epsilon_{ij} = \mu_i + \epsilon_{ij}$, where $\mu_i = \mu + \tau_i$ is the mean value of Y under treatment level i. The values ϵ_{ij} are called residuals or errors and represent the random element in the responses. In the model we are considering, we will assume a homoskedastic model, with $E(\epsilon_{ij}) = 0$ and $Var(\epsilon_{ij}) = \sigma^2$ — that is, we assume the variance is constant for all treatment levels. This is an important element in our assumptions, and often, is not justified by the data. However, we will not concern ourselves with heteroskedastic models here.

We have some things to estimate: the global mean μ, the treatment effects τ_i, and the variance of the errors, σ^2. Our main concern is deciding if there is a **treatment effect**: Do the different treatment levels have no effect on the average response of the variable Y, or is there a statistically verifiable effect of at least one of the treatment levels on the mean response in Y? This can be restated as: the treatment does not impact the average of the response variable Y. Our test can be formalized as: $H_0 : \tau_1 = \tau_2 = \cdots = \tau_k = 0$ versus $H_1 :$ at least one τ_i is nonzero. Note that if one τ_i isn't zero, then there must be at least one more that isn't zero, since the global average is μ. Generally, it may be that all τ_i are zero; some are zero, and the rest are nonzero; or all are nonzero. Our test will only provide evidence that all τ_i values are zero, or statistical evidence that at least one is not. There are methods for identifying which are nonzero (or, at least, ones we have sufficient evidence to infer they are nonzero), but we will not look at these here. The null and alternative hypotheses can also be expressed as: $H_0 : \mu_1 = \cdots = \mu_k$ versus $H_1 : \mu_i \neq \mu_j$ for at least one pair (i,j), $j \neq i$.

In our model, we assume $Y_{ij} \sim N(\mu + \tau_i, \sigma^2) = N(\mu_i, \sigma^2)$. Thus $\overline{Y}_{i.} \sim N(\mu_i, \sigma^2/n_i)$, where $\overline{Y}_{i.} = \sum_{j=1}^{n_i} Y_{ij}/n_i$. The dot notation to indicate one of the subscripts has been summed over will be used throughout. This avoids confusion about what, e.g., $\overline{Y}_i$ might mean. Because of the normality assumption, our best point estimate of μ_i will be $\overline{Y}_{i.}$, and one estimate of σ^2 is the sample variance from the data from level i:

$$S_i^2 = \sum_{j=1}^{n_i}(Y_{ij} - \overline{Y}_{i.})^2/(n_i - 1), \text{ with } (n_i - 1)S_i^2/\sigma^2 \sim \chi^2(n_i - 1).$$

Since σ^2 is common to all levels, we can get a better estimate by using the other data as well, as will be demonstrated shortly. However, it is useful to know that $E(\overline{Y}_{i.}) = \mu_i$ and $E(S_i^2) = \sigma^2$.

Note that, in general, $E(\overline{Y}_{i.}) = \mu + \tau_i$, and $E(\frac{1}{N}\sum_{i=1}^{k}\sum_{j=1}^{n_i} Y_{ij}) = \frac{1}{N}\sum_{i=1}^{k} n_i(\mu + \tau_i) = \mu + \frac{1}{N}\sum_{i=1}^{k} n_i\tau_i$. This means that unless the design is balanced, the global sample mean is not unbiased for the global mean, at least in general. However, if the null hypothesis is correct and all the τ_is are 0, this is unbiased for μ, even if the design is unbalanced.

A balanced design has aesthetic appeal, as well as some distributional implications. Beyond the observation in the last paragraph, it can be shown that if a design is unbalanced, then the statistical results can be greatly affected by heteroskedasticity, that is, unequal variances corresponding to different treatment levels. We cannot expect that homoskedasticity holds perfectly, but in balanced designs, the impact of different variances is not severe until the difference is quite significant. Basically, this is because no one distribution is overrepresented compared to the others in the data. We formally assume all variances are the same, but the methods produce very good results even if the variances differ somewhat.

Let $Y_{..}$ denote the grand sum of the Y values and $\frac{1}{N}Y_{..} = \overline{Y}_{..}$ the overall sample mean. If we define $Y' = \frac{1}{k}\sum_{i=1}^{k}\overline{Y}_{i.} = \frac{1}{k}\sum_{i=1}^{k}\frac{1}{n_i}Y_{i.}$, then $E(Y') = \mu$ regardless of whether the design is balanced or the null hypothesis is true or false (since $\sum_{i=1}^{k}\tau_i = 0$ is assumed true), so this weighted average of the different averages within the treatment levels is unbiased for μ. Thus, we do have a method to estimate the global mean with an unbiased, normally distributed random variable. The variance of this estimator of μ is left to the reader.

However, at the moment, we are only interested in whether there is a treatment effect, not the effect of the individual treatment levels or the grand mean. Thus, in some sense, they are nuisance parameters. In the following, we will determine a method for testing our hypotheses in the balanced and unbalanced cases simultaneously. Set:

$$SST = \text{Total Sum of Squares} = \sum_{i=1}^{k}\sum_{j=1}^{n_i}(Y_{ij} - \overline{Y}_{..})^2$$

also called the sum of squares total. This is a measure of the total variability seen in the response variable Y. It can be shown by introducing the term $\overline{Y}_{i.} - \overline{Y}_{i.}$ into each summand and squaring and simplifying that:

$$SST = \sum_{i=1}^{k}n_i(\overline{Y}_{i.} - \overline{Y}_{..})^2 + \sum_{i=1}^{k}\sum_{j=1}^{n_i}(Y_{ij} - \overline{Y}_{i.})^2 + 2\sum_{i=1}^{k}\sum_{j=1}^{n_i}(\overline{Y}_{i.} - \overline{Y}_{..})(Y_{ij} - \overline{Y}_{i.}).$$

The last sum involving products is in fact 0 algebraically, and the proof is left to the reader. We define:

$$SSTr = \sum_{i=1}^{k} n_i (\overline{Y}_{i.} - \overline{Y}_{..})^2 \quad \text{and} \quad SSE = \sum_{i=1}^{k} \sum_{j=1}^{n_i} (Y_{ij} - \overline{Y}_{i.})^2.$$

Thus, we can write $SST = SSTr + SSE$. We also write $SSTr = SS(between)$ and $SSE = SS(within)$, where $SSTr$ is the **sum of squares for treatments** (or treatment sum of squares) and SSE is the **sum of squares for error** (or error sum of squares). Note that $SSTr$ is a measure of variability between the treatment levels as it measures the deviation of the treatment means from the global mean, while SSE is a measure of variability found within each treatment level, then aggregated. SSE is crucial to the analysis: It is a measure of the global variance and is fundamentally linked to σ^2 and not to any of the means. It is in fact related to the remaining variability after the means has been fitted. Each summand is an estimate of the corresponding ϵ_{ij} term (the residual or error) in our model, then squared. It cannot be emphasized enough that the independence of the residuals is fundamental to the statistical analysis to come.

The partitioning of a global sum of squares into separate sums of squares is used in many more complicated designs, so it is important to understand how we proceed to get the distributional results needed in this relatively simple case. First, SST can be viewed as a sample variance from a normal distribution, except it is not divided by N−1. If we now divide this by σ^2, we would have a statistic that is $\chi^2(N-1)$, where we have lost a degree of freedom from estimating the global mean. An application of Cochran's Theorem establishes that $SSTr$ and SSE are independently distributed under our assumptions and are related to specific χ^2 variables. The sum of squares for error can also be viewed as a pooled estimate of the variance, generalizing the results from testing the difference of means from two independent populations with a common variance. We can also argue in the SSE term that we need to estimate k sample means, and hence, the number of degrees of freedom for this term should be $N-k$. For the $SSTr$ term, there are k summands, but we need a global estimate of the mean, and so there are $k-1$ degrees of freedom for these terms. Note that $N-1 = N-k+k-1$, so the two sides of the degrees-of-freedom calculations balance.

More formally, SSE is also the sum of independent $\chi^2(n_i - 1)$ random variables (except they are not divided by σ^2), so $SSE/\sigma^2 \sim \chi^2(N-k)$. Under the null hypothesis, $SSTr$ must be related to a $\chi^2(k-1)$ variable by Cochran's Theorem since it is a sum of squares. Because $\sum_{i=1}^{k}(\overline{Y}_{i.} - \overline{Y}_{..}) = 0$, then there are only $k-1$ linearly independent expressions in $SSTr$, which means it is related to a $\chi^2(k-1)$ random variable.

Define $MSE = SSE/(N-k)$ and $MSTr = SSTr/(k-1)$, called the **mean square error** and **mean square for treatments**, respectively. Under the null or alternate hypothesis, and balanced or unbalanced designs, we can show MSE is unbiased for σ^2. It can be shown that $E(MSTr) = \sigma^2 + \frac{1}{k-1}\sum_{i=1}^{k} n_i(\tau_i - \overline{\tau})^2$, where $\overline{\tau} = \sum_{i=1}^{k} n_i \tau_i / N$. Under the null hypothesis, $E(MSTr) = \sigma^2$. We suspect that the null has evidence against it if MSTr/MSE is significantly larger than 1. Basically, all that is left is to find a reasonable pivotal statistic to use. Cochran's Theorem has shown that under the null hypothesis $(k-1)MSTr/\sigma^2 \sim \chi^2(k-1)$ and is independent of $(N-k)MSE/\sigma^2 \sim \chi^2(N-k)$. Recall that if $U_i \sim \chi^2(\nu_i)$, $i = 1, 2$ are independent, then $[U_1/\nu_1]/[U_2/\nu_2] \sim F(\nu_1, \nu_2)$; that is, this particular ratio has Snedecor's F distribution. Note the numerator degrees of freedom come first in the statement of the F distribution. Thus, under the null hypothesis, balanced design or not, we would have MSTr/MSE has an F distribution with $k-1, N-k$ degrees of freedom. The rest of the testing procedure is the same as for any one-tail test.

The information associated with this completely randomized design (CRD) is summarized in an ANOVA table as follows:

Source	d.o.f.	SS	MS	F
Treatment	k−1	SSTr	SSTr/(k−1)	MSTr/MSE
Error	N−k	SSE	SSE/(N−k)	
Total	N−1	SST		

We use a table of the F distribution or use a computer program to generate percentage points and find the appropriate upper percentage points. Normally, tables of the F distribution are organized first by the

Chapter 10 — Analysis of Variance

numerator degrees of freedom and then within that table by the denominator degrees of freedom. For denominator degrees of freedom not listed, use the degrees of freedom closest to the ones in the problem.

The ANOVA table above is sometimes called the brief ANOVA table. A full ANOVA table includes a line related to the mean, but does add to the current analysis.

To determine SST, SS(between), and SSE=SS(within), we first note that:

$$SST = \sum\sum Y_{ij}^2 - \frac{1}{N}\left(\sum\sum Y_{ij}\right)^2$$

and

$$SSTr = SS(between) = \sum_{i=1}^{k} n_i \overline{Y}_{i.}^2 - \frac{1}{N}\left(\sum\sum Y_{ij}\right)^2.$$

Then $SSE = SS(within) = SST - SSTr$. The sum of squares total is relatively easy to compute, since it involves the sum of the observations and the sum of the squares of the observations. SSTr involves the sum of the observations and the k values $\overline{Y}_{i.}$. Spreadsheet programs generate these numbers, of course, but these forms can be useful in determining expected values and variances.

Consider the balanced design in the case $k = 2$. Then $\overline{Y}_{..} = \frac{1}{2}(\overline{Y}_{1.} + \overline{Y}_{2.})$. Thus,

$$\sum_{i=1}^{k}(\overline{Y}_{i.} - \overline{Y}_{..})^2 = \sum_{i=1}^{2}(\overline{Y}_{i.} - \overline{Y}_{..})^2 = (\overline{Y}_{1.} - \overline{Y}_{..})^2 + (\overline{Y}_{2.} - \overline{Y}_{..})^2$$

$$= \left(\frac{1}{2}(\overline{Y}_{1.} - \overline{Y}_{2.})\right)^2 + \left(\frac{1}{2}(\overline{Y}_{2.} - \overline{Y}_{1.})\right)^2 = \frac{1}{2}(\overline{Y}_{1.} - \overline{Y}_{2.})^2.$$

This calculation shows that the t test for testing the equality of means in two independent populations (versus they are different) is equivalent to the F test because the square of the pivotal statistic for the t test is exactly the same as the calculated F value.

If we have concluded that there are differences among the treatment effects, we may wish to determine where those differences are. Typically, this will involve some linear combination of the means, such as $\mu_1 - \mu_3 = 0$, or $(\mu_5 + \mu_4)/2 - \mu_3 = 0$, etc. In general, any expression of the form $L = \sum_{i=1}^{k} c_i \mu_i$ will be called a linear contrast of the means, and is called an orthogonal contrast if $\sum_{i=1}^{k} c_i = 0$. If we want an estimate of this contrast, it would be the corresponding linear combination of sample means, namely $\sum_{i=1}^{k} c_i \overline{Y}_{i.}$.

To determine how to test whether a linear contrast is 0, versus it isn't, we consider the linear combination of sample averages $C = \sum_{i=1}^{k} c_i \overline{Y}_{i.}$. If we take the expected value of C, we get $E(C) = \sum_{i=1}^{k} c_i \mu_i$. If this is an orthogonal contrast, i.e. we have $\sum_{i=1}^{k} c_i = 0$, then $E(C) = \sum_{i=1}^{k} c_i \tau_i$, which means we can view questions about combinations of means to be questions about the treatment effects as well.

Because a linear combination of normal variables is normal, a linear contrast of the sample means is normally distributed. As a normal variable, we need to know its mean and variance, but that is straightforward. By linearity of expectation, $E(C) = \sum_{i=1}^{k} c_i \mu_i$ and $var(C) = \sum_{i=1}^{k} c_i^2 \sigma^2 / n_i$. Under the null hypothesis, $H_0 : L = 0$, $C \sim N(0, \sum_{i=1}^{k} c_i^2 \sigma^2 / n_i)$. That means:

$$U = C^2 / \left(\sum_{i=1}^{k} c_i^2 \sigma^2 / n_i\right) \sim \chi^2(1)$$

and, since it is made up of the sample means, is independent of SSE. We call $U \times \sigma^2$, denoted SSC, the contrast sum of squares. If we now take SSC/MSE, this has an $F(1, N - k)$ distribution under the null hypothesis. The null hypothesis is rejected if the calculated score is greater than some critical value, expressed as the upper $\alpha \times 100\%$ percentage point of $F(1, N - k)$, the usual right-tail procedure.

Sometimes, this process is written as follows: Replace c_i by $w_i n_i$. Then we can write $C = \sum_{i=1}^{k} c_i \overline{Y}_{i.} = \sum_{i=1}^{k} w_i Y_{i.}$, so C is a linear combination of sample totals. It is still normally distributed, so the only remaining change is the expression of the variance. It now takes the form $\sum_{i=1}^{k} w_i^2 n_i \sigma^2$. The rest of developing the pivotal statistic is unchanged.

In the balanced design, we usually require $\sum_{i=1}^{k} c_i = 0$, but in the unbalanced case, we typically require $\sum_{i=1}^{k} c_i n_i = 0$. Contrasts can arise quite naturally. If an experiment involves two or more new treatments, and one or more traditional treatments for a particular disease, it may be of interest to see if there is an overall difference between the old treatments and the new.

10.2.2 ANOVA for Model 2: Random Effects

The major difference in the setup of a random effects model is that only a small random selection of possible values of the design variable are selected in advance, and then observations of the response variable on the experimental units under the assigned values of the design variable are taken. The design can be balanced or unbalanced, but as in Model 1, it is best to choose a balanced design. The point is that another researcher may randomly select a completely different set of values of the design variable. However, the procedure for testing should lead to the same inferences, within the same level of significance.

Let Y be the response variable. Our design variable X has a very large number of possible values (typically, assumed infinite), and k of these values are selected at random; each of these values is applied to a certain number of experimental units, and the responses are observed. Again, it is useful to have a balanced design, that is, we apply each value of X selected to the same number of experimental units. However, we will set up the analysis in the general unbalanced case, and the balanced design becomes just one of these.

We assume a linear model, which means each observed value of Y can be written $Y_{ij} = \mu + \tau_i + \epsilon_{ij}$, where the subscript i represents the selected treatment levels, $1 \leq i \leq k$, j represents which observation of Y under a given value of i is listed, μ is the global mean response of Y, τ_i is the effect of treatment level i, and ϵ_{ij} is the random element in observing Y_{ij}. Since τ is now a random variable, our assumptions change. In the fixed effects model, we could assume $\sum_{i=1}^{k} \tau_i = 0$, but that is not possible here, since the τ_is are simply the effects of a random selection of values from the design variable. The most typical assumption is that $\tau \sim N(0, \sigma_\tau^2)$, that is, the τ_i's form a simple random sample from the specified normal distribution. This means $E(\sum_{i=1}^{k} \tau_i) = 0$, and $var(\tau_i) = var(\tau) = \sigma_\tau^2$. It is further assumed that τ and the τ_i's are independent of the error terms. We still assume $\epsilon_{ij} \sim N(0, \sigma^2)$.

Based on our assumptions, $var(Y) = \sigma^2 + \sigma_\tau^2$. Model 2 is sometimes called the components of variance model, with σ^2 and σ_τ^2 the components of variance. Our goal is to find estimates, hopefully independent of one another, of these two variances. Our goal is to test $H_0 : \tau_i = 0$ over all i, including the values of X we did not select. Thus, our null really is $H_0 : \tau = 0$, i.e., the effect of any treatment observed or not is 0. This the same thing as saying τ is a constant, specifically 0. How can a random variable be identically 0 (or any constant)? Its variance must be 0. We now have the null hypothesis in a form we can try to test, namely $H_0 : \sigma_\tau^2 = 0$. The alternative would be expressed as $H_1 : \sigma_\tau^2 > 0$.

To develop an estimate of σ_τ^2, consider the MSTr and MSE as derived for the fixed effects model. These values come from an algebraic calculation breaking down SST. If we compute $E(MSE)$, this will still be σ^2, regardless of whether the null is true or false. After some lengthy calculation, $E(MSTr) = \sigma^2 + n\sigma_\tau^2$ if the design is balanced, and $= \sigma^2 + M\sigma_\tau^2$ where $M = (N - \sum_{i=1}^{k} n_i^2/N)/(k-1)$, where N is the total sample size, $\sum_{i=1}^{k} n_i$. The argument about the distributions of the sums of squares still holds, so that $MSTr/MSE$ will have an $F(k-1, N-k)$ distribution under the null. Note that under the null, $E(MSTr) = \sigma^2$ so we would reject the null only if this ratio is much greater than one, so this is a right-tail test, exactly as before. Basically, the methodologies of analyzing the two models is the same; it is the inferences drawn from large values of the pivotal statistic that change.

This analysis also provides a way to estimate σ_τ^2: By using the expected values of the mean squares as guide, we have $\hat{\sigma}_\tau^2 = (MSTr - MSE)/n$ in the balanced case and $= (MSTr - MSE)/M$ in the unbalanced case. It is then possible to derive confidence intervals for σ^2 and for $\sigma_\tau^2/(\sigma^2 + \sigma_\tau^2)$. This last ratio is useful because it gives a measure of what fraction of the variability seen in Y is due to the treatment effect. It is less obvious about how to find confidence intervals for σ_τ^2 since its estimate is the difference of variables related to χ^2 variables, and there is no recognizable distribution for this case.

Chapter 10 — Analysis of Variance

Analysis of variance is used in more complicated designs. We will outline two such experimental designs and the basics of their analysis, and leave finding and analyzing real data to the reader.

10.3 ANOVA IN COMPLETE BLOCK DESIGNS

Up to this point, we have been assuming our experimental units are more or less uniform in how they will react to application of a particular treatment level. We assume that if we had assigned the experimental units to different treatment levels, the analysis would not be affected. However, it is frequently the case that there are extraneous factors that differentiate the experimental units, and we want to control their impact on the treatment levels as much as possible; so our results are about the effect of the treatment level, not about the experimental units chosen.

In the **complete block design**, experimental units are grouped into homogeneous units of sufficient size that each of the treatment levels can be applied within the block or group. This process is called **blocking**. The purpose is to keep the variability due to the experimental units themselves to a minimum. At the same time, it would be useful to know if it was actually necessary to apply blocking, so we would also like to determine if there is a blocking effect. This isolation of the blocking effect is much more efficient than applying a CRD, that is, we get better information for fewer experimental units.

Blocking is often used in experiments involving men and women and health issues; different research assistants preparing materials for an experiment; different age groups and different order of birth in families; and different sections of land for agricultural experiments. Again, the goal of this kind of experiment is to determine treatment effects separate from the possible effects of the type of experimental units available.

We assume our experimental units are now grouped into essentially homogeneous blocks of size k or such that we can select k at a time from the groups to form blocks, where as usual k is the number of treatment levels. We randomly select b blocks and randomly assign the treatment levels in those blocks. The sample then has a total of $N = kb$ observations, and this is clearly a balanced design. Our model is $Y_{ij} = \mu + \tau_i + \beta_j + \epsilon_{ij}$, where τ_i is the effect of treatment level i, β_j is the effect of block j, and ϵ_{ij} is the random error independent of blocks and treatments. This is an additive model, in that we assume the treatment and block effects are represented by additive terms. It is possible that there is an **interaction** between blocks and treatments, and another term may have to be included. We will not study such models here, but note that the fundamental methods developed in this chapter provide the basis for analyzing more complicated models.

We will consider the fixed effects model here, and note that random effects for blocks and treatments can be analyzed. We assume $\sum_{i=1}^{k} \tau_i = 0$ and $\sum_{j=1}^{b} \beta_j = 0$, which means we are observing all different types of blocks available. The null hypothesis of interest is $H_0 : \mu_1 = \cdots = \mu_k = \mu$ where $\mu_i = \mu + \tau_i$. Note that $\mu_i = (\sum_{j=1}^{b} (\mu + \tau_i + \beta_j))/b$ and that the null can be restated as $H_0 : \tau_1 = \cdots = \tau_k = 0$.

One of the assumptions is that $E(\epsilon_{ij}) = 0$ and another is that the model is homoskedastic, meaning $var(\epsilon_{ij}) = \sigma^2$, a constant over all (i,j). Thus, $E(Y_{ij}) = \mu + \tau_i + \beta_j$. Using the usual dot notation, we have $E(\overline{Y}_{i.}) = \sum_{j=1}^{b} E(Y_{ij})/b = \mu + \tau_i$ so we have an estimate of μ_i. Similarly, $E(\overline{Y}_{.j}) = \sum_{i=1}^{k} E(Y_{ij})/k = \mu + \beta_j$. We also have $E(\overline{Y}_{..}) = \mu$ regardless of the validity of the null. That means $\overline{Y}_{i.} - \overline{Y}_{..}$ is unbiased for τ_i and $\overline{Y}_{.j} - \overline{Y}_{..}$ is unbiased for β_j. This motivates breaking up the total sum of squares in a way that isolates terms related to the error, treatment, and blocking elements. In fact, it can be shown by a bit of algebra that:

$$SST = \sum_{i=1}^{k}\sum_{j=1}^{b}(Y_{ij} - \overline{Y}_{..})^2 = \sum_{i=1}^{k}\sum_{j=1}^{b}(\overline{Y}_{i.} - \overline{Y}_{..})^2 + \sum_{i=1}^{k}\sum_{j=1}^{b}(\overline{Y}_{.j} - \overline{Y}_{..})^2 + \sum_{i=1}^{k}\sum_{j=1}^{b}(Y_{ij} - \overline{Y}_{i.} - \overline{Y}_{.j} + \overline{Y}_{..})^2$$

$$= b\sum_{i=1}^{k}(\overline{Y}_{i.} - \overline{Y}_{..})^2 + k\sum_{j=1}^{b}(\overline{Y}_{.j} - \overline{Y}_{..})^2 + \sum_{i=1}^{k}\sum_{j=1}^{b}(Y_{ij} - \overline{Y}_{i.} - \overline{Y}_{.j} + \overline{Y}_{..})^2$$

$$= SSTr + SSB + SSE.$$

This breaks the total sum of squares into a set of sums of squares, one summand related to treatments, one to blocking, and one to error. It is at this point that we now assume normality of the errors, that is, $\epsilon_{ij} \sim N(0, \sigma^2)$, which is equivalent to assuming all Y_{ij} random variables are distributed as $N(\mu + \tau_i + \beta_j, \sigma^2)$. Since the errors are assumed independent, so are the Y_{ij}s. Under this assumption, we can apply Cochran's theorem to find related statistics that have independent χ^2 distributions. SST has $N-1$ degrees of freedom, $SSTr$ has $k-1$ degrees of freedom, and SSB has $b-1$ degrees of freedom. Now SSE has $N-1-k+1-b+1 = N-k-b+1 = ab-k-b+1 = (k-1)(b-1)$ degrees of freedom. We can also show $MSE = SSE/(k-1)(b-1)$ is unbiased for σ^2, regardless of the validity of the null hypothesis or the presence or not of a block effect. In comparison to the completely randomized design, the SSE term has lost $b-1$ degrees of freedom in the complete block design to help remove the effect of the blocks, if there is one. This produces a better estimate of the variance term σ^2 since the new SSE is the result of removing both the treatment effects and blocking effects from the total sum of squares.

Define the following mean squares:

$$MSE = \frac{SSE}{(k-1)(b-1)}; \quad MSTr = \frac{SSTr}{k-1}; \quad \text{and} \quad MSB = \frac{SSB}{b-1}.$$

By extending the methods for the CRD (completely randomized design) in section 10.2, it can be shown that:

$$E(MSE) = \sigma^2; \quad E(MSTr) = \sigma^2 + \frac{b\sum_{i=1}^{k}\tau_i^2}{k-1}; \quad \text{and} \quad E(MSB) = \sigma^2 + \frac{k\sum_{j=1}^{b}\beta_j^2}{b-1}.$$

Because of the independence of the sums of squares, then under the null hypothesis $H_0: \tau_i = 0$ for all $1 \leq i \leq k$, we must have $MSTr/MSE \sim F(k-1, (k-1)(b-1))$. Given the form of the expected values above, we would reject the null when this ratio is too large, quantified by an appropriate value from the right tail of the given F distribution.

It is tempting to test the existence of a blocking effect by examining MSB/MSE as we did with treatments. Large values may indeed make us suspect there is a real effect of the blocks, and values closer to 1 would indicate otherwise. The problem is that the process of blocking has introduced a restriction on randomization — we randomly assign the treatment levels within the blocks, not across all experimental units. Thus, the terms associated with blocks also include the effect of this restriction on randomization. For this reason, any inference about the blocking effect based on assuming $MSB/MSE \sim F(b-1, (k-1)(b-1))$ will not be valid. The calculations of the sums of squares and the mean squares are included in the ANOVA table, but only the F value related to treatments is included.

The standard from of the ANOVA table in the complete block experiment (fixed effects model) is:

Source	d.o.f.	SS	MS	F
Treatment	k−1	SSTr	SSTr/(k−1)	MSTr/MSE
Blocks	b−1	SSB	SSB/(b−1)	
Error	(k−1)(b−1)	SSE	SSE/((k−1)(b−1))	
Total	kb−1	SST		

Sometimes, the value MSB/MSE is included, but caution should be used in interpreting the value.

Other considerations in blocking include replication and random effects. Replication is the complete repetition of the basic random experiment. In the case of blocking, we have each treatment level in each block. The assumption is that treatment effects and blocking effects are simply additive. If, however, we wanted to determine if there is an interaction between blocks and treatments, we do not have information to determine this without repeating the experiment over several blocks of each type. Also, we have assumed the blocking is a fixed effect, but if, for example, we have human subjects do each of three tasks, and our main concern is with time to complete the task, then the block or human subject can be viewed as a random effect. We will not examine such extensions of the methodology, but note that the fundamental methods already explored provide the basis for the analyses in these different cases.

10.4 ANOVA IN TWO-WAY CLASSIFICATIONS

The methods outlined in sections 10.2 and 10.3 extend to more complicated experimental designs. We will look at one extension and illustrate how the analysis is a relatively simple extension of analysis of the one-way classification.

The problem we consider is the analysis of the data in a two-way classification. This means that we have two factors, A and B, each with a finite number of levels. This type of model is called the fixed effects model, because the number of levels is finite and we will observe all levels. We will consider a random effects model, in which the number of levels is very large or even infinite and a random selection of levels is taken, when we study linear regression in Chapter 11.

We have a response variable Y, and we suspect the response depends on the levels of A and B. Let the levels of A be $1 \leq i \leq r$ and those of B be $1 \leq j \leq c$. We are assuming a linear model of the form $Y_{ij} = \mu + \tau_i + \beta_j + \epsilon_{ij}$, where μ is a global mean response, τ_i is the effect of level i of factor A on the response variable, and β_j is the effect of level j of factor B on Y. Because μ is the average of Y over all levels of the factors, we assume $\sum_{i=1}^{r} \tau_i = 0$ and $\sum_{j=1}^{c} \beta_j = 0$. As in the earlier case, we assume $E(\epsilon_{ij}) = 0$ and $var(\epsilon_{ij}) = \sigma^2$, the homoskedastic model. In order to complete our analysis, we will add the assumptions that the errors (or residuals after fitting the mean and the effects to the observed Y value) are independent, and that in fact $\epsilon_{ij} \sim N(0, \sigma^2)$ and so $Y_{ij} \sim N(\mu + \tau_i + \beta_j, \sigma^2)$.

The structure of the Two-Way Classification is the same as that in the Complete Block design. The fundamental difference is that the two-way classification is a CRD; that is, the experimental units are assigned at random to the different combinations of treatment levels. This means that we can include both factors in the statistical analysis. Recall that the blocking effect was not amenable to statistical analysis because of the restriction on randomization.

We observe a value of Y for each cell (i, j) and record the results in contingency table format, namely:

		B		
		1	$\cdots$	c	
	1	Y_{11}	$\cdots$	Y_{1c}	$Y_{1.}$
A	$\vdots$	$\vdots$	$\vdots$	$\vdots$	$\vdots$
	r	Y_{r1}	$\cdots$	Y_{rc}	$Y_{r.}$
		$Y_{.1}$	$\cdots$	$Y_{.c}$	$Y_{..}$

In accordance with the notation introduced earlier, $Y_{i.} = \sum_{j=1}^{c} Y_{ij}$, $Y_{.j} = \sum_{i=1}^{r} Y_{ij}$ and $Y_{..} = \sum_{i=1}^{r} \sum_{j=1}^{c} Y_{ij}$. Denote the averages as $\overline{Y}_{i.} = Y_{i.}/c$, $\overline{Y}_{.j} = Y_{.j}/r$ and $\overline{Y}_{..} = Y_{..}/(rc)$. Note that the total sample size is rc. In this case, we observe Y once per cell, but the procedure generalizes to more than one observation per cell.

We proceed as before: We start with the total sum of squares,

$$SST = \sum_{i=1}^{r} \sum_{j=1}^{c} (Y_{ij} - \overline{Y}_{..})^2$$

and partition it into parts that can be related to just A, B and error. The decomposition is:

$$SST = c \sum_{i=1}^{r} (\overline{Y}_{i.} - \overline{Y}_{..})^2 + r \sum_{j=1}^{c} (\overline{Y}_{.j} - \overline{Y}_{..})^2 + \sum_{i=1}^{r} \sum_{j=1}^{c} (Y_{ij} - \overline{Y}_{i.} - \overline{Y}_{.j} + \overline{Y}_{..})^2$$

which is then written as:

$$SST = SSA + SSB + SSE.$$

SSE is the sum of squares for error, SSA is the sum of squares for factor A, and SSB is the sum of squares for factor B. Under our assumptions, these summands are independent (this uses Cochran's Theorem). SST has $rc - 1$ degrees of freedom, while SSA has $r - 1$ degrees of freedom, and SSB has $c - 1$ degrees of freedom (why?). Since the degrees of freedom must balance, then SSE has $rc - 1 - (c - 1) - (r - 1) = (r - 1)(c - 1)$ degrees of freedom. If we take each of these sums of squares, divide by their respective degrees of freedom, and call the result the mean square, we can show:

$$E(MSA) = \sigma^2 + \frac{c}{r-1} \sum_{i=1}^{r} \tau_i^2$$

$$E(MSB) = \sigma^2 + \frac{r}{c-1} \sum_{ij=1}^{c} \beta_j^2$$

and

$$E(MSE) = \sigma^2$$

where σ^2 is the common variance of the errors. If we wanted to determine if factor A's effects are 0, that is, all τ_is are 0, then we can look at MSA/MSE. If this is near 1, we would not have much evidence that even one τ_i is nonzero. Similarly, if we look at MSB/MSE and this is near 1, we would not have much evidence that even one β_j is nonzero. On the other hand, if the ratio is substantially larger than 1, we would have evidence that the appropriate factor does impact the observed value of Y. Under all of our assumptions, $MSA/MSE \sim F(r - 1, (r - 1)(c - 1))$ under $H_0 : \tau_1 = \cdots = \tau_r = 0$ and $MSB/MSE \sim F(c - 1, (r - 1)(c - 1))$ under $H_0 : \beta_1 = \cdots = \beta_c = 0$. The alternative in each case is that at least one of the specified parameters is nonzero.

We can summarize the values in an expanded ANOVA table:

Source	d.o.f.	SS	MS	F
A	r−1	SSA	SSA/(r−1)	MSA/MSE
B	c−1	SSB	SSB/(c−1)	MSB/MSE
Error	(r−1)(c−1)	SSE	SSE/((r−1)(c−1))	
Total	rc−1	SST		

With the null distributions of our pivotal statistics known, we can carry out the two tests of hypotheses described above.

10.5 ANOVA IN OTHER DESIGNS

The method of ANOVA extends to many other designs, including **incomplete blocks**, **Latin squares**, and **linear regression**. An extension to cases where there are continuous covariates is called Analysis of Covariance, abbreviated as ANCOVA. One point to stress here is that the methodology in these different models is fundamentally the same as those models considered in this chapter, with some changes to reflect the particular experimental design and the interpretation of the results of the data analysis.

In the last section of this chapter, some of the fundamental concerns, notation, and concepts in experimental design are developed. This is meant as an introduction to the subject, not a formal development. Detailed analyses can be found in, for example, D. C. Montgomery (1991), R. M. Bethea, R. S. Duran and T.L. Boullion (1985).

10.6 ASPECTS OF EXPERIMENTAL DESIGN

An experimenter with the government is to investigate the effects of lead on lake trout in Canada. One issue of concern is the length of adult trout: Does the presence of lead in the lakes reduce the size of such fish? In trying to answer this question, there are many different factors that may be considered: the latitude and the longitude of the lakes in the study; the proximity of pollution sources; the time of year under consideration; the surface area of the lake; the depth of the lake; the amount of fresh water coming into the lake; how much fishing is done; the proximity to settled areas; other pollutants in the lake; the presence (or absence) of predators; and other influences.

In setting up a procedure to get the data necessary, our researcher must know how much money is available for carrying out the experiment. The length of time available to complete the study is a factor. Unless a full population study is undertaken, it is necessary to know the precision needed. She needs to know the experience level of any person involved in the experiment. Measuring devices have limited accuracy, and so the best choice given the resources must be made. What information is available concerning lake trout in non-polluted lakes?

Before trying to gather data, our researcher must decide on the exact question(s) to be answered. Is it really the length of the fish, or is it the weight and general health of the fish? Is it only this type of fish, or is it any game fish that is of interest? Assuming the question has been determined, as well as the population under study, then the methodology for answering the question can be developed. This is the subject of experimental design.

The following is a list of steps to be used in developing an appropriate experimental procedure:

1. Determine precisely the nature of the problem under consideration, the questions that are to be answered, and the precision required. This includes determining the actual population involved.
2. Determine the factors that are to considered, including those to be manipulated and those that are to be held fixed.
3. Determine the variable(s) of interest, including the possible values of those variables.
4. Determine the likely distribution of the variables involved.
5. Choose an experimental design that will generate the data needed to be able to answer the question(s) with the required precision.
6. Set out explicit protocols to follow to carry out the experiment, and then gather the data.
7. Analyze the data, through descriptive statistics and through statistical methodology pertinent to the experiment's design.
8. Draw appropriate inferences based on the analysis.
9. Critique the experiment, including possible sources of error, unexpected results, possible factors for further investigation, and whether the goals have been met.

There are different types of experiments. For example, the **observational study** is frequently used in biology. In this instance, the researcher does not manipulate the factors that might influence the variables of interest. An **experiment** is a study in which one or more of the factors is manipulated or set by the researcher. We may be interested in changes over a period of time, and so a **longitudinal study** may be most appropriate. In this case, **repeated measures**, that is, the same type(s) of measurements taken on the same individuals at different points along some time line, may be necessary. There may be two or more populations with a particular characteristic, and a **comparative study** might be used. In drug trials, for example, the **placebo effect** needs to be considered, as well as the effect of no treatment at all. The type of experiment determines the statistical methods available for the quantification of results.

Other considerations include reducing or eliminating **bias**. Generally, bias is the difference between what is measured and the actual value of a particular quantity. For example, if a doctor is aware of which patients are receiving a drug and which are receiving a placebo, then the level of improvement for those receiving the drug may be influenced by the fact that the doctor is actually expecting better results for that group. The doctor's attitude can be encouraging or discouraging, and this may not be a conscious act on the part of the doctor. Such reasons have led to the **single blind** and **double blind** experiments.

We might also have the problem that the experimental units are not **homogeneous**. If there are differences, we call the units **heterogeneous**. This becomes a source of variability and is essentially a **nuisance factor**.

One of the main concerns in developing an appropriate design is **randomization**. This is fundamental to the statistical procedures used in drawing inferences with given precision. Typically, there is one (possibly more) variable (the response variable) we are observing or measuring, and we have some kind of model in mind about how the factors and levels help explain the differences in values of the response variable. These factors are sometimes called **explanatory variables**. After we have taken into account the effects of the factors, invariably there will be some part of the response variable's values "left over" or unexplained by the model. We call these **errors** or **residuals**. These are random variables and are assumed to have a particular distribution. In order to generate statistically reliable results, these must be **independent** of each other. This is one of the most important aspects of experimental design — if these are not independent, the statistical procedures discussed in this text are not appropriate. This is because the procedures, including determining distributions of estimators or pivots, are based on the assumption of independence. For example, the likelihood function in maximum likelihood is based on assuming the sample is a simple random one. The application of Cochran's Theorem to decompositions of the total sum of squares relies heavily on the independence assumption, as well as other distributional assumptions on the errors.

When we want to draw inferences, generally, we will not be in a position to say with 100% certainty some condition holds. However, we will want to have some level of assurance that the results represent the population, so we want a level of confidence in our results. This is related to the idea of **precision** and **confidence**. In the experiments under consideration, some value is being estimated (a mean, a variance, a proportion, etc.). The estimate is a number based on the data generated by a function of the observations, but we need to quantify just how far from the actual or true population value we might be. Even with the best experiments, we may get information that is not representative of the population, so there is a chance what we observe is far away from the true value. The confidence that we have is measured by how often our procedure will give us a value close to the true value, or, alternately, how often the value is far away from the true value. (This idea was developed in the chapter on confidence intervals and tests of hypothesis.) However, to attain a given level of precision and confidence, we will need a large enough sample. This is often determined from theoretical considerations related to distributions and based on the assumption of independence of the experimental errors. Precision is also dependent on the amount of **replication** in an experiment. Replication is the complete reproduction of the basic experiment, so in a completely randomized design, replication is the number of observations to be taken. In a complete block design, one replication would a selection of one of each type of block with all treatments applied within each block. A second replication would be another random selection of one of each type of block and then the random assignment of the treatments inside each block. Replication can be costly, but if it is important to isolate and estimate the blocking effect, it is the procedure that needs to be followed.

Another consideration is **power**. Power is the ability to detect differences in magnitude of effects. Often, we have an historical record of a population, and we are trying to determine if there has been some change. Power considerations are related to the sensitivity of our statistical methods to change. It is often the case that there can be different tests available, and, other factors being equal, the test with the highest power is preferred.

In the next subsection, we will discuss in broad terms some of the designs commonly used in the life sciences. This is by no means an exhaustive list, and details are given in broad form only.

10.6.1 Principles in Experimental Design

In the last section, we gave a list of nine steps in developing an experiment. In this section, we continue the development and reasons for these steps. Not all agree on this specific list, but generally similar considerations are made by all researchers and theoreticians.

1. **The question(s) or hypotheses and the population under study:**
   Consider our researcher and the problem of the health of adult fish in lakes in Canada. The resources available to her will allow for samples of fish to be taken from a small number of lakes, and that the underlying concern is the effect of new mining operations in Northern Ontario. In particular, the government is interested in any effects a new gold mine, opened five years ago, has had on the environment. Trout fishing is popular in the area and has been important to the tourism industry, something that the local inhabitants depend on. There are several lakes, mostly small, but there is a lake very close the the

mining operation, one that is a few miles upstream, and one that is a few miles downstream that are of similar size. Our researcher decides to take samples of adult trout from each of these lakes. Past studies indicate that the average length of such trout in such lakes is 52 cm., so the question she wants to try and answer is, is the length of adult lake trout in these lakes 52 cm. or is it significantly less? She may also want to know if the position of the lake in relation to the mine has an effect.

Note how the population under study has changed as the question becomes clearer. The resources available, the accessibility of experimental units and other factors may limit the study, and then in turn, the results of the study may have limited applicability. We also have to decide what it is we are measuring or observing to answer the questions.

2. **Factors:**

Unless an experiment is very controlled, a number of factors can impact the outcomes. If we want to study the effect of a new fertilizer on growth rates of plants, we may have access to greenhouses with very effective environmental controls, or we may have to run the experiment on plots of land out in the open. Even in a greenhouse, the amount of sunlight/cloud cover can have an influence, as can the training of the people maintaining the seed beds. Clearly, outdoor plots are subject to many more factors that are beyond our control: rainfall, sunlight, temperature, insects, previous plantings, and animals are all potential sources of variability. We can observe some or all of these factors, but we cannot set a level for them in advance.

Some factors have a relatively small number of discrete levels, and some have a large (usually assumed infinite) number of levels. If we can control a factor's level, then usually our experiment will involve having information from all of its levels. If there are too many, we randomly select some levels to observe. These cases have some similarity in analysis, but there will be differences. When there are a small number of levels to a factor, and we observe something about each level, we say the factor is a **fixed effect**. When we randomly select some of the levels from a large number for a factor, we say the factor is a **random effect**.

A **treatment** is the experimental procedure as it is applied to an individual experimental unit. Thus, a treatment may be the measurement of an adult fish from a specific lake, at a specific time of year. A treatment may be the application of a particular pain remedy to a male, age 25–30, with a particular condition. A treatment may be the administration of a memory test to a female, over 70 years of age, with early Alzheimer's disease. If a factor has a number of levels, the application of one of the factor levels to an experimental unit is a treatment. The separation of experimental units into categories is a treatment — the factor could be sex, and the treatment is the categorization of experimental units into male or female. Factors may affect the outcome of the response variable. Often, there is one factor and its effect on the response variable under consideration, and other factors that are not of particular interest that we try to isolate from the main factor.

Sometimes, treatments and factors are used interchangeably in describing experiments. This should be avoided to ensure minimum confusion.

3. **Variable(s):**

In trying to get an answer to the questions of interest, we will have a variable, say Y, that we want to use to reflect the quantity or quality of interest. Y is called the **response variable**. As we have seen, if we are performing a measurement, then Y's values fill out at least one interval, and we call it a continuous random variable. If its possible values are finite or countably infinite, we call it discrete. Sometimes, we simply want to count how often something happens, and then this is a categorical variable. Other times, we have discrete states and a rank ordering between the states (first, second, third, other, for example). Such cases are called rank ordered. In order to answer our question, another variable, typically a function of our response variable, may be used. For example, in the fish study, it may be the average length of fish that is of interest, and often, the sample average:

$$\overline{Y} = \frac{1}{n} \sum_{i=1}^{n} Y_i,$$

where n is the number of fish (or sample size) used in the study, is the variable considered. Other variables may be of interest: the variability in the length of the adult fish may indicate a problem that the average length does not reflect. The proportion of fish that exceed a particular length can be important.

In trying to determine the sources of variability in the response variable Y, there may be factors we think influence Y's values, and we want to build a model that relates Y to these other factors. The effect of individual factors are expressed as **explanatory variables**, and we often want to know how these variables affect our response variable. In this text, we concentrate on the **linear model**. **Non-linear models** are also used, but we will only touch on this briefly.

A common model in the linear case and with two fixed effects and interaction is of the form:

$$Y_{ijk} = \mu + \tau_i + \beta_j + (\tau\beta)_{ij} + \epsilon_{ijk}, \quad 1 \le i \le a, 1 \le j \le b, 1 \le k \le c.$$

Here our first factor A has a levels, and the level is indicated by i. The second factor B has b levels and is indicated by j. We are considering a balanced design, so each pair of levels (i, j) is observed equally often, i.e., c times. The number of times each treatment pair is observed is, therefore c, and which one of these observations is indicated by k.

The specific formulation of our model of the response variable reflects the questions to be answered and the overall experimental design. In the above, implicit in the model is that we think the response variable Y is affected by the level of A and B, denoted by τ_i and β_j, respectively. In many problems, there can be an **interaction** between the two factors, and this is denoted by the $(\tau\beta)_{ij}$ term. It is possible a particular combination of levels of A and levels of B can reinforce their individual effects, or one could work counter to the effect of the other. If there is no interaction, then the term $(\tau\beta)_{ij}$ would not be part of the model.

It is assumed there are certain constants, called parameters, that describe the probability distribution of the response variable. Some of these may be known, and some may have to be estimated. These include, in the linear model, the global mean, treatment effects, σ^2, and interaction effects. Once these have been accounted for, it is very likely that a part of the variability in the observations on Y has not been explained by the factors and any interaction, and that residual value is ϵ_{ijk}. Two problems emerge: We need to estimate values such as τ_i, β_j, and $(\tau\beta)_{ij}$, and we need to analyze the behaviour of the residuals (or errors).

4. Distributions:

Once we have decided on our variables, we need to determine the general distributional nature of these variables. Past studies may allow us to make assumptions, such as the average or expected value of the error terms is 0, and that the variability within the different factor levels of the errors is a constant. It may also be the case that we are comfortable assuming a very specific distributional form. For count data, with a fixed number of trials and the assumptions of independence of trials and constant probability of success, the sum of the successes is assumed binomial. If our population is finite and has two types of experimental units, the count of the number in our sample with one of the characteristics can be assumed hypergeometric (if the sample is taken correctly).

In the case of the linear model described earlier, the distribution of the errors is fundamental to being able to quantify our confidence in any inferences drawn from the data about the population. Thus, we may believe they follow a normal distribution: the errors have an average value of 0, are symmetric about 0, and when graphed, suggest a bell shape. Or, as is often the case, there may be thicker tails than the normal (suggesting perhaps a Student's t would be a better fit), or there may be sufficient skewness to make the assumption of normality questionable. The distribution selected needs to be verified as a reasonable fit once the data is collected. Significant departures from model assumptions call into question any inference drawn.

The quantification of our results depends heavily on the assumption(s) about the distribution of the variable. As noted earlier, much greater precision can be achieved for the same number of experimental units if the data is bell-shaped, rather than when no specific shape is assumed. The functions of the data almost invariably have different distributions than the original variable that we record. For this reason, it is important to have a working knowledge of an array of finite and continuous distributions.

5. **Experimental design:**
Once we have decided on the question, the factors, and the variable of interest, we need to determine how to get our information in the most efficient way. Experimental design is the method of incorporating our factors into treatments to apply to the individual experimental units. There can be different choices, and the one that requires fewest resources is often the best choice. Specific designs are discussed in the next section.

6. **Protocols and data collection:**
A perfectly well-designed experiment can be ruined by inconsistent data collection. A set of protocols or rules about how to carry out the experiment should be laid out, and everyone involved trained in following the protocols. In a recent incident, a plane was refuelled before heading to its next destination. However, it ran out of fuel and had to divert to an alternate airport. What happened? The ground crew filled the tanks with X liters, which should have been X gallons. Note that a similar problem would occur if the tanks were filled with a certain number of U.S. gallons, rather than the same number of imperial gallons.

Example 1: Suppose we are administering a questionnaire, and our procedure is to assign to each of five assistants one district, and they are to randomly sample adults in their district, with the total sampled in each district set at 200. One assistant selects 100 families and interviews the two oldest members. Another stands outside a grocery store and interviews the first 200 adults that will stop and talk. Another goes to a large park in the middle of the afternoon on a Wednesday and seeks out 200 adults. Another sends out 1,000 of the questionnaires by mail and gets 200 responses. The fifth assistant gets access to the most recent voters list and randomly selects 200 names, and then interviews those individuals. Discuss these different approaches.

In large and/or complicated experiments, it may be very useful to run a **pilot project**. Essentially, this is a miniature version of the experiment and allows us to work out any bugs, perhaps determine if there are some irrelevant factors, or factors that are important that we didn't consider, etc. This may also be necessary so that we can determine the sample size: the sample size depends on many elements, and in particular, on the distributional assumptions. The variance is a critical value, and getting an estimate of this value helps determine the sample size needed to achieve the precision required.

7. **Data analysis:**
Our data will typically be a set of numbers, and it may be difficult to get a sense of how well our assumptions are reflected in the data. Various procedures include descriptive statistics (summary values and pictorial displays), tests of model assumptions, and calculation of the values of the random variables of interest. Estimates of parameters (values that determine the specific member of a given distributional family) and confidence intervals or tests of hypotheses are carried out.

8. **Inferences:**
With our analysis of the data, we can then return to the original question(s) and determine what the evidence supports. In this part, we also indicate our level of confidence in these inferences.

9. **Summary and observations:**
After the experiment is completed, it is important to evaluate the procedures, the implementation, and the limitations encountered. There may be questionable observations (although values thought of as outliers may in fact be the most interesting values). It may emerge from this study that specific factors should be explored in more detail.

10.6.2 Specific Experimental Designs

As noted earlier, there are different types of variables and different types of factors. We have a level of precision required and questions to be answered. We need to get data to allow for appropriate inferences in the most efficient manner possible. Different combinations of these considerations lead to specific designs. In this section, we describe some of these designs.

10.6.2.1 Completely Randomized Design

The most common design is the completely randomized design, CRD. In this, we have a single factor to be applied to a homogeneous population. A fixed sample size n of experimental units is randomly selected. If the factor has only one level, then we apply that one treatment to all of the randomly selected experimental units. If the factor has several levels, then the number of units n_i to receive factor level i as treatment is determined in advance, and the experimental units are assigned the different treatments randomly so that n_i receive factor level i. Values of the random variable of interest are then observed for the sample. Typically, it is one of: the population mean μ; or the population variance σ^2; or the population proportion π that we want to draw inferences about. Normally, a Greek letter is used to denote the population value, an uppercase English letter is used to denote a variable, and the corresponding lowercase letter is used to denote a realization of the variable. Once the data has been gathered in a manner consistent with the design, we will be able to estimate such population values, and we often denote the estimate to be used in our analysis by placing a ˆ over the population value. Such estimates come from applying methods such as least squares, maximum likelihood, or Bayesian analysis.

10.6.2.2 Block Designs

Block designs can significantly increase efficiency, which means that we get better information for fewer experimental units. A standard problem is to have different treatments (such as placebo, half dose, full dose) and different subpopulations among the experimental units (sex, age, place of origin). Basically, we want to separate out the effects of the different characteristics of the population from the reaction to the treatments.

In the simplest form, one characteristic with a small number of variants forms the blocking.

10.6.2.3 Blocking

At its simplest, blocking is the segmentation of the population into homogeneous parts. In most problems, the number of types of blocks is small enough so that all types can be considered. In other cases, there can be a large number of types, and so we randomly select some types and then randomly select blocks within the individual types. One way to view blocking is that it allows us to separate the block effect, often viewed as a nuisance effect, from the effect of interest.

10.6.2.4 Complete Blocks

Complete block designs are such that all treatments can be applied to each individual block. In such cases, we may also need to consider **replication**. Replication is the repetition of the experiment under the same conditions, and this allows us to generate more precise estimates of the effects of interest. A **balanced design** is one where all treatments are observed equally often, under the different block types.

10.6.2.5 Latin Squares

If we have two factors, and a small number of treatments, a Latin square design can again increase efficiency over simpler designs. A specific form is to have k treatments, and then we design the experiment so that block type A has k levels, as does block type B. We label the treatments with letters $a, b, \cdots$ (there are k of these letters). In a matrix of size $k \times k$, the rows represent A's levels, and the columns represent B's levels. We then assign treatments to each row and column so that each letter $a, b, \cdots$ appears once and only once in each row and in each column.

Example 2: A biologist has four different methods available to clone certain DNA fragments. He is interested in the time required for each one to generate a certain amount of usable material. He selects four of his students to prepare the genetic samples; the preparation of one batch takes two hours, so four batches can be prepared in a day. The researcher has noted that there can be differences in the quality of preparation as the day goes on, and so he uses a Latin square design to isolate the effect of the preparer and the effect of the time of day. The letters a, b, c, d represent the different methods of preparation. His design is:

Student	Time of Day			
	1	2	3	4
1	b	c	d	a
2	a	b	c	d
3	d	a	b	c
4	c	d	a	b

There are many different Latin squares possible, and for different values of k, they have been tabulated. We randomly select one of these formulations for our design.

This method generalizes to more than two nuisance factors. As an exercise, find information on **Graeco-Latin square designs**.

10.6.2.6 Incomplete Blocks

In some instances, we cannot apply all treatments to an individual block. The incomplete block design allows us to block various factors, and in such a way that every **pair** of treatments occur equally often over the blocks. Thus, if each batch of genetic material allows for only three lots, but there are four treatments, then we would choose say four batches, and run three treatments expressed as (a,b,c), (a,b,d), (a,c,d), or (b,c,d) on each block. In this way, we see all treatments equally often. This also means that pairs of treatments (factor levels) are spread equally across the blocks, so, for example, we don't just apply a to one block, b to another, etc.

10.6.2.7 Factorial Designs

Factorial designs allow for several different factors while retaining the efficiency needed to remain cost effective. Various types of factorial designs reflect many different experimental needs. The simplest is the 2^k factorial design: There are k different factors, each with two levels. These levels may be "high"; "low"; No, Yes; male, female; etc. If k is relatively small, our experiment consists of applying the treatment to all 2^k combinations. Replications are then possible. If, however, k is large, we may need to choose a modification of this design to get information about the effect(s) we are really interested in.

Note that there can be interaction terms, that is, terms associated with combinations of levels of the factors. It is often assumed that higher-level interactions (interactions between three or more factors) are negligible, and we confound these effects with lower order interactions. **Confounding** is the result of a design in which an interaction term cannot be distinguished from the main effects.

10.6.2.8 Nested, or Hierarchical, Designs

Suppose that we are interested in how well patients with a particular physical disability respond to a new rehabilitation program. There are three hospitals in the area, each with a rehabilitation department with four physiotherapists. Each will be assigned five patients with the physical disability, and the range of motion after a month of rehabilitation is to be measured.

In such problems, we have (at least) two factors to consider: each hospital may have different quality of resources available; and the physiotherapists may have different attitudes to the rehabilitation program, for example. However, we cannot assign a doctor from one hospital to another (we will assume the patients are more or less uniform in the degree of physical disability and other characteristics).

Exercises

1. Show in the unbalanced, completely randomized design that if there are two treatment levels, the F test and the t test for testing $H_0: \mu_1 = \mu_2$ vs $H_1: \mu_1 \neq \mu_2$ are numerically identical.

2. From the internet, textbooks, researchers, and other sources, find the basic ANOVA in at least one of the following designs: Latin square; factorial designs; split-plot; hierarchical designs. Determine the conditions of experiments leading to these designs, and examine the partitioning of the total sums of squares that leads to the formal statistical analysis of the components of the model.

3. One approach to determining if distributional assumptions are reasonably valid is to plot the **estimated residuals**. These are the values that remain after the various effects, including the global mean, have been fitted. In a completely randomized design involving one factor with several levels, the response variable is assumed to take the form $Y_{ij} = \mu + \tau_i + \epsilon_{ij}$. We have an estimate of $\mu + \tau_i$, namely $\overline{Y}_{i.}$ so the estimated residual is $\hat{\epsilon}_{ij} = Y_{ij} - \overline{Y}_{i.}$. In the complete block design, the estimated residuals are $\hat{\epsilon}_{ij} = Y_{ij} - \overline{Y}_{i.} - \overline{Y}_{.j} + \overline{Y}_{..}$. These will be normal if the original distribution of (estimated) residuals is normal. Residuals can be used to examine the normality assumption and the assumption of homogeneity of variances.

    (a) In a plot of the estimated residuals, how would a plot of these values versus the different populations (coded 1 to k) help verify the normal assumption for the residuals, regardless of the design?

    (b) In the plot in (a), what would indicate that the variance of the residuals is not constant across the different populations, regardless of the design?

    (c) In the case of the completely randomized design, sometimes the residuals for the different treatment levels are plotted against $\overline{Y}_{i.}$ on the same axes. What would the distribution of points for each i look like if there is in fact a difference in variances across the different populations?

4. An agriculturist is interested in the effect of combinations of direct sunlight and the use of artificial light on the growth of two-week-old maple saplings. In particular, she wants to determine the effect, if any, of using: two hours of sunlight and six hours of artificial light; four hours of sunlight and four hours of artificial light; and eight hours of sunlight in 24-hour periods. The experiment is set up so that the experimenter can control the exposure of light on the saplings. She will measure the change in height of each sapling in millimeters and measured to two decimals of accuracy, after two days of the given exposures, with the exposures coded 1, 2, 3, respectively. The saplings were otherwise grown under the same conditions up to the start of the experiment, and the two-day period had sun for the entire day. The sunlight exposure was set to be symmetric about midday, and the artificial light was applied equally before and after the direct sunlight exposure. Ten saplings were selected for each combination of light, with the results:

		Growth in Millimeters									
Light	1	2.99	1.97	1.34	1.98	1.57	3.46	0.16	1.65	1.86	3.08
Exposure	2	0.86	2.34	1.77	3.43	2.95	6.46	0.27	6.14	3.34	5.01
Type	3	6.47	2.53	5.44	5.22	4.02	4.73	3.00	1.99	2.45	3.64

    (a) Use ANOVA of a completely randomized design to determine if this data is evidence that there is an effect of the length of sunlight and artificial light on the growth of these saplings. Summarize your results in an appropriate ANOVA table.

(b) Calculate the estimated residuals (see exercise 3) and use appropriate plots to determine if there is significant departure from the normality and homoskedastic assumptions.

5. An executive in a large company that replants deforested areas wants to determine if experience of workers planting young trees affects the number of trees they can plant in a standard work day. He breaks the workers into four categories, namely those with up to five years experience; those with from five up to 10 years; those with from 10 up to 15 years; and those with 15 up to 20 years experience (the company is 20 years old). These types of workers are coded 1 to 4. He randomly selects a number of each type of employee, and on a given day, the following number of trees planted by each of the workers was recorded:

		Trees Planted									
	1	133	89	68	139	115	63	104	101			
Experience	2	133	83	104	136	134	104	111	114	99	91	
	3	91	96	112	98	87	143	99	129	104	92	
	4	101	112	85	95	100	118	96	101	117	97	110	104

(a) Use analysis of variance to determine if the average number of trees a worker can plant in a day is affected by the number of years of experience. Summarize your results in an appropriate ANOVA table.

(b) Is there evidence that the variability in the number of trees planted in a day is not the same for the different types of workers? If so, what impact would this have on your answer to (a)?

6. One of the steps in dating ice cores from glaciers is the separation of carbon isotopes from the cores. There are four different methods to distill carbon from a core, and a researcher wants to know if there is any difference in these methods. In a particular region in Norway, there are three glaciers in close proximity. Since the chemical composition of cores taken at 10 meters below the surface of a glacier at a particular altitude may be impacted by the specific glacier tested, the lead analyst selects a core from each glacier at the requisite depth and altitude, and then separates the core into four equal parts and applies at random each of the extraction methods to one of the core quarters. The percentage purity of the resulting samples are:

		Glacier		
		1	2	3
	1	84.89	88.39	90.08
Distillation	2	87.24	88.48	90.74
Method	3	89.73	90.89	93.94
	4	90.25	91.52	94.33

(a) Determine if there is a difference in mean purity of the carbon extraction by the different methods. Summarize your results in an appropriate ANOVA table.

(b) Is there evidence of an effect of the specific glacier from which the core samples are taken?

(c) Examine the estimated residuals for this design, and use graphical methods to decide if the distributional assumptions are valid.

7. A researcher is studying effects of temperature and aeration on the weight gain of young trout. There are four levels of aeration determined by the speed of a pump (levels 1 (lowest), 2 , 3, 4 (highest)) and three temperatures of water that can be maintained in any holding tank (65°, 67.5°, and 70° in Fahrenheit, coded 1, 2 and 3). Three tanks are available, and the researcher establishes the different possible temperatures in each, and then sets the aeration level at the same value for each and changes that level at the end of a week. An individual trout is selected from a large source of young, similar-sized fish of approximately the same age to be put into each tank at the start of a week, and the difference in weight (in grams) over the week recorded. The amount and type of food is the same at all times. Then the aeration level is changed, and a

new trout is selected for each tank; this continues until all four aeration levels have been implemented and the change in weights recorded as follows:

		Temperature Level		
		1	2	3
Aeration Level	1	1.76	1.95	1.61
	2	2.31	1.87	1.84
	3	2.55	2.16	1.42
	4	2.44	2.38	1.81

(a) Use the two factor ANOVA to determine if there is an effect from temperature levels, and from aeration rates. Summarize your results in an appropriate ANOVA table.

(b) Calculate the estimated residuals, and use them to explore the model assumptions.

(c) Are there any other steps that should be taken to try to ensure the fish are subjected to the same environment, regardless of the temperature and aeration levels?

CHAPTER 11

LINEAR REGRESSION

Synopsis
The fundamental theory and methods of linear regression under fixed effects and random effects are developed. The use of regression in prediction is discussed.

11.1 PREDICTION

In this chapter we will discuss some of the basic ideas associated with fitting curves to data. In particular, we will assume that we have two variables X, often called a regressor or predictor variable, and Y, usually called the response variable. Either both are measured or observed simultaneously on a random selection of experimental units, or we set the X values in advance and observe or measure Y one or more times for each different X value. The information is organized in ordered pairs, where the first entry is the X value and the second the corresponding Y value. The ordered pairs are then a random sample of all such pairs. If we select the Xs, then the distribution under consideration is univariate, related to Y. If both X and Y are random, then we are considering a bivariate distribution and trying to establish if X and Y are statistically related. In both cases, we call the study of statistical relationships between variables **regression analysis**. The main goals of regression include determining if knowing X helps in predicting what Y value is likely, what form the relationship between X and Y takes, and measuring the strength of such relationships.

For example, we may believe the quality of the product of a chemical process is dependent on the temperature at a particular point in the process. We may then choose a selection of temperatures and run the process several times at each such temperature. In another case, we may believe that a child's height at age 2 is a good predictor of that person's adult height (at age 18, say). Clearly, we cannot set the child's height at age 2, but it is a constant for that child. In this case, there are many possible age 2 heights, and if we take a random sample of 18 year olds with records including their age 2 height, the predictor variable is assumed measured without error, but we will only see a selection of them. A third scenario might be that we measure X and Y simultaneously on experimental units, and both X and Y are random variables. An example may be measuring an adult male's blood pressure and plaque build up in the aorta (measured by the thickness at a particular point). In this case, blood pressure is affected by many factors, so it is a random variable.

Regardless of the type of problem, we have some reason to believe the variables are not independent, and we want to determine whether there is evidence to support this assumption. If the evidence is in favor of the hypothesis that the variables are dependent, then we would like a measure of the strength of that dependence. To fix notation, we will let Y be a response variable of interest, and also call it the dependent variable, and X is the independent variable or regressor or predictor variable. We are interested in determining Y as a function of X, up to random error. Another way to think of this is that we are interested in determining the average response in Y for each given value of X. There are many possibilities for the function describing the relationship, but we will assume a linear relationship, namely $Y = \beta_0 + \beta_1 X + \epsilon_X$, where β_0 and β_1 are called **regression coefficients**. We also call β_0 the Y-intercept and β_1 the slope of the regression line. The value ϵ_X is random error, or the residual after the intercept and slope of the regression line has been fitted.

When a linear relationship of the form $Y = \beta_0 + \beta_1 X + \epsilon_X$ is assumed, we call the result the **linear regression of Y on X**. The value β_1 is also called the true rate of change of Y with respect to X, and if

X changes by one unit, we expect to see **on average** β_1 units of change in Y. The equation is also written $E(Y \mid X = x) = \beta_0 + \beta_1 x$ under the standard assumption $E(\epsilon_X) = 0$, indicating the fact that the mean value of Y for a given value of X lies on the **regression line** in the $X - Y$ plane. As will be discussed, the variance σ_X^2 of the errors ϵ_X may be constant across all values of X, and we call this the homoskedastic model, or the variance of the errors may depend on the specific value of X; and we call this the heteroskedastic model. The heteroskedastic case is left for more advanced studies in regression. Under the assumption of homoskedasticity, we write $\sigma_X^2 = \sigma^2$. Also in this model we will write ϵ instead of ϵ_X to further emphasize that the distribution of the errors does not depend on the value of X.

Nonlinear regression assumes the relationship between X and Y can be expressed as $Y = H(\beta, \mathbf{X}) + \epsilon$, where β is a vector of unknown constants, called regression coefficients. It is also possible we have two or more regressor variables, but we will assume there is just one. The function H can take on many forms. If H is linear in the coefficients, we will call such models linear: If, for example, $Y = \beta_0 + \beta_1 X + \beta_2 e^X + \epsilon$, this will be called linear, since the methodology for estimating the coefficients is a straightforward generalization of estimating the coefficients in $Y = \beta_0 + \beta_1 X + \epsilon$. There are many important nonlinear models in biology, such as:

$$Y = \frac{A}{1 + B\exp(X)} + \epsilon$$

called the logistic model, and:

$$Y = A + Be^{-X} + \epsilon$$

sometimes called asymptotic growth or von Bertalanffy growth model (see L. von Bertalanffy, (1938)). These are examples of important models that cannot be linearized, that is, transformed into a related linear model. For example, if $Y = Ae^{BX}$ (the exponential growth model), then $\ln Y = \beta_0 + \beta_1 X + \epsilon$ is a related linear model resulting from taking natural logarithms and setting $\ln B = \beta_0$ and $\beta_1 = B$. This makes estimating coefficients much easier, but then caution has to be taken in the interpretation of results in terms of the original variable Y.

Throughout this chapter, we will assume linear regression means that the regression equation is linear in the coefficients, and no transformation was needed to get to that form.

It is often assumed that the X values are constants, either set in advance, or observed at the same time as Y. This is referred to as the **fixed effects model** or conditional model, and the analysis of this model is called **regression analysis**. Frequently, we think there is a direct link between X and Y, that is, X causes Y. Such models are called **causal models**. If we instead assume the X values are also random so that X and Y have a joint probability distribution, it is called the **random effects model**, and the analysis of the random effects model is called **correlation analysis**. In the random effects model, it is not necessarily the case that we think X causes Y. It may very well be that there is an unobserved variable Z which does directly cause Y and is highly correlated with X. In this case, we are looking for the strength of statistical relationship, assumed linear, between X and Y. The basic calculations in the causal and correlational models are the same, but the interpretation of the results is different.

We want to use X to predict Y, so we call X the independent variable and Y the dependent variable. X is also called a **regressor** variable, or a **covariate**. We need to have a procedure for estimating the parameters, as well as information about what kind of distributions for the errors are appropriate. From these, we can compute confidence intervals and carry out tests of hypotheses on the parameters. Recall also that $\rho = cov(X, Y)/\sqrt{\sigma_X^2 \sigma_Y^2}$ is the correlation coefficient and measures the strength of a statistical linear relationship between two variables, so we will want to estimate it also.

One way to decide if there is any level of linear relationship is to draw a **scatter plot**. We take a random sample of size n of observations (X_i, Y_i), $1 \leq i \leq n$, regardless of the type of model, and then plot these points on an $X - Y$ coordinate plane. We are looking for a general trend to the data, specifically linear in appearance. This does not mean the points must all fit on a line, but that a line goes through the "middle" of the observations. If the line has a positive slope, it indicates that on average as X increases so does Y, while if the slope is negative, then as X increases, Y decreases. These statements are about the trend of the average value of Y for given

Chapter 11 — Linear Regression

X values. From the form of the regression line, it is interpreted to mean that for each fixed X, there is a mean response to Y, and all these means fit on a line. Sometimes, if there isn't a linear relationship, a transformation may produce one and then we do linear regression on the transformed variables. Sometimes, we will need to use a more complex regression model, but we will leave details to more advanced courses on regression. We will assume that in fact there is a linear trend, and now we want to find the line of best overall fit to the data.

To determine the parameter estimates, we have to have a principle concerning goodness of fit of a line or trend line to the data. In general, we want to minimize how far the totality of points is away from the line, and the method typically used is **least squares**. Define:

$$L = \sum_{i=1}^{n}(Y_i - \beta_0 - \beta_1 x_i)^2.$$

Note that this is essentially the square of the error terms, or in other words, the residuals squared, after fitting the regression line. The **Principle of Least Squares** is: Choose β_0 and β_1 to minimize the function L. By using partial derivatives, we can readily show that this means:

$$\hat{\beta}_1 = S_{XY}/S_{XX} \text{ and } \hat{\beta}_0 = \bar{Y} - \hat{\beta}_1 \bar{X}$$

where:

$$S_{XX} = \sum_{i=1}^{n}(X_i - \bar{X})^2 = \sum_{i=1}^{n}X_i^2 - n\bar{X}^2, \quad S_{YY} = \sum_{i=1}^{n}(Y_i - \bar{Y})^2 = \sum_{i=1}^{n}Y_i^2 - n\bar{Y}^2,$$

and:

$$S_{XY} = \sum_{i=1}^{n}(X_i - \bar{X})(Y_i - \bar{Y}) = \sum_{i=1}^{n}X_i Y_i - n\bar{X}\bar{Y}.$$

We have included the computing formulae for the sum of squares in the above equations. Such formulae help avoid round-off error in calculations done by hand. They also prove useful in determining expected values and variances of the estimators.

In the fixed effects case (the X values are constants), it is readily shown that $E(\hat{\beta}_1) = \beta_1$ and $E(\hat{\beta}_0) = \beta_0$, and so these are unbiased. Also recall that the correlation coefficient is $\rho = cov(X, Y)/\sqrt{\sigma_X^2 \sigma_Y^2}$. Its estimate is:

$$r = S_{XY}/\sqrt{S_{XX}S_{YY}} = \hat{\beta}_1 \sqrt{S_{XX}/S_{YY}}.$$

Numerically, then, there is relationship between r and $\hat{\beta}_1$. However, when X is a random effect, tests should be about the correlation since its distribution depends on the assumption of a joint distribution between X and Y (typically, that they have a bivariate normal distribution), and hence, about ρ. When X is a fixed effect, then we make hypotheses about β_1 based on the specific values of X chosen and from that determine confidence intervals and carry out tests of hypotheses about β_1.

Define $SST = \sum_{i=1}^{n}(Y_i - \bar{Y})^2 \equiv S_{YY}$, the total sum of squares. Let $\hat{Y}_i = \hat{\beta}_0 + \hat{\beta}_1 X_i$, and then define $SSR = \sum_{i=1}^{n}(\hat{Y}_i - \bar{Y})^2$, the sum of squares due to regression. Further, set $SSE = \sum_{i=1}^{n}(Y_i - \hat{Y}_i)^2$, the sum of squares for error (or residual sum of squares). By using the algebraic methods in this text, $SST = S_{YY} = SSR + SSE$. The **coefficient of determination** is $R^2 = SSR/SST = 1 - SSE/SST$. The value $R^2 100\%$ can be interpreted as the percentage of the variability seen in Y explained by the regression equation. A rule of thumb is that, for practical purposes, $R^2 \geq 0.5$ is necessary for the regression to be of value. It can be shown that $R^2 = r^2$, the square of the correlation coefficient.

Various terms are used: We define $\widehat{E}(Y_i \mid X = x_i) \equiv \hat{Y}_i = \hat{\beta}_0 + \hat{\beta}_1 x_i$ and call this the fitted Y value, although it really is the estimated mean value of Y for $X = x_i$. We then set $\hat{\epsilon}_i = Y_i - \hat{Y}_i$ and call this the estimated residual, or estimated error. Note that $SSE = S_{YY} - (S_{XY})^2/S_{XX}$, and hence, we can compute this from the previously defined values.

Example 1: Let X be extra phosphorous in a fertilizer mixture over the standard, with X being 1%, 2%, or 3%, and let Y be the increased height of geranium cuttings two weeks after the cuttings have been planted, measured in mm. Nine cuttings of the same size are selected at random. Three are planted in soil mixed with one of the three fertilizer mixtures, and this is repeated for the other fertilizer mixtures. The results after two weeks were:

X	1	1	1	2	2	2	3	3	3
Y	25.32	26.05	23.42	30.49	30.19	29.51	34.87	34.30	34.44

Some summary values include: $\sum x_i y_i = 566.00$, $\sum x_i^2 = 42$, $\sum y_i^2 = 8158.64$, $\sum x_i = 18$, and $\sum y_i = 268.59$. Thus, $S_{XX} = 6$, $S_{YY} = 143.02$, $S_{XY} = 28.82$, and so $\hat{\beta}_1 = 28.82/6 = 4.80$ and $\hat{\beta}_0 = 20.24$. Hence, the least squares regression line is $\hat{Y} = 20.24 + 4.80X$. Thus, for each % increase in phosphorus, the height increased by 4.8 mm, on average. Note that the intercept in this case can be viewed as the average height of the cuttings after two weeks in soil prepared from the standard fertilizer mixture. Sometimes the intercept has meaning, and sometimes it doesn't. We should not use the regression equation for prediction too far away from the range of the X values observed: A fertilizer can become destructive if concentrations of constituents are too high.

In our problem, $r^2 = .98^2 = .97$ so that about 97% of Y's variability is explained by the regression of Y on X.

Now that we have some fitted values, we would like to determine how good this regression line is in fitting the data. The first thing we want to test is whether there is an actual relationship between X and Y, or if X and Y are independent. This is the same as determining if the true rate of change of Y with respect to X is 0. For this purpose, we assume the errors are normally distributed and have a common variance for the different values of X in the **regression analysis**, i.e., when we assume the X values are constants. We can express this as $Y \sim N(\beta_0 + \beta_1 x, \sigma_e^2)$ for all values $X = x$, or equivalently, $\epsilon \sim N(0, \sigma_e^2)$, where σ_e^2 is the (constant) variance of the errors.

Under the assumption of independence and normality of the errors, and common variance, we have the following results:

Under the assumption that X and Y have a bivariate normal distribution (the assumption in the random effects model), if $\rho = 0$, then $r\sqrt{(n-2)/(1-r^2)} \sim t(n-2)$. If $\rho \neq 0$, then we have approximately:

$$\frac{1}{2}\ln\left(\frac{1+r}{1-r}\right) \sim N\left(\frac{1}{2}\ln\left(\frac{1+\rho}{1-\rho}\right), \frac{1}{n-3}\right].$$

Analyzing the relationship between X and Y in this way is called **correlation analysis**. The transform of r above is called the Fisher z-transformation.

If we assume the X values are fixed (the conditional model or regression analysis), then:

$$t = (\hat{\beta}_1 - \beta_1)/S_{\hat{\beta}_1} \sim t(n-2)$$

where $S_{\hat{\beta}_1}^2 = S_e^2/S_{XX}$ and $S_e^2 = SSE/(n-2)$. Further, $SSE/\sigma_e^2 \sim \chi^2(n-2)$, and $E(SSE/(n-2)) = \sigma_e^2$. The value σ_e^2 is the variance of the errors, assumed constant across all values of X. We will leave the derivation of these results to the reader, noting these come from partitioning sums of squares, and, under the assumption of independence and normality of the errors, the application of Cochran's Theorem. It is easy to show that $var(\hat{\beta}_1) = \sigma_e^2/S_{XX}$, which is why the distribution of the pivot for β is as stated.

We can develop tests for β_0, but it can also be shown that the distributions of $\hat{\beta}_0$ and $\hat{\beta}_1$ are not independent. A broader discussion of this would be taken up in a more detailed course on regression.

Example 1 (Continued): Regression Analysis: The values of X are fixed in this problem. The test of interest is $H_0: \beta_1 = 0$ vs $H_1: \beta_1 \neq 0$ at the $\alpha = 0.05$ significance level. Directional tests can be carried out if this is given in advance. $SST = S_{YY} = 143.02$ and $SSR = \hat{\beta}_1^2 S_{XX} = 4.80^2(6) = 138.24$. Thus, $SSE = SST - SSR = 4.78$ and so

Chapter 11 — Linear Regression

$S_e^2 = 0.114$. From this, $t_{calc} = 14.23$. Since this is a two-sided test, $t_{crit} = \pm 2.365$, recalling the degrees of freedom to use is 7 here, so we can reject H_0. A confidence interval for β_1 can be found as usual: A 95% confidence interval would be (4.80-2.365(0.114), 4.80 +2.365(0.114)) =(4.53, 5.07).

Correlation Analysis: If it is the correlation we want to test, then the hypotheses will be $H_0 : \rho = 0$ vs $H_1 : \rho \neq 0$, again at the $\alpha = 0.05$ significance level. The alternative can also be a directional alternative. The assumption of bivariate normality or nearly so is essential for the distributional assumption about the sample correlation under the null hypothesis.

Example 2: A biologist is studying various characteristics of mature groundhogs and is interested to know if there is a statistical relationship between chest span (in inches) and weight (in ounces) of groundhogs at the first week of August, in the year they are born. She randomly selects 20 of such groundhogs through humane traps and records the measurements in question for each as:

X	9.02	11.17	11.39	9.80	11.00	9.84	8.91	10.47	9.31	10.26
Y	95.70	103.56	102.58	99.38	100.42	97.71	98.25	102.75	98.77	98.57

X	10.31	9.91	10.68	10.29	10.20	10.78	9.63	10.42	8.66	8.24
Y	101.87	99.98	103.67	101.79	99.02	105.47	99.29	100.23	97.88	98.24

Summary values include:

$$\sum x_i = 200.28 \quad \sum x_i^2 = 2019.43 \quad \sum y_i = 2005.15 \quad \sum y_i^2 = 201146.60 \quad \sum x_i y_i = 20106.50.$$

From these values we have:

$$S_{XX} = 13.83, \quad S_{YY} = 115.27 = SST, \quad S_{XY} = 26.93, \quad SSR = 52.27, \quad SSE = 62.83.$$

Also,

$$\hat{\beta}_1 = 26.93/13.83 = 1.95, \quad \hat{\beta}_0 = \bar{Y} - \hat{\beta}_1 \bar{X} = 80.76, \quad r = \hat{\beta}_1 \sqrt{S_{XX}/S_{YY}} = 0.67.$$

Since the assumption is that X and Y have a bivariate normal distribution, so this is a random effects model, we will test $H_0 : \rho = 0$ vs $H_1 : \rho > 0$. Under the null, $r\sqrt{18/(1-r^2)} \sim t(18)$. The critical t score is 1.734 when the significance level is $\alpha = 0.05$. The calculated t value is 3.83, we conclude that the null should be rejected.

To find a confidence interval for ρ, we use $\frac{1}{2} \ln((1 + r)/(1 - r)) \sim N(\frac{1}{2} \ln((1 + \rho)/(1 - \rho)), 1/17)$. Based on this approximation, the 95% symmetric confidence interval for $\frac{1}{2} \ln((1 + \rho)/(1 - \rho)$ is (0.335, 1.286), leading to (-0.023, 0.847) as the approximate confidence interval for ρ. Note that 0 is in this confidence interval. Does this contradict the result of the test of hypothesis for ρ?

A couple of checks can be run to ensure the calculations are correct. The **centroid** $(\bar{X}, \bar{Y})$ must be on the regression line. Also, the estimated residuals must add up to 0 (this is because of the application of the Principle of Least Squares method). If one of these conditions is not true, then a calculation error must have occurred.

The calculations in these examples illustrate the fact that the numerical results of correlation analysis versus regression analysis are identical as far as the pivotal statistic is concerned. That does not mean we can just use one for both types of designs, because the interpretation of what is under consideration is different, as are the assumptions. In the fixed effects model, we treat the X values as constants, and when we compute expected values, the fact that they are constants is critical to the calculations. In the random effects model, we are assuming a bivariate distribution, in particular a bivariate normal distribution. Calculating expected values is more complicated because we have two random variables. The value ρ characterizes independence, and that is why tests of ρ are the correct ones in the correlation model.

The test of whether there is evidence of a statistical linear relationship between X and Y can be expressed in an ANOVA table, as follows:

Source	d.o.f.	SS	MS	F
Regression	1	SSR	SSR/1	MSR/MSE
Error	n−2	SSE	SSE/(n−2)	
Total	n−1	SST		

This comes from partitioning the total sum of squares into a component related to error (essentially, what variability is left after fitting the estimated regression equation) and a sum of squares related to the regression. The total sum of squares has $n - 1$ degrees of freedom, since we estimate a global mean; the sum of squares for error has $n - 2$ degrees of freedom, since there are two parameters being estimated; and since the degrees of freedom must balance, there is 1 degree of freedom for regression. In particular,

$$SSE = \sum_{i=1}^{n} \hat{\epsilon}_i^2 \text{ and } SST = \sum_{i=1}^{n} (Y_i - \bar{Y})^2 = S_{YY}$$

and then,

$$SSR = SST - SSE.$$

Normally, if this calculation is done by hand, we find SSR by subtraction. It can be shown by algebraic manipulation that in fact $SSR = \hat{\beta}_1^2 S_{XX}$. [Note: $S_{XX} = \sum_{i=1}^{n}(X_i - \bar{X})^2$, $S_{YY} = \sum_{i=1}^{n}(Y_i - \bar{Y})^2$ and $S_{XY} = \sum_{i=1}^{n}(X_i - \bar{X})(Y_i - \bar{Y})$.] Also, in the fixed effects model, it can be shown that $E(MSE) = \sigma_e^2$ and $E(MSR) = \sigma_e^2 + \beta_1^2 S_{XX}$. Further, MSR and MSE are, except for certain multiplicative constants, independent χ^2 variables with the degrees of freedom indicated, so their ratio is $F(1, n - 2)$ under the assumption of independence. We can see that large values of the pivot F will lead to rejection of the null hypothesis, and so we will use the right tail of the F distribution to calculate p-values and critical values.

There are some plots used to help determine if our analytic procedures are valid. For example, we can look at the estimated residuals as a random sample of the error distribution. We know that MSE is unbiased for σ_e^2, so we can, at least approximately, standardize the residuals and the result should be close to a standard normal random sample. On ordering the estimated standardized residuals, and then plotting them versus the appropriate quantiles of the standard normal, the values should fit a straight line. This is sometimes called a Q-Q plot. Another visual display that can be helpful is the plot of the standardized residuals versus the fitted Y values. A possibility is that the variance in the residuals gets larger as the Y values get larger, which would mean the variance is not constant. There are many other plots, and in a full regression analysis, some of the model verification plots should be undertaken to see if the assumptions hold, or at least are reasonable.

As an exercise, the reader should go through the ANOVA analysis and verify the same conclusion is reached for the groundhog data in Example 2.

It is possible that we have in mind a certain level for ρ already, i.e., $H_0 : \rho = \rho_0 \neq 0$, and then we would use the Fisher z transformation and the corresponding normal distribution approximation. Even for relatively small samples, this approximation is fairly good. As an exercise, the interested reader could run the test $H_0 : \rho = 0.8$ vs $H_1 : \rho \neq 0.8$ for Example 2.

In Example 1, we have taken into account one aspect of what might affect Y. We could equally want to incorporate the local field conditions, the compactness of the soil, weather factors during the spring, etc. We can then build into the model multiple regression of Y on several regressor variables, including discrete and continuous variables simultaneously.

One final note in this section: We have presented the information as if there was a single observation on Y for a given level of X in using the notation (X_i, Y_i) to represent the sample. This need not be the case. We may have, say, four levels of X and then make several observations of Y for these specific X values. It is only for notational simplicity that we have used the particular notation for samples. If we in fact do have several observations on Y

Chapter 11 — Linear Regression

for one or more X values, we can run a test called the **lack of fit test**. In this case, we can partition SSE further to within-factor error (that is, within each level of X), and then the remainder is between the different levels of X. If the model fits the data well, these two terms should represent the same thing, namely estimates of the variance of the error terms. As an exercise, the reader should find information on this procedure and use the general concepts introduced in this section to understand how it is run.

11.2 PREDICTION: NEW VALUES FROM OLD

We will briefly describe three problems in this area. First, suppose we want to estimate the mean value of Y for a new value of X, call it x_0. This is assumed a value for which we haven't observed Y in the regression analysis. Thus, we want $\hat{\mu}_{Y|X=x_0}$. Since all the means are supposed to fit on the regression line, our best point estimate must be $\hat{\mu}_{Y|X=x_0} = \hat{\beta}_0 + \hat{\beta}_1 x_0$. This is a normally distributed variable, and it is readily apparent that it is unbiased. The only other quantity we need is the variance to completely specify the distribution, and in this case,

$$var(\hat{\mu}_{Y|X=x_0}) = \frac{\sigma_e^2}{n} + (x_0 - \bar{X})^2 \frac{\sigma_e^2}{S_{XX}}.$$

As $n \to \infty$ this tends to 0, because $S_{XX} \to \infty$. For finite n, the smallest variance occurs when $x_0 = \bar{X}$ and then gets progressively larger in each direction away from the mean. One general observation here is that we should not use the regression line in prediction or estimation beyond the extreme X values observed. The variability increases, as indicated by the above, but also we may very well have a reasonably straight regression curve between the extremes of X, but very different shapes beyond the extremes. Since we don't have information beyond the limits of X, we should refrain from using the regression equation for prediction beyond the X extremes.

The above method gives us a way to estimate what we would see on average for a new value of X, along with a way to measure the variability. Instead, suppose that we plan to run the experiment at $X = x_0$, but just once. We would like to have a prediction of what we would see, and also how variable this observation will be. In this case, our best point estimate is still the fitted mean, so:

$$\hat{Y}_0 = \hat{\beta}_0 + \hat{\beta}_1 x_0.$$

The variance calculation, however, is different from estimating the mean, because we have to incorporate the variance of a single observation together with the variance in the fitted mean. What we really should use as the expression for the predicted value is:

$$\hat{Y}_0 = \hat{\beta}_0 + \hat{\beta}_1 x_0 + \epsilon_0$$

where ϵ_0 is the residual, just as in the general regression model. We have $E(\epsilon_0) = 0$ and $Var(\epsilon_0) = \sigma_e^2$ since this is just another residual, and they are all assumed to have the same distribution. Further, this new residual will be independent of all previous residuals (if the experiment is carried out properly), so:

$$var(\hat{Y}_0) = var(\hat{\beta}_0 + \hat{\beta}_1 x_0 + \epsilon_0) = var(\hat{\beta}_0 + \hat{\beta}_1 x_0) + var(\epsilon_0) = \sigma_e^2 \left(1 + \frac{1}{n} + \frac{(x_0 - \bar{X})^2}{S_{XX}}\right).$$

Again, under our assumptions, we know the distribution of $\hat{Y}_0$, and can relate it to a t variable when we use MSE as our estimate of σ_e^2. Thus, we can find confidence intervals for a predicted new value for $X = x_0$.

The last problem is to determine a prediction interval for a sample mean $\bar{Y}_0$ if we were to run the experiment at $X = x_0$, say $k \geq 1$ times. This really is just a generalization of the problem above, and it is left to the reader to show that the best point estimate is still:

$$\hat{\bar{Y}}_0 = \hat{\beta}_0 + \hat{\beta}_1 x_0$$

and the variance of this estimator is:

$$\text{var}(\hat{Y}_0) = \sigma_e^2 \left(\frac{1}{k} + \frac{1}{n} + \frac{(x_0 - \bar{X})^2}{S_{XX}} \right).$$

Exercises

1. A particular bacteria culture and its rate of growth is studied under four different temperatures: 40°, 50°, 60°, and 70° Fahrenheit. Similar-sized cultures are selected, 16 separate cultures are prepared, and four are randomly selected to be placed in temperature-controlled environments at the different temperatures. The percentage growth in each culture is recorded after five days in their environment, with the following results:

Temperature Level							
1	2	3	4	1	2	3	4
4.78	5.25	9.83	14.34	4.54	6.84	8.84	17.00
5.15	7.29	9.07	14.78	4.80	5.81	9.16	14.96
6.84	5.73	8.67	16.26	8.30	6.98	8.40	14.06

(a) Determine if this data should be analyzed as a fixed effects or a random effects model.

(b) Determine the estimated regression equation.

(c) Carry out the appropriate test of independence of ambient temperature and percentage growth in the bacteria cultures.

(d) Calculate the residuals, and use them to give visual evidence for or against the assumption of normality and the assumption of homoskedasticity in the residuals.

(e) Find a 95% confidence interval for the percentage growth in a culture if the temperature is set at 65°.

2. Four diets to help overweight adult Siamese cats reduce have 0%, 1%, 2%, and 3% of an extra ingredient. It is thought that increasing the special ingredient will increase weight loss on average. To determine the effects of the formulae, six overweight (but otherwise healthy) adult Siamese cats were selected to be given one of the formulae, with six different cats selected for each of the other formulae. The weight loss in grams was recorded for the 24 cats as:

Diet Formula							
16.6	19.7	15.8	20.0	18.7	15.9	17.2	21.0
15.8	17.7	20.8	15.4	11.2	13.8	21.1	21.2
17.1	15.6	14.9	18.7	18.1	18.9	21.3	15.0

The first two columns are the weight losses under the first formula, the second two columns are under the second formula, etc.

(a) Determine if this data should be analyzed as a fixed effects or a random effects model.

(b) Determine the estimated regression equation.

(c) Carry out the appropriate test of independence of diet and weight loss in the Siamese cats under study.

Chapter 11 — Linear Regression

(d) Calculate the residuals, and use them to give visual evidence for or against the assumption of normality, and the assumption of homoskedasticity.

(e) Find a 95% confidence interval for the average weight loss in 10 Siamese cats if a mixture including 2.5% of the extra ingredient is used.

3. It is believed that the bigger the flower, the longer a honeybee will stay with the flower. In a study, 10 flowers in a bed of petunias were selected at random, then observed for an hour on an early summer morning. The total number of seconds of visits by honeybees to each were recorded, along with the radius of the flowers in centimeters, with the results:

Diameter	10.4	8.6	10.2	9.5	10.1	13.0	13.3	11.3	11.3	8.7
Time	23.1	25.0	26.1	28.0	24.1	27.0	25.1	24.7	25.9	24.4

Diameter	8.6	8.7	9.7	9.5	10.8	8.2	8.6	8.0	8.3	11.9
Time	26.3	24.2	25.8	24.3	23.7	30.3	24.9	25.3	24.7	23.9

(a) Determine if this data should be analyzed as a fixed effects or a random effects model.
(b) Determine the estimated regression equation.
(c) Carry out the appropriate test of independence of size of flower and length of visit.
(d) Calculate the residuals, and use them to give visual evidence for or against the assumption of normality, and the assumption of homoskedasticity.
(e) Carry out the analysis of variance related to the null hypothesis of independence of length of visit and size of flower.

4. The weight of eight-month-old blue jays is thought to be related to their wingspan. Twenty-five such blue jays, observed from birth, are selected and their wingspans in cm and weight in grams recorded as:

X	39.75	40.21	41.22	41.58	39.52	40.09	41.32	43.36	40.88	43.19
Y	68.9	69.1	69.6	69.8	68.8	69.0	69.8	70.7	69.4	70.6
X	43.3	39.9	38.7	43.2	42.2	42.7	40.4	49.2	44.2	41.7
Y	70.6	68.9	68.3	70.6	70.1	70.4	69.2	69.6	71.1	69.8
X	41.8	42.7	40.5	43.0	42.8					
Y	69.9	70.4	69.3	70.5	70.4					

(a) Determine if this data should be analyzed as a fixed effects or a random effects model.
(b) Determine the estimated regression equation.
(c) Carry out the appropriate test of independence of wingspan and weight of eight-month-old blue jays.
(d) Calculate the residuals, and use them to give visual evidence for or against the assumption of normality, and the assumption of homoskedasticity.
(e) Find a 95% confidence interval for the correlation coefficient between weight and wingspan.
(f) Carry out the analysis of variance in testing the independence of weight and wingspan.

5. The mass of eggs, measured in hundreds of eggs in the mass, laid by a type of frog is thought to be adversely affected by the oxygen level in the local pond where the eggs are laid. The oxygen level is measured in parts per million, and 20 random observations of egg masses laid in a pond and the oxygen level of the pond were recorded as follows:

X	6.5	5.4	6.0	5.9	6.7	6.7	6.8	7.0	8.7	5.8
Y	15.0	16.3	15.6	15.6	14.8	15.9	15.8	14.4	12.5	15.8
X	6.0	7.7	7.1	7.2	7.1	7.3	5.4	6.7	6.0	6.6
Y	15.5	13.7	14.4	14.3	14.4	14.1	16.2	14.8	15.5	14.9

(a) Determine if this data should be analyzed as a fixed effects or a random effects model.

(b) Determine the estimated regression equation.

(c) Carry out the appropriate test of independence of pond oxygen concentration and mass of eggs laid by this type of frog.

(d) Calculate the residuals, and use them to give visual evidence for or against the assumption of normality, and the assumption of homoskedasticity.

(e) Determine a 95% symmetric confidence interval for the correlation coefficient between the mass of eggs and the oxygen concentration.

(f) Carry out the analysis of variance for testing independence of oxygen concentration and mass of eggs laid.

Appendix A: Notation

The Greek Alphabet

Lower Case	Upper Case	Name	Lower Case	Upper Case	Name
α	A	alpha	ν	N	nu
β	B	beta	ξ	Ξ	xi
γ	Γ	gamma	o	O	omicron
δ	Δ	delta	π	Π	pi
ϵ	E	epsilon	ρ	P	rho
ζ	Z	zeta	σ	Σ	sigma
η	H	eta	τ	T	tau
θ	Θ	theta	υ	Υ	upsilon
ι	I	iota	ϕ	Φ	phi
κ	K	kappa	χ	X	chi
λ	Λ	lambda	ψ	Ψ	psi
μ	M	mu	ω	Ω	omega

Sigma Notation

Let f be a function of some set of consecutive integers, say $n_0, \ldots, n_1$, where $n_0 \leq n_1$. The notation for summing up the values of f on this set of integers is:

$$\sum_{i=n_0}^{n_1} f(i) = \sum_{i=n_0}^{n_1} f_i = \sum_{j=n_0}^{n_1} f_j.$$

We often replace the formula $f(i)$ by f_i. In some cases, there is no convenient formula, and we are using the subscript to indicate the place in the sum the value f_i occurs. The Greek letter Σ is used to indicate we are forming a sum. The value i is called the index of summation, n_0 is the lower limit of summation, and n_1 is the upper limit of summation. Any sum with the lower limit of summation greater than the upper limit of summation is not defined. It does not matter the name used for the index of summation: we will get exactly the same value if we replace i by j, or by k, etc. The main consideration is to choose a letter that is not otherwise being used in the sum. This compact notation replaces the more cumbersome:

$$f_{n_0} + f_{n_0+1} + \cdots + f_{n_1-1} + f_{n_1}.$$

It is most often the case that n_0 is either 0 or 1, but sums can start from convenient values other than 0 or 1.

Since n_0 and n_1 are integers, they are finite, and so the above sums are called finite sums. f_i is called a summand. Sometimes, we want to sum an unbounded number of summands. For example, in the negative binomial distribution, where the random variable X counts the number of trials needed to get the kth success, with each trial assumed independent of the other trials, and the probability of a success for all trials is the constant p, the values of X start at k, but there is no upper bound. The population is assumed infinite, so we may need a large number of trials. As seen in Chapter 3, the probability distribution function is:

$$P(X = x) = \binom{x-1}{k-1} p^k (1-p)^{x-k}, \quad x \geq k.$$

The expected value of X uses an infinite sum of the form:

$$E(X) = \sum_{x=k}^{\infty} x \binom{x-1}{k-1} p^k (1-p)^{x-k}.$$

There are situations where doubly infinite sums are necessary and where iterated sums are used.

Common Notation and Terms

$\forall$	means for all.
$\exists$	means there exists.
iff	is the abbreviation for "if and only if".
$\Leftrightarrow$	connects two statements that are equivalent, and is read "if and only if".
$\Rightarrow$	is used between two statements a and b such that knowing a is true implies b is true. $a \Rightarrow b$, and is read "a implies b".
$E()$	is the expectation operator.
$\sum$	is the symbol used to indicate a sum.
$\prod$	is the symbol used to indicate a product.
$\emptyset$	denotes the empty set.
$\cup$	is the symbol for the union of sets.
$\cap$	is the symbol for the intersection of sets.
$\overline{A}$	is the notation for the complement of the set A in a given universal set.
$\in$	is the symbol for an object being an element of a set.
$\{\ |\ \}$	is called set builder notation.
$\subset$	is used to denote that one set's elements are in a second set.
$\subseteq$	is used to denote the case that one set is either a proper subset or exactly equal to another set.
pdf or p.d.f.	is short form for probability distribution function for discrete variables and probability density function for continuous variables.
pmf	is short for probability mass function, and is sometimes used in place of pdf for discrete variables.
cdf	is short for cumulative distribution function.
mgf	is the abbreviation for moment generating function.

Discrete Distributions

The following is a summary of discrete distributions named in this text, along with basic properties such as notation, parameters, domain, pdf, mean = μ, variance = σ^2, and mgf. Throughout, the variable in the mgf is t, and the mgf is denoted $M(t)$. Also, the pdf is assumed 0 at all reals other than the numbers specified. For the multivariate and the multinomial distributions, only the joint pdfs are provided

Uniform: $X \sim U(\{1, 2, .., n\})$ (other sets of integers are possible); pdf: $u(x) = 1/n, x \in \{1, 2, \ldots, n\}$; $\mu = (n+1)/2; \sigma^2 = (n^2 - 1)/12; M(t) = e^t(1 - e^{nt})/[n(1 - e^t)]$.

Bernoulli: $X \sim B(1, \pi)$; pdf: $b(x; 1, \pi) = \binom{1}{x}\pi^x(1-\pi)^{1-x} = \pi^x(1-\pi)^{1-x}, x = 0, 1; \mu = \pi$; $\sigma^2 = \pi(1-\pi); M(t) = 1 - \pi + e^t\pi$.

Binomial: $X \sim B(n, \pi)$; pdf: $b(x; n, \pi) = \binom{n}{x}\pi^x(1-\pi)^{n-x}, x = 0, 1, \ldots, n; \mu = n\pi; \sigma^2 = n\pi(1-\pi)$; $M(t) = (\pi e^t + 1 - \pi)^n$.

Negative Binomial: $X \sim B^*(k, \pi)$; pdf: $b^*(x; k, \pi) = \binom{x-1}{k-1}\pi^k(1-\pi)^{x-k}, x \geq k$ and integer; $\mu = k/\pi$; $\sigma^2 = k(1-\pi)/\pi^2; M(t) = \pi^k e^{kt}(1 - (1-\pi)e^t)^{-k}$.

Appendix A — Notation

Geometric: $X \sim B^*(1, \pi)$; pdf: $b^*(x; 1, \pi) = \pi(1-\pi)^{x-1}$, $x \geq 1$ and integer; $\mu = 1/\pi$; $\sigma^2 = (1-\pi)/\pi^2$; $M(t) = \pi e^t (1-(1-\pi)e^t)^{-1}$.

Hypergeometric: $X \sim H(N, K, n)$; pdf: $h(x; N, K, n) = \binom{K}{x}\binom{N-K}{n-x}/\binom{N}{n}$, whenever all of the binomial coefficients are defined; $\mu = nK/N$; $\sigma^2 = nK(N-K)(N-n)/[N^2(N-1)]$; $M(t)$ exists, but is complicated.

Multivariate Hypergeometric: $(X_1, \ldots, X_k) \sim MH(N, K_1, \ldots, K_k, n)$;
pdf: $mh(x_1, \ldots, x_k; N, K_1, \ldots, K_k, n) = \binom{K_1}{x_1} \cdots \binom{K_k}{x_k}/\binom{N}{n}$.

Multinomial: $(X_1, \ldots, X_k) \sim M(n, \pi_1, \ldots, \pi_k)$; $X_1 + \cdots + X_k = n$; $\pi_1 + \cdots + \pi_k = 1$; $\pi_j \geq 0$;
pdf: $m(x_1, \ldots, x_k; n, \pi_1, \ldots, \pi_k) = \frac{n!}{x_1! \cdots x_k!} \pi_1^{x_1} \cdots \pi_k^{x_k}$.

Poisson: $X \sim P_o(\lambda)$; pdf: $p(x; \lambda) = \lambda^x e^{-\lambda}/x!$, $x \geq 0$ and integer; $\mu = \lambda$; $\sigma^2 = \lambda$; $M(t) = \exp(\lambda(e^t - 1))$.

Continuous Distributions

The following is a summary of special continuous distributions used in this text. It includes: notation, pdf and domain, mean, variance, and moment-generating function, denoted $M(t)$, when it exists. Note that because of continuity, domains may or may not include endpoints of intervals. Also, the density function is defined to be 0 at all other real numbers than those specified for the density. Some fundamental properties of the bivariate normal are given at the end.

Uniform: $X \sim U(\alpha, \beta)$, $\alpha < \beta$; $f(x) = 1/(\beta - \alpha)$, $\alpha < x < \beta$; mean $= (\alpha + \beta)/2$; variance $= (\beta - \alpha)^2/12$; $M(t) = (e^{\beta t} - e^{\alpha t})/[t(\beta - \alpha)]$.

Gamma: $X \sim \Gamma(\alpha, \beta)$, $\alpha, \beta > 0$; $f(x) = \frac{1}{\Gamma(\alpha)\beta^\alpha} x^{\alpha-1} \exp(-x/\beta)$, $x > 0$; mean $= \alpha\beta$; variance $= \alpha\beta^2$; $M(t) = (1 - \beta t)^{-\alpha}$.

Exponential: $X \sim E(\theta)$, $\theta > 0$; $f(x) = \frac{1}{\theta} \exp(-x/\theta)$, $x > 0$; mean $= \theta$; variance $= \theta^2$; $M(t) = (1 - \theta t)^{-1}$.

χ^2 (Chi-Square): $X \sim \chi^2(\nu)$, $\nu > 0$; $f(x) = \frac{1}{2^{\nu/2}\Gamma(\nu/2)} x^{(\nu-2)/2} \exp(-x/\nu)$, $x > 0$; mean $= \nu$; variance $= 2\nu$; $M(t) = (1 - 2t)^{-\nu/2}$.

Beta: $X \sim \text{Beta}(\alpha, \beta)$, $\alpha, \beta > 0$; $f(x) = \frac{1}{B(\alpha, \beta)} x^{\alpha-1}(1-x)^{\beta-1}$, $0 < x < 1$; mean $= \alpha/(\alpha + \beta)$; variance $= \alpha\beta/[(\alpha + \beta)^2(\alpha + \beta + 1)]$; $M(t)$ exists, but is complicated.

Normal: $X \sim N(\mu, \sigma^2)$, $\mu \in \mathbb{R}$, $\sigma > 0$; $f(x) = \frac{1}{\sigma\sqrt{2\pi}} \exp(-(x-\mu)^2/[2\sigma^2])$, $x \in \mathbb{R}$; mean $= \mu$; variance $= \sigma^2$; $M(t) = \exp(\mu t + \sigma^2 t^2/2)$.

Standard Normal: $Z \sim N(0, 1)$; $\phi(z) = \frac{1}{\sqrt{2\pi}} \exp(-z^2/2)$, $z \in \mathbb{R}$; mean $= 0$; variance $= 1$; $M(t) = \exp(t^2/2)$.

Log Normal: $X \sim LN(\mu, \sigma^2)$, $\mu \in \mathbb{R}$, $\sigma > 0$; $f(x) = \frac{1}{x\sigma\sqrt{2\pi}} \exp(-(\ln x - \mu)^2/[2\sigma^2])$, $x > 0$; mean $= \exp(\mu + \sigma^2)$; variance $= (\exp(\sigma^2) - 1)\exp(2\mu + \sigma^2)$; $M(t)$ does not exist.

Student's t: $T \sim t(\nu)$ or $T(\nu)$, $\nu > 0$; $f(t) = \frac{\Gamma((\nu+1)/2)}{\Gamma(\nu/2)\sqrt{\pi\nu}} \left(1 + \frac{t^2}{\nu}\right)^{-(\nu+1)/2}$, $t \in \mathbb{R}$; mean $= 0$ for $\nu > 1$; variance $= \nu/(\nu - 2)$ for $\nu > 2$; $M(t)$ does not exist.

F, or Snedecor's F: $X \sim F(\nu_1, \nu_2)$, $\nu_1, \nu_2 > 0$;

$$f(x) = \frac{(\nu_1/\nu_2)^{\nu_1/2}}{B(\nu_1/2, \nu_2/2)} \frac{x^{\nu_1/2 - 1}}{(1 + \nu_1 x/\nu_2)^{(\nu_1+\nu_2)/2}}, \quad x > 0;$$

mean $= \nu_2/(\nu_2 - 2)$, $\nu_2 > 2$; variance $= \frac{2\nu_2^2(\nu_1 + \nu_2 - 2)}{\nu_1(\nu_2 - 2)^2(\nu_2 - 4)}$, $\nu_2 > 4$; $M(t)$ does not exist.

Bivariate Normal: $(X, Y) \sim BN(\mu_X, \mu_Y, \sigma_X^2, \sigma_Y^2, \rho)$, $\mu_X, \mu_Y \in \mathbb{R}, \sigma_X, \sigma_Y > 0, -1 < \rho < 1$;

$$f(x,y) = \frac{1}{2\pi\sigma_X\sigma_Y\sqrt{1-\rho^2}} \exp(-Q/[2(1-\rho^2)]), \text{ where}$$

$$Q = \left(\frac{x-\mu_X}{\sigma_X}\right)^2 - 2\rho\left(\frac{x-\mu_X}{\sigma_X}\right)\left(\frac{y-\mu_Y}{\sigma_Y}\right) + \left(\frac{y-\mu_Y}{\sigma_Y}\right)^2$$

The conditional distribution of $X|Y = y$ is $N(\mu_X + \rho\frac{\sigma_X}{\sigma_Y}(y - \mu_Y), \sigma_X^2(1 - \rho^2))$.

The marginal distribution of X is $N(\mu_X, \sigma_X^2)$, and of Y is $N(\mu_Y, \sigma_Y^2)$.

The correlation coefficient is ρ with $\rho = cov(X, Y)/\sqrt{\sigma_X\sigma_Y}$. In the bivariate normal, $\rho = 0$ iff X and Y are independent.

Appendix B: Calculus Review

Note: The material in this appendix includes a very brief summary of the ideas of integration, and is limited in scope and application to the essentials for this text. There are many very good texts on univariate and multivariate calculus that include a detailed development of integration and the methods used to evaluate integrals. Similarly, the material on optimization for two or more variables is a summary of the basic techniques used in maximum likelihood and least squares.

It is assumed that students have had a good grounding in differential calculus. Fundamental formulae for calculating derivatives of powers, exponentials, logarithms, products and quotients are provided in the table, **Fundamental Formulae**, just after Example 1 of this Appendix. It is good practice to review results related to derivatives.

Our main interest in integrals is their use in calculating areas under curves over a given range. The general integration problem, however, involves a function f defined on an interval (a, b), $a < b$. The general approach is to break the interval (a, b) into n parts, which for our purposes, can be viewed as of equal length, and introduce the values $x_i = a + (b - a)i/n$, where $0 \leq i \leq n$ are whole numbers. We then select values in each interval, $[x_{i-1}, x_i]$, $1 \leq i \leq n$, call them x_i^*, $1 \leq i \leq n$. We call the value:

$$\sum_{i=1}^{n} f(x_i^*)(x_i - x_{i-1}) = \sum_{i=1}^{n} f(x_i^*) \frac{(b-a)}{n}$$

a Riemann sum. As we let n get larger and larger, and if this sum tends to the same finite value, regardless of how the x_i^* values are chosen in the subintervals, the value is called the integral, or Riemann integral, of f over (a, b). If f is non-negative on the interval, the integral is defined to be the area under f over $[a, b]$. Note that, because the area under a single point is 0, it does not matter whether a or b or both are included in the interval or not. This limiting value is referred to as the **definite integral** of f over (a, b), and we denote it as

$$\int_a^b f(t)dt$$

We call f the **integrand**, b the **upper limit of integration**, a the **lower limit of integration**, $\int$ is the **integral sign**, and dt is called a **differential**. We will also define:

$$\int_b^a f(t)dt = -\int_a^b f(t)dt.$$

The letter t is a dummy variable, in that it can be any other letter not already used, since the value of the integral does not depend on what name we give the variable.

When do Riemann sums for a given function f and interval $[a, b]$ have a limit as described above? As long as f is either continuous or has at most a finite number of jump discontinuities, then the limit of the Riemann sums will be unique, i.e., the Riemann integral exists. For the continuous distributions considered in this text, integrals of densities will exist. The problem is, How are they computed? Fortunately, the Fundamental Theorem of Calculus answers this question.

Finding areas or definite integrals is not complicated in many cases because of the Fundamental Theorem of Calculus (stated below). This theorem links the definite integral problem to finding derivatives. Let f be a function defined on some interval. We call a function F such that $F'(x) = f(x)$ on the interval an **antiderivative** of f on that interval. Thus, an antiderivative of $f(x) = x$ is $F(x) = x^2/2$. We also use the term

indefinite integral: If F is an antiderivative of f, then so is $F(x) + C$ for any constant C, and we call $F(x) + C$ the indefinite integral of f. C is called the **constant of integration**, and we often write $\int f(x)dx = F(x) + C$ for the indefinite integral. With these preliminaries, we can state:

Fundamental Theorem of Calculus: Let f be a continuous function on $[a, b]$. Then:

$$\text{(a)} \quad \frac{d}{dx}\int_a^x f(t)dt = f(x)$$

and if F is any antiderivative of f on the interval,

$$\text{(b)} \quad \int_a^b f(t)dt = F(b) - F(a).$$

The first part says that the derivative of an integral of the given form is the integrand, and also that such integrals are in fact antiderivatives. The second part says that once you have found one antiderivative of f, then it can be used to evaluate the definite integral as per the formula.

Example 1: Find the area under $f(x) = x^2$ over $[2,3]$.

Solution: Basically, we want to solve $\int_2^3 x^2 dx$. By the Fundamental Theorem of Calculus, we need to find a function $F(x)$ such that $F'(x) = x^2$. From differential calculus, we know $\frac{dx^3}{dx} = 3x^2$, so we suspect that an antiderivative will have x^3 as a factor. From the rules for derivatives, a function of the form ax^3 might work, if we choose a correctly. If we form $\frac{d(ax^3)}{dx} = 3ax^2$, we see that taking $a = 1/3$ will do the trick. Thus, an antiderivative is $F(x) = x^3/3$, and again by the Fundamental Theorem, $\int_2^3 x^2 dx = F(3) - F(2) = 3^3/3 - 2^3/3 = 19/3$.

Some rules from differentiation help in solving integrals. Basically, we read the rules "backwards" to get an integration rule:

Fundamental Formulae

Derivative Rule	Integral Rule
$(cf(x))' = cf'(x)$, c a constant	$\int cf(x)dx = c\int f(x)dx$
$(f(x) + g(x))' = f'(x) + g'(x)$	$\int (f(x) + g(x))dx = \int f(x)dx + \int g(x)dx$
$(f(x) - g(x))' = f'(x) - g'(x)$	$\int (f(x) - g(x))dx = \int f(x)dx - \int g(x)dx$
Chain Rule:	**Substitution:**
$(f(g(x)))' = f'(y)g'(x)$, where $y = g(x)$ $= f'(g(x))g'(x)$	$\int f(g(x))g'(x)dx = F(g(x)) + C$, F an antiderivative of f
Product Rule:	**Integration by Parts:**
$(f(x)g(x))' = f'(x)g(x) + f(x)g'(x)$	$\int f(x)g'(x)dx = f(x)g(x) - \int f'(x)g(x)dx$
Quotient Rule:	
$(f(x)/g(x))' = \frac{f'(x)g(x) - f(x)g'(x)}{g^2(x)}$	
Power Rule:	**Power Rule:**
$(x^r)' = rx^{r-1}$, $r \in \mathbb{R}\setminus\{0\}$	$\int x^r dx = (1/(1+r))x^{r+1} + C$, $r \neq -1$
Exponentials and Logarithms	**Exponentials and Logarithms**
$(\ln(x))' = 1/x$, $x > 0$	$\int (1/x)dx = \ln(x) + C$, $x > 0$
$(a^x)' = a^x \ln(a)$	$\int a^x dx = a^x/\ln(a) + C$
$(e^{ax})' = ae^{ax}$	$\int e^{ax}dx = (1/a)e^{ax} + C$

Appendix B — Calculus Review

Note that the derivative of a constant will be 0, and if we interpret $x^0 = 1$, then the power rule can include $r = 0$. When the base of an exponential function is 1, we take $1^x \equiv 1$. Otherwise, we assume the base is positive.

Another important fact concerning definite integrals is this: Let f be **integrable** over an interval containing a, b, c. This means the definite integral exists as a finite number on this interval. Then we can break the integral up as:

$$\int_a^b f(t)dt = \int_a^c f(t)dt + \int_c^b f(t)dt.$$

Note that no assumption is made on which of a, b, c is the largest. In probability and statistics, important functions are often defined in pieces, and this formula allows us to break an integral into pieces corresponding to the changes of formula on different intervals.

Piecewise defined functions occur frequently. Typical examples include:

$$f(x) = \begin{cases} 0 & x < 0 \\ 2x & 0 \leq x \leq 1 \\ 0 & x > 1 \end{cases}$$

$$g(x) = \exp(-|x|) = \begin{cases} \exp(x) & x < 0 \\ \exp(-x) & x \geq 0 \end{cases}$$

and

$$h(x) = \begin{cases} x & 0 \leq x \leq 1 \\ 2 - x & 1 \leq x \leq 2 \\ 0 & \text{otherwise} \end{cases}$$

When we want to evaluate integrals involving such functions, we need to remember which piece of the domain is involved, which in turn tells us what formula to use.

Example 2: For the function h above, evaluate $\int_{1/2}^{3/2} xh(x)dx$.

Solution: $\int_{1/2}^{3/2} xh(x)dx = \int_{1/2}^{1} xh(x)dx + \int_{1}^{3/2} xh(x)dx$, because the function changes definition. Now that we have broken this up into the different pieces, we can use the formula for h on those pieces to complete the problem. $\int_{1/2}^{3/2} xh(x)dx = \int_{1/2}^{1} xh(x)dx + \int_{1}^{3/2} xh(x)dx = \int_{1/2}^{1} x^2 dx + \int_{1}^{3/2} x(2-x)dx = \frac{x^3}{3}\big|_{1/2}^{1} + (x^2 - \frac{x^3}{3})\big|_{1}^{3/2} = 7/12$.

Improper Integration

In probability, we frequently work with a function f that is defined on the entire real line. Because of this, we need to be able to evaluate integrals of functions over infinite ranges, rather than finite ranges. There are also some functions that are not bounded on the interval of interest, yet we want to integrate them as well over a region containing this point of infinite discontinuity.

Examples of Functions on Infinite Regions or Becoming Unbounded at a Point

$$f(x) = \begin{cases} \exp(-ax), & x \geq 0 \\ 0, & \text{otherwise} \end{cases}$$

where a is a positive constant;

$$g(x) = \exp(-x^2), \quad x \in \mathbb{R};$$

$$h(x) = \begin{cases} x^{a-1}(1-x)^{b-1}, & 0 \leq x \leq 1 \\ 0, & \text{otherwise} \end{cases}$$

where a and b are a positive constants; and

$$k(x) = \begin{cases} x^{a-1} \exp(-x), & x \geq 0 \\ 0, & \text{otherwise} \end{cases}$$

where a is a positive constant.

The functions f, g, k have infinite domains, while h may have (depending on the values of a and b) an infinite discontinuity at $x = 0$ or $x = 1$, or both. Note that k may also have an infinite discontinuity at $x = 0$, depending on the size of a.

How do we handle them? Basically, we convert these into limits to infinity or negative infinity (or both), or limits to the finite value(s) where there is an infinite discontinuity. Once this is done, we will have an expression involving an integral over a finite region where the function is bounded. Assuming a reasonable function, we can then apply our usual integration methods and then take the appropriate limits.

There are only a few limit rules that are needed. We have the usual arithmetic of limits, but we also need to know what happens in expressions of the form $u(x)v(x)$ as x gets large, or as x approaches a finite value. This generally comes down to the relative size of divergent factors and factors tending to 0.

Example 3: $\lim_{x \to 0^+} x^2 \ln(x) = 0$.

Note that $\lim_{x \to 0^+} x^2 = 0$ but $\lim_{x \to 0^+} \ln(x) = -\infty$, so we cannot simply substitute 0 into the formula. There is a rule, L'Hospital's Rule, which allows us to evaluate this result in a formal way. However, we will need some simple consequences only. For our purposes, it is enough to know the following domination rules. (1) Assume we have an exponential function of the form $\exp(ax)$ where a is a positive constant, a polynomial, and a logarithm function of the form $\ln(x)$. Then as $x \to \infty$: exponential functions grow faster than polynomials, and polynomials grow faster than logarithms. This means, for example, with $n > 0$:

$$\lim_{x \to \infty} x^n / \exp(ax) = 0, \quad \lim_{x \to \infty} \exp(ax)/x^n = \infty, \quad \lim_{x \to \infty} \ln x / x^n = 0, \text{ etc.}$$

(2) for any $a > 0$, $\lim_{x \to 0^+} x^a \ln(x) = 0$, that is, positive powers of x go to 0 faster than logarithms go to infinity.

Example 4:

$$\int_0^\infty \exp(-x)\, dx = \lim_{w \to \infty} \int_0^w \exp(-x)\, dx = \lim_{w \to \infty} \exp(-x) \big|_0^w = \lim_{w \to \infty} (1 - \exp(-w)) = 1.$$

$$\int_0^1 x^{-1/2}(1-x)^2\, dx = \int_0^1 (x^{-1/2} - 2x^{1/2} + x^{3/2})\, dx = \lim_{w \to 0^+} \int_w^1 (x^{-1/2} - 2x^{1/2} + x^{3/2})\, dx$$

$$= \lim_{w \to 0^+} \left(2x^{1/2} - \frac{4}{3}x^{3/2} + \frac{2}{5}x^{5/2}\right) \Big|_w^1 = \lim_{w \to 0^+} \left[2 - \frac{4}{3} + \frac{2}{5} - \left(2w^{1/2} - \frac{4}{3}w^{3/2} + \frac{2}{5}w^{5/2}\right)\right] = \frac{16}{15}.$$

$$\int_0^\infty x \exp(-2x)\, dx = \lim_{w \to \infty} \int_0^w x \exp(-2x)\, dx = \lim_{w \to \infty} \left(-\frac{1}{2}x \exp(-2x) - \frac{1}{4}\exp(-2x)\right) \Big|_0^w$$

$$= \lim_{w \to \infty} \left[-\frac{1}{2}w \exp(-2w) - \frac{1}{4}\exp(-2w) + \frac{1}{4}\right] = \frac{1}{4}.$$

Appendix B — Calculus Review

A couple of observations may be useful: First, we made use of integration by parts in solving the last integral. Second, there can be combinations of limits needed. And third, there are some integrals we simply cannot solve by antidifferentiation. In particular,

$$\int_b^a \exp(-x^2)dx, \quad \text{where } a \text{ and } b \text{ are real numbers.}$$

It is not obvious, but there is no simple antiderivative we can use, and yet, this particular type of integrand shows up in many applications. It is related to what is called the normal distribution, and the normal is common to many scientific experiments, especially where random errors are mostly small and can be positive or negative. This is not the only case of what is called an intractable integral. A common situation is expressed in the following integral:

$$\int_0^\infty x^{a-1} \exp(-x)dx = \Gamma(a).$$

The Γ function is defined to be the value of this integral in general. Particular cases are well known: $\Gamma(n) = (n-1)!$ when n is a natural number. We also have $\Gamma(a+1) = a\Gamma(a)$, and it can be shown that $\Gamma(1/2) = \sqrt{\pi}$, for example. Because this function and the generating integral occur so frequently, highly accurate tables of values have been compiled and algorithms are built into most statistical and mathematical software.

Some Problems

To be able to work effectively with integrals, it is important to practice. Students should review their first-year course notes on calculus and attempt several problems to remind themselves of the basic methods. Often, algebra is used to manipulate expressions, and a review of factoring, simplifying, exponents, etc., is essential. The following problems are meant as a refresher, and if the reader has difficulties completing them, you should spend more time on review and practice.

Exercises with an * denote problems involving improper integrals.

1. Let $f(x) = e^{2x}$. Evaluate each of the following:

   (a) $\int_0^2 f(x)dx$

   (b)* $\int_{-\infty}^0 f(x)dx$. Note: $\lim_{x \to -\infty} e^{ax} = 0$ for any $a > 0$.

   (c) $\int_{-3}^3 f(x)dx$

2. Determine k in each of the following so that the value of the definite integral over the region where the function is non-negative is 1:

   (a) $f(x) = kx^2(1-x)$ for $0 \le x \le 1$, and $f(x) = 0$ otherwise.

   (b) $f(x) = kx(1-x)^2$ for $0 \le x \le 1$, and $f(x) = 0$ otherwise.

   (c)* $f(x) = ke^{-3x}$ for $x \ge 0$, and $f(x) = 0$ otherwise.

3. Let $f(x) = k$, with $k > 0$, for $a \le x \le b$ ($a < b$), and $f(x) = 0$ otherwise. Determine (in terms of a and b) the value of k that makes the definite integral of f from a to b equal to 1.

4. Let $f(x) = 1/2$, $0 \le x \le 2$, and $f(x) = 0$ otherwise. Determine $\int_{-\infty}^\infty xf(x)dx = \mu$ and $\int_{-\infty}^\infty (x-\mu)^2 f(x)dx$, where μ is the value of the first integral.

5*. Let $f(x) = 5e^{-5x}$, $x \ge 0$, and $f(x) = 0$ otherwise. Determine $\int_{-\infty}^\infty xf(x)dx = \mu$ and $\int_{-\infty}^\infty (x-\mu)^2 f(x)dx$, where μ is the value of the first integral.

6. (a) Let $f(x) = x$, $0 \leq x \leq 1$, $f(x) = 2 - x$, $1 \leq x \leq 2$, and $f(x) = 0$ otherwise. Determine $\int_{-\infty}^{\infty} xf(x)dx = \mu$ and $\int_{-\infty}^{\infty}(x - \mu)^2 f(x)dx$, where μ is the value of the first integral.

   (b) Let $f(x) = 20x^3(1 - x)$, $0 \leq x \leq 1$, and $f(x) = 0$ otherwise. Determine $\int_{-\infty}^{\infty} xf(x)dx = \mu$ and $\int_{-\infty}^{\infty}(x - \mu)^2 f(x)dx$, where μ is the value of the first integral.

7. (a) Let $f(x) = k/x$, $1 \leq x \leq 5$, and 0 otherwise. Determine the value of k that makes $\int_{-\infty}^{\infty} f(x)dx = 1$

   (b) Determine $\int_{-\infty}^{\infty} xf(x)dx = \mu$ and $\int_{-\infty}^{\infty}(x - \mu)^2 f(x)dx$, where μ is the value of the first integral.

   (c) Evaluate $\int_2^3 f(x)dx$, to two decimals of accuracy.

8. Let $f(x) = 4x^3$, $0 \leq x \leq 1$ and 0 otherwise.

   (a) Determine the value M_2 so that $\int_{-\infty}^{M_2} f(x) = 1/2$.

   (b) Determine the value P_{80} so that $\int_{-\infty}^{P_{80}} f(x) = 0.80$.

   (c) Determine the value M_1 so that $\int_{-\infty}^{M_1} f(x) = 1/4$.

9. Let $f(x) = 5e^{-5x}$, $x \geq 0$, and $f(x) = 0$ otherwise.

   (a) Determine the value M_2 so that $\int_{-\infty}^{M_2} f(x) = 1/2$.

   (b) Determine M_1 and M_3 so that $\int_{-\infty}^{M_1} f(x) = 1/4$ and $\int_{-\infty}^{M_3} f(x) = 3/4$, respectively. Use two decimals of accuracy.

10. Define for a given function f:

$$M(t) = \int_{-\infty}^{\infty} e^{tx} f(x) dx$$

   (a) Using integration by parts, find $M(t)$ when $f(x) = 2x$, $0 \leq x \leq 1$, and 0 otherwise.

   (b)* Find $M(t)$ when $f(x) = 3e^{-3x}$, $x \geq 0$ and 0 otherwise.

Optimization

Univariate Optimization

Let f be a function of a single variable x on a closed (and finite) interval. Let the interval be $I = [a,b]$. The Extreme Value Theorem says that a continuous function on a closed interval must attain its maximum and must attain its minimum. The procedure to find the max/min values is to solve $f'(x) = 0$ on (a, b) and determine any values $x \in (a, b)$ for which $f'(x)$ is undefined. These values of x are called critical numbers. Then the largest in the set $\{f(\text{critical numbers}), f(\text{end points})\}$ is the maximum value, and the smallest is the minimum. The Second Derivative Test for relative extrema when the second derivative exists is: if $f''(c.n.) < 0$, then f has a relative maximum at the critical number; if $f''(c.n.) > 0$, then f has a relative minimum at the critical number; and if $f''(c.n.) = 0$ the test is inconclusive.

Bivariate Optimization

The ideas of optimization in one variable extend to functions of more than one variable. In this book, optimization of functions of two variables is all that is needed. Thus, let $z = f(x,y)$ be a function of two variables on some domain. We assume this function has all second partial derivatives continuous on the given domain.

Partial Derivatives: For a function of two variables $z = f(x,y)$, if we treat y as a constant and then differentiate with respect to x, the result is called the partial derivative of z with respect to x. If we hold x fixed, we can partially differentiate with respect to y as well. Once we have computed these first-order partials, denoted z_x and z_y, these two functions can also be partially differentiated with respect to x or y. These are called the second

Appendix B — Calculus Review

partials, or second-order partials, and are denoted z_{xx}, z_{xy}, z_{yx}, z_{yy}. If these are all continuous functions, as stated in the assumption above, then $z_{xy} = z_{yx}$. All of the standard univariate differentiation rules apply.

Example 5: Let $z = x\exp(xy) + \sin(y)$. Then: $z_x = \exp(xy) + xy\exp(xy)$; $z_y = x^2\exp(xy) + \cos(y)$; and $z_{xy} = 2x\exp(xy) + x^2 y\exp(xy) = z_{yx}$.

For any function $z = f(x, y)$ that has continuous second partial derivatives on an appropriate domain (details are omitted — this involves determining what is a closed domain in two dimensions), then to determine relative extrema:

1. Calculate z_x and z_y.
2. Solve the equations $z_x = 0$ and $z_y = 0$ simultaneously. Denote the results as ordered pairs (a, b), and call them critical points.
3. Define the discriminant Δ by $\Delta = z_{xx} z_{yy} - z_{xy}^2$.
4. (a) If $\Delta(a, b) > 0$ and $z_{xx}(a, b) < 0$ (that is, the values of these functions at the point (a, b)), then z has a relative maximum at (a, b). (b) If $\Delta(a, b) > 0$ and $z_{xx}(a, b) > 0$, then z has a relative minimum at (a, b). (c) If $\Delta(a, b) < 0$, then z has a saddle point (neither a maximum nor a minimum) at (a, b). (d) If $\Delta(a, b) = 0$, the test fails and other methods must be used to determine the nature of (a, b).

Example 6: Let $z = 7x + 5y + x^2 + y^2$. Then $z_x = 7 + 2x$, $z_y = 5 + 2y$, $z_{xx} = 2$, $z_{yy} = 2$, and $z_{xy} = 0$. The simultaneous solution of $z_x = 0$ and $z_y = 0$ is the single point $(-7/2, -5/2)$. Note that $\Delta = 4 > 0$, so by the second derivative test, this must yield a relative minimum.

Exercises

For each of the following functions, determine all critical points, and apply the second derivative test.

1. $z = 2xy + 100/x + 100/y$.
2. $z = -(x + y)\exp(-(x^2 + y^2)/2)$.
3. $z = (x - 2)^2 + 3(y - 1)^2 + (x - y + 1)^2$.

Appendix C: Statistical Tables

1. Values of the Cumulative Distribution Function of the Standard Normal Distribution
2. Quantiles of Student's $t(\nu)$
3. Quantiles of $\chi^2(\nu)$
4. Quantiles of $F(\nu_1, \nu_2)$

Cumulative Distribution Function $\Phi(z)$ of the Normal, $z \geq 0.00$

z	0.00	0.01	0.02	0.03	0.04	0.05	0.06	0.07	0.08	0.09
0.0	0.5000	0.5040	0.5080	0.5120	0.5160	0.5199	0.5239	0.5279	0.5319	0.5359
0.1	0.5398	0.5438	0.5478	0.5517	0.5557	0.5596	0.5636	0.5675	0.5714	0.5753
0.2	0.5793	0.5832	0.5871	0.5910	0.5948	0.5987	0.6026	0.6064	0.6103	0.6141
0.3	0.6179	0.6217	0.6255	0.6293	0.6331	0.6368	0.6406	0.6443	0.6480	0.6517
0.4	0.6554	0.6591	0.6628	0.6664	0.6700	0.6736	0.6772	0.6808	0.6844	0.6879
0.5	0.6915	0.6950	0.6985	0.7019	0.7054	0.7088	0.7123	0.7157	0.7190	0.7224
0.6	0.7257	0.7291	0.7324	0.7357	0.7389	0.7422	0.7454	0.7486	0.7517	0.7549
0.7	0.7580	0.7611	0.7642	0.7673	0.7704	0.7734	0.7764	0.7794	0.7823	0.7852
0.8	0.7881	0.7910	0.7939	0.7967	0.7995	0.8023	0.8051	0.8078	0.8106	0.8133
0.9	0.8159	0.8186	0.8212	0.8238	0.8264	0.8289	0.8315	0.8340	0.8365	0.8389
1.0	0.8413	0.8438	0.8461	0.8485	0.8508	0.8531	0.8554	0.8577	0.8599	0.8621
1.1	0.8643	0.8665	0.8686	0.8708	0.8729	0.8749	0.8770	0.8790	0.8810	0.8830
1.2	0.8849	0.8869	0.8888	0.8907	0.8925	0.8944	0.8962	0.8980	0.8997	0.9015
1.3	0.9032	0.9049	0.9066	0.9082	0.9099	0.9115	0.9131	0.9147	0.9162	0.9177
1.4	0.9192	0.9207	0.9222	0.9236	0.9251	0.9265	0.9279	0.9292	0.9306	0.9319
1.5	0.9332	0.9345	0.9357	0.9370	0.9382	0.9394	0.9406	0.9418	0.9429	0.9441
1.6	0.9452	0.9463	0.9474	0.9484	0.9495	0.9505	0.9515	0.9525	0.9535	0.9545
1.7	0.9554	0.9564	0.9573	0.9582	0.9591	0.9599	0.9608	0.9616	0.9625	0.9633
1.8	0.9641	0.9649	0.9656	0.9664	0.9671	0.9678	0.9686	0.9693	0.9699	0.9706
1.9	0.9713	0.9719	0.9726	0.9732	0.9738	0.9744	0.9750	0.9756	0.9761	0.9767
2.0	0.9772	0.9778	0.9783	0.9788	0.9793	0.9798	0.9803	0.9808	0.9812	0.9817
2.1	0.9821	0.9826	0.9830	0.9834	0.9838	0.9842	0.9846	0.9850	0.9854	0.9857
2.2	0.9861	0.9864	0.9868	0.9871	0.9875	0.9878	0.9881	0.9884	0.9887	0.9890
2.3	0.9893	0.9896	0.9898	0.9901	0.9904	0.9906	0.9909	0.9911	0.9913	0.9916
2.4	0.9918	0.9920	0.9922	0.9925	0.9927	0.9929	0.9931	0.9932	0.9934	0.9936

z	0.00	0.01	0.02	0.03	0.04	0.05	0.06	0.07	0.08	0.09
2.5	0.9938	0.9940	0.9941	0.9943	0.9945	0.9946	0.9948	0.9949	0.9951	0.9952
2.6	0.9953	0.9955	0.9956	0.9957	0.9959	0.9960	0.9961	0.9962	0.9963	0.9964
2.7	0.9965	0.9966	0.9967	0.9968	0.9969	0.9970	0.9971	0.9972	0.9973	0.9974
2.8	0.9974	0.9975	0.9976	0.9977	0.9977	0.9978	0.9979	0.9979	0.9980	0.9981
2.9	0.9981	0.9982	0.9982	0.9983	0.9984	0.9984	0.9985	0.9985	0.9986	0.9986
3.0	0.9987	0.9987	0.9987	0.9988	0.9988	0.9989	0.9989	0.9989	0.9990	0.9990
3.1	0.9990	0.9991	0.9991	0.9991	0.9992	0.9992	0.9992	0.9992	0.9993	0.9993
3.2	0.9993	0.9993	0.9994	0.9994	0.9994	0.9994	0.9994	0.9995	0.9995	0.9995
3.3	0.9995	0.9995	0.9995	0.9996	0.9996	0.9996	0.9996	0.9996	0.9996	0.9997
3.4	0.9997	0.9997	0.9997	0.9997	0.9997	0.9997	0.9997	0.9997	0.9997	0.9998
3.5	0.9998	0.9998	0.9998	0.9998	0.9998	0.9998	0.9998	0.9998	0.9998	0.9998
3.6	0.9998	0.9998	0.9999	0.9999	0.9999	0.9999	0.9999	0.9999	0.9999	0.9999
3.7	0.9999	0.9999	0.9999	0.9999	0.9999	0.9999	0.9999	0.9999	0.9999	0.9999
3.8	0.9999	0.9999	0.9999	0.9999	0.9999	0.9999	0.9999	0.9999	0.9999	0.9999
3.9	1.000	1.000	1.000	1.000	1.000	1.000	1.000	1.000	1.000	1.000

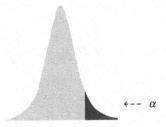

Quantiles of Student's t Distribution
The Upper $\alpha \times 100$ Percentage Points of Student's t Distribution

DoF	$\alpha=0.10$	0.05	0.025	0.01	0.005
1	3.078	6.314	12.71	31.82	63.66
2	1.886	2.920	4.303	6.965	9.925
3	1.638	2.353	3.182	4.541	5.841
4	1.533	2.132	2.776	3.747	4.604
5	1.476	2.015	2.571	3.365	4.032
6	1.440	1.943	2.447	3.143	3.707
7	1.415	1.895	2.365	2.998	3.499
8	1.397	1.860	2.306	2.896	3.355
9	1.383	1.833	2.262	2.821	3.250
10	1.372	1.812	2.228	2.764	3.169
11	1.363	1.796	2.201	2.718	3.106
12	1.356	1.782	2.179	2.681	3.055
13	1.350	1.771	2.160	2.650	3.012
14	1.345	1.761	2.145	2.624	2.977
15	1.341	1.753	2.131	2.602	2.947
16	1.337	1.746	2.120	2.583	2.921
17	1.333	1.740	2.110	2.567	2.898
18	1.330	1.734	2.101	2.552	2.878
19	1.328	1.729	2.093	2.539	2.861
20	1.325	1.725	2.086	2.528	2.845

DoF	$\alpha=0.10$	0.05	0.025	0.01	0.005
21	1.323	1.721	2.080	2.518	2.831
22	1.321	1.717	2.074	2.508	2.819
23	1.319	1.714	2.069	2.500	2.807
24	1.318	1.711	2.064	2.492	2.797
25	1.316	1.708	2.060	2.485	2.787
26	1.315	1.706	2.056	2.479	2.779
27	1.314	1.703	2.052	2.473	2.771
28	1.313	1.701	2.048	2.467	2.763
29	1.311	1.699	2.045	2.462	2.756
30	1.310	1.697	2.042	2.457	2.750
40	1.303	1.684	2.021	2.423	2.704
50	1.299	1.676	2.008	2.403	2.678
100	1.290	1.661	1.984	2.364	2.626
∞	1.22	1.645	1.960	3.326	2.576

Appendix C — Statistical Tables

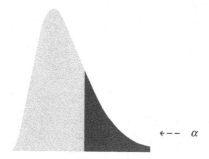

The Upper $\alpha \times 100$ Percentage Points of $\chi^2(\nu)$

ν	$\alpha =$ 0.995	0.99	0.975	0.95	0.90	0.10	0.05	0.025	0.01	0.005
1	0.000093	0.000157	0.000982	0.00393	0.0158	2.71	3.84	5.02	6.64	7.88
2	0.0100	0.0201	0.0506	0.103	0.211	4.61	5.99	7.38	9.21	10.60
3	0.0717	0.115	0.216	0.352	0.584	6.251	7.815	9.348	11.34	12.84
4	0.0207	0.297	0.484	0.711	1.064	7.779	9.488	11.14	13.28	14.86
5	0.412	0.554	0.831	1.145	1.610	9.236	11.07	12.83	15.09	16.75
6	0.676	0.872	1.237	1.635	2.204	10.64	12.59	14.45	16.81	18.55
7	0.989	1.239	1.690	2.167	2.833	12.02	14.07	16.01	18.48	20.28
8	1.344	1.646	2.180	2.733	3.490	13.36	15.51	17.53	20.09	21.95
9	1.735	2.088	2.700	3.325	4.168	14.68	16.92	19.02	21.67	23.59
10	2.156	2.558	3.247	3.940	4.865	15.99	18.31	20.48	23.21	25.19
11	2.603	3.053	3.816	4.575	5.578	17.28	19.68	21.92	24.72	26.76
12	3.074	3.571	4.404	5.226	6.304	18.55	21.03	23.34	26.22	28.30
13	3.565	4.107	5.009	5.892	7.042	19.81	22.36	24.74	27.69	29.82
14	4.075	4.660	5.629	6.571	7.790	21.06	23.68	26.12	29.14	31.32
15	4.601	5.229	6.262	7.261	8.547	22.31	25.00	27.49	30.58	32.80
16	5.142	5.812	6.908	7.962	9.312	23.54	26.30	28.85	32.00	34.27
17	5.697	6.408	7.564	8.672	10.09	24.77	27.59	30.19	33.41	35.72
18	6.265	7.015	8.231	9.390	10.86	25.99	28.87	31.53	34.81	37.16
19	6.844	7.633	8.907	10.12	11.65	27.20	30.14	32.85	36.19	38.58
20	7.434	8.260	9.591	10.85	12.44	28.41	31.41	34.17	37.57	40.00
21	8.034	8.897	10.28	11.59	13.24	29.62	32.67	35.48	38.93	41.40
22	8.643	9.542	10.98	12.34	14.04	30.81	33.92	36.78	40.29	42.80
23	9.260	10.20	11.69	13.09	14.85	32.01	35.17	38.08	41.64	44.18
24	9.886	10.86	12.40	13.85	15.66	33.20	36.42	39.36	42.98	45.56
25	10.52	11.52	13.12	14.61	16.47	34.38	37.65	40.65	44.31	46.93
26	11.16	12.20	13.84	15.38	17.29	35.56	38.89	41.92	45.64	48.29
27	11.81	12.88	14.57	16.15	18.11	36.74	40.11	43.19	46.96	49.64
28	12.46	13.56	15.31	16.93	18.94	37.92	41.34	44.46	48.28	50.99
29	13.12	14.26	16.05	17.71	19.77	39.09	42.56	45.72	49.59	52.34
30	13.79	14.95	16.79	18.49	20.60	40.26	43.77	46.98	50.89	53.67

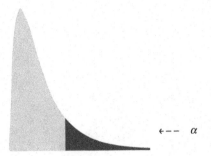

The Upper $\alpha \times 100$ Percentage Points of $F(v_1, v_2)$: $\alpha = 0.10$

Note: It can be shown that $F_\alpha(v_1, v_2) = 1/F_{1-\alpha}(v_2, v_1)$ so that lower percentages points can be determined from upper percentage points.

v_1	1	2	3	4	5	6	7	8	9	10
v_2										
1	39.86	49.50	53.59	55.83	57.24	58.20	58.91	59.44	59.86	60.19
2	8.526	9.000	9.162	9.243	9.293	9.326	9.349	9.367	9.381	9.392
3	5.538	5.462	5.391	5.343	5.309	5.285	5.266	5.252	5.240	5.230
4	4.545	4.325	4.191	4.107	4.051	4.010	3.979	3.955	3.936	3.920
5	4.060	3.780	3.619	3.520	3.453	3.405	3.368	3.339	3.316	3.297
6	3.776	3.463	3.289	3.181	3.108	3.055	3.014	2.983	2.958	2.937
7	3.589	3.257	3.074	2.961	2.883	2.827	2.785	2.752	2.725	2.703
8	3.458	3.113	2.924	2.806	2.726	2.668	2.624	2.589	2.561	2.538
9	3.360	3.006	2.813	2.693	2.611	2.551	2.505	2.469	2.440	2.416
10	3.285	2.924	2.728	2.605	2.522	2.461	2.414	2.377	2.347	2.323
11	3.225	2.860	2.660	2.536	2.451	2.389	2.342	2.304	2.274	2.248
12	3.177	2.807	2.606	2.480	2.394	2.331	2.283	2.245	2.214	2.188
13	3.136	2.763	2.560	2.434	2.347	2.283	2.234	2.195	2.164	2.138
14	3.102	2.726	2.522	2.395	2.307	2.243	2.193	2.154	2.122	2.095
15	3.073	2.695	2.490	2.361	2.273	2.208	2.158	2.119	2.086	2.059
16	3.048	2.668	2.462	2.333	2.244	2.178	2.128	2.088	2.055	2.028
16	3.026	2.645	2.437	2.308	2.218	2.152	2.102	2.061	2.028	2.001
18	3.007	2.624	2.416	2.286	2.196	2.130	2.079	2.038	2.005	1.977
19	2.990	2.606	2.397	2.266	2.176	2.109	2.058	2.017	1.984	1.956
20	2.975	2.589	2.380	2.249	2.158	2.091	2.040	1.999	1.965	1.937
21	2.961	2.575	2.365	2.233	2.142	2.075	2.023	1.982	1.948	1.920
22	2.949	2.561	2.351	2.219	2.128	2.060	2.008	1.967	1.933	1.904
23	2.937	2.549	2.339	2.207	2.115	2.047	1.995	1.953	1.919	1.890
24	2.927	2.538	2.327	2.195	2.103	2.035	1.983	1.941	1.906	1.877
25	2.918	2.528	2.317	2.184	2.092	2.024	1.971	1.929	1.895	1.866
26	2.909	2.519	2.307	2.174	2.082	2.014	1.961	1.919	1.884	1.855
27	2.901	2.511	2.299	2.165	2.073	2.005	1.952	1.909	1.874	1.845
28	2.894	2.503	2.291	2.157	2.064	1.996	1.943	1.900	1.865	1.836
29	2.887	2.495	2.283	2.149	2.057	1.988	1.935	1.892	1.857	1.827
30	2.881	2.489	2.276	2.142	2.049	1.980	1.927	1.884	1.849	1.819

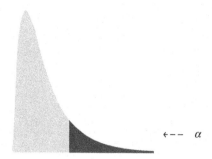

The Upper $\alpha \times 100$ Percentage Points of $F(v_1, v_2): \alpha = 0.10$

v_1	11	12	13	14	15	16	17	18	19	20
v_2										
1	60.47	60.71	60.90	61.07	61.22	61.35	61.46	61.57	61.66	61.74
2	9.401	9.408	9.415	9.420	9.425	9.429	9.433	9.436	9.439	9.441
3	5.222	5.216	5.210	5.205	5.200	5.196	5.193	5.190	5.187	5.184
4	3.907	3.896	3.886	3.878	3.870	3.864	3.858	3.853	3.849	3.844
5	3.282	3.268	3.257	3.247	3.238	3.230	3.223	3.217	3.212	3.207
6	2.920	2.905	2.892	2.881	2.871	2.863	2.855	2.848	2.842	2.836
7	2.684	2.668	2.654	2.643	2.632	2.623	2.615	2.607	2.601	2.595
8	2.519	2.502	2.488	2.475	2.464	2.455	2.446	2.438	2.431	2.425
9	2.396	2.379	2.364	2.351	2.340	2.329	2.320	2.312	2.305	2.298
10	2.302	2.284	2.269	2.255	2.244	2.233	2.224	2.215	2.208	2.201
11	2.227	2.209	2.193	2.179	2.167	2.156	2.147	2.138	2.130	2.123
12	2.166	2.147	2.131	2.117	2.105	2.094	2.084	2.075	2.067	2.060
13	2.116	2.097	2.080	2.066	2.053	2.042	2.032	2.023	2.014	2.007
14	2.073	2.054	2.037	2.022	2.010	1.998	1.988	1.978	1.970	1.962
15	2.037	2.017	2.000	1.985	1.972	1.961	1.950	1.941	1.932	1.924
16	2.005	1.985	1.968	1.953	1.940	1.928	1.917	1.908	1.899	1.891
17	1.978	1.958	1.940	1.925	1.912	1.900	1.889	1.879	1.870	1.862
18	1.954	1.933	1.916	1.900	1.887	1.875	1.864	1.854	1.845	1.837
19	1.932	1.912	1.894	1.878	1.865	1.852	1.841	1.831	1.822	1.814
20	1.913	1.892	1.875	1.859	1.845	1.833	1.821	1.811	1.802	1.794
21	1.896	1.875	1.857	1.841	1.827	1.815	1.803	1.793	1.784	1.776
22	1.880	1.859	1.841	1.825	1.811	1.798	1.787	1.777	1.768	1.759
23	1.866	1.845	1.827	1.811	1.796	1.784	1.772	1.762	1.753	1.744
24	1.853	1.832	1.814	1.797	1.783	1.770	1.759	1.748	1.739	1.730
25	1.841	1.820	1.802	1.785	1.771	1.758	1.746	1.736	1.726	1.718
26	1.830	1.809	1.790	1.774	1.760	1.747	1.735	1.724	1.715	1.706
27	1.820	1.799	1.780	1.764	1.749	1.736	1.724	1.714	1.704	1.695
28	1.811	1.790	1.771	1.754	1.740	1.726	1.715	1.704	1.694	1.685
29	1.802	1.781	1.762	1.745	1.731	1.717	1.705	1.695	1.685	1.676
30	1.794	1.773	1.754	1.737	1.722	1.709	1.697	1.686	1.676	1.667

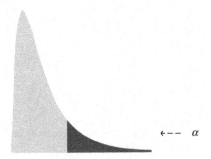

The Upper $\alpha \times 100$ Percentage Points of $F(\nu_1, \nu_2): \alpha = 0.05$

ν_1	1	2	3	4	5	6	7	8	9	10
ν_2										
1	161.4	199.5	215.7	224.6	230.2	234.0	236.8	238.9	240.5	241.9
2	18.51	19.00	19.16	19.25	19.30	19.33	19.35	19.37	19.38	19.40
3	10.13	9.552	9.277	9.117	9.013	8.941	8.887	8.845	8.812	8.786
4	7.709	6.944	6.591	6.388	6.256	6.163	6.094	6.041	5.999	5.964
5	6.608	5.786	5.409	5.192	5.050	4.950	4.876	4.818	4.772	4.735
6	5.987	5.143	4.757	4.534	4.387	4.284	4.207	4.147	4.099	4.060
7	5.591	4.737	4.347	4.120	3.972	3.866	3.787	3.726	3.677	3.637
8	5.318	4.459	4.066	3.838	3.687	3.581	3.500	3.438	3.388	3.347
9	5.117	4.256	3.863	3.633	3.482	3.374	3.293	3.230	3.179	3.137
10	4.965	4.103	3.708	3.478	3.326	3.217	3.135	3.072	3.020	2.978
11	4.844	3.982	3.587	3.357	3.204	3.095	3.012	2.948	2.896	2.854
12	4.747	3.885	3.490	3.259	3.106	2.996	2.913	2.849	2.796	2.753
13	4.667	3.806	3.411	3.179	3.025	2.915	2.832	2.767	2.714	2.671
14	4.600	3.739	3.344	3.112	2.958	2.848	2.764	2.699	2.646	2.602
15	4.543	3.682	3.287	3.056	2.901	2.790	2.707	2.641	2.588	2.544
16	4.494	3.634	3.239	3.007	2.852	2.741	2.657	2.591	2.538	2.494
17	4.451	3.592	3.197	2.965	2.810	2.699	2.614	2.548	2.494	2.450
18	4.414	3.555	3.160	2.928	2.773	2.661	2.577	2.510	2.456	2.412
19	4.381	3.522	3.127	2.895	2.740	2.628	2.544	2.477	2.423	2.378
20	4.351	3.493	3.098	2.866	2.711	2.599	2.514	2.447	2.393	2.348
21	4.325	3.467	3.072	2.840	2.685	2.573	2.488	2.420	2.366	2.321
22	4.301	3.443	3.049	2.817	2.661	2.549	2.464	2.397	2.342	2.297
23	4.279	3.422	3.028	2.796	2.640	2.528	2.442	2.375	2.320	2.275
24	4.260	3.403	3.009	2.776	2.621	2.508	2.423	2.355	2.300	2.255
25	4.242	3.385	2.991	2.759	2.603	2.490	2.405	2.337	2.282	2.236
26	4.225	3.369	2.975	2.743	2.587	2.474	2.388	2.321	2.265	2.220
27	4.210	3.354	2.960	2.728	2.572	2.459	2.373	2.305	2.250	2.204
28	4.196	3.340	2.947	2.714	2.558	2.445	2.359	2.291	2.236	2.190
29	4.183	3.328	2.934	2.701	2.545	2.432	2.346	2.278	2.223	2.177
30	4.171	3.316	2.922	2.690	2.534	2.421	2.334	2.266	2.211	2.165

Appendix C — Statistical Tables

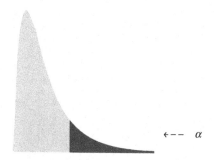

The Upper $\alpha \times 100$ Percentage Points of $F(v_1, v_2): \alpha = 0.05$

v_1	11	12	13	14	15	16	17	18	19	20
v_2										
1	243.0	243.9	244.7	245.4	245.9	246.5	246.9	247.3	247.7	248.0
2	19.40	19.41	19.42	19.42	19.43	19.43	19.44	19.44	19.44	19.45
3	8.763	8.745	8.729	8.715	8.703	8.692	8.683	8.675	8.667	8.660
4	5.936	5.912	5.891	5.873	5.858	5.844	5.832	5.821	5.811	5.803
5	4.704	4.678	4.655	4.636	4.619	4.604	4.590	4.579	4.568	4.558
6	4.027	4.000	3.976	3.956	3.938	3.922	3.908	3.896	3.884	3.874
7	3.603	3.575	3.550	3.529	3.511	3.494	3.480	3.467	3.455	3.445
8	3.313	3.284	3.259	3.237	3.218	3.202	3.187	3.173	3.161	3.150
9	3.102	3.073	3.048	3.025	3.006	2.989	2.974	2.960	2.948	2.936
10	2.943	2.913	2.887	2.865	2.845	2.828	2.812	2.798	2.785	2.774
11	2.818	2.788	2.761	2.739	2.719	2.701	2.685	2.671	2.658	2.646
12	2.717	2.687	2.660	2.637	2.617	2.599	2.583	2.568	2.555	2.544
13	2.635	2.604	2.577	2.554	2.533	2.515	2.499	2.484	2.471	2.459
14	2.565	2.534	2.507	2.484	2.463	2.445	2.428	2.413	2.400	2.388
15	2.507	2.475	2.448	2.424	2.403	2.385	2.368	2.353	2.340	2.328
16	2.456	2.425	2.397	2.373	2.352	2.333	2.317	2.302	2.288	2.276
17	2.413	2.381	2.353	2.329	2.308	2.289	2.272	2.257	2.243	2.230
18	2.374	2.342	2.314	2.290	2.269	2.250	2.233	2.217	2.203	2.191
19	2.340	2.308	2.280	2.256	2.234	2.215	2.198	2.182	2.168	2.155
20	2.310	2.278	2.250	2.225	2.203	2.184	2.167	2.151	2.137	2.124
21	2.283	2.250	2.222	2.197	2.176	2.156	2.139	2.123	2.109	2.096
22	2.259	2.226	2.198	2.173	2.151	2.131	2.114	2.098	2.084	2.071
23	2.236	2.204	2.175	2.150	2.128	2.109	2.091	2.075	2.061	2.048
24	2.216	2.183	2.155	2.130	2.108	2.088	2.070	2.054	2.040	2.027
25	2.198	2.165	2.136	2.111	2.089	2.069	2.051	2.035	2.021	2.007
26	2.181	2.148	2.119	2.094	2.072	2.052	2.034	2.018	2.003	1.990
27	2.166	2.132	2.103	2.078	2.056	2.036	2.018	2.002	1.987	1.974
28	2.151	2.118	2.089	2.064	2.041	2.021	2.003	1.987	1.972	1.959
29	2.138	2.104	2.075	2.050	2.027	2.007	1.989	1.973	1.958	1.945
30	2.126	2.092	2.063	2.037	2.015	1.995	1.976	1.960	1.945	1.932

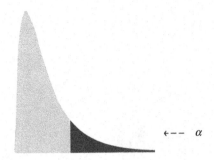

The Upper $\alpha \times 100$ Percentage Points of $F(v_1, v_2): \alpha = 0.025$

v_1 \ v_2	1	2	3	4	5	6	7	8	9	10
1	647.8	799.5	864.2	899.6	921.8	937.1	948.2	956.7	963.3	968.6
2	38.51	39.00	39.17	39.25	39.30	39.33	39.36	39.37	39.39	39.40
3	17.44	16.04	15.44	15.10	14.88	14.73	14.62	14.54	14.47	14.42
4	12.22	10.65	9.979	9.605	9.364	9.197	9.074	8.980	8.905	8.844
5	10.01	8.434	7.764	7.388	7.146	6.978	6.853	6.757	6.681	6.619
6	8.813	7.260	6.599	6.227	5.988	5.820	5.695	5.600	5.523	5.461
7	8.073	6.542	5.890	5.523	5.285	5.119	4.995	4.899	4.823	4.761
8	7.571	6.059	5.416	5.053	4.817	4.652	4.529	4.433	4.357	4.295
9	7.209	5.715	5.078	4.718	4.484	4.320	4.197	4.102	4.026	3.964
10	6.937	5.456	4.826	4.468	4.236	4.072	3.950	3.855	3.779	3.717
11	6.724	5.256	4.630	4.275	4.044	3.881	3.759	3.664	3.588	3.526
12	6.554	5.096	4.474	4.121	3.891	3.728	3.607	3.512	3.436	3.374
13	6.414	4.965	4.347	3.996	3.767	3.604	3.483	3.388	3.312	3.250
14	6.298	4.857	4.242	3.892	3.663	3.501	3.380	3.285	3.209	3.147
15	6.200	4.765	4.153	3.804	3.576	3.415	3.293	3.199	3.123	3.060
16	6.115	4.687	4.077	3.729	3.502	3.341	3.219	3.125	3.049	2.986
17	6.042	4.619	4.011	3.665	3.438	3.277	3.156	3.061	2.985	2.922
18	5.978	4.560	3.954	3.608	3.382	3.221	3.100	3.005	2.929	2.866
19	5.922	4.508	3.903	3.559	3.333	3.172	3.051	2.956	2.880	2.817
20	5.871	4.461	3.859	3.515	3.289	3.128	3.007	2.913	2.837	2.774
21	5.827	4.420	3.819	3.475	3.250	3.090	2.969	2.874	2.798	2.735
22	5.786	4.383	3.783	3.440	3.215	3.055	2.934	2.839	2.763	2.700
23	5.750	4.349	3.750	3.408	3.183	3.023	2.902	2.808	2.731	2.668
24	5.717	4.319	3.721	3.379	3.155	2.995	2.874	2.779	2.703	2.640
25	5.686	4.291	3.694	3.353	3.129	2.969	2.848	2.753	2.677	2.613
26	5.659	4.265	3.670	3.329	3.105	2.945	2.824	2.729	2.653	2.590
27	5.633	4.242	3.647	3.307	3.083	2.923	2.802	2.707	2.631	2.568
28	5.610	4.221	3.626	3.286	3.063	2.903	2.782	2.687	2.611	2.547
29	5.588	4.201	3.607	3.267	3.044	2.884	2.763	2.669	2.592	2.529
30	5.568	4.182	3.589	3.250	3.026	2.867	2.746	2.651	2.575	2.511

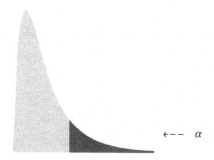

The Upper $\alpha \times 100$ Percentage Points of $F(v_1, v_2)$: $\alpha = 0.025$

v_1	11	12	13	14	15	16	17	18	19	20
v_2										
1	973.0	976.7	979.8	982.5	984.9	986.9	988.7	990.3	991.8	993.1
2	39.41	39.41	39.42	39.43	39.43	39.44	39.44	39.44	39.45	39.45
3	14.37	14.34	14.30	14.28	14.25	14.23	14.21	14.20	14.18	14.17
4	8.794	8.751	8.715	8.684	8.657	8.633	8.611	8.592	8.575	8.560
5	6.568	6.525	6.488	6.456	6.428	6.403	6.381	6.362	6.344	6.329
6	5.410	5.366	5.329	5.297	5.269	5.244	5.222	5.202	5.184	5.168
7	4.709	4.666	4.628	4.596	4.568	4.543	4.521	4.501	4.483	4.467
8	4.243	4.200	4.162	4.130	4.101	4.076	4.054	4.034	4.016	3.999
9	3.912	3.868	3.831	3.798	3.769	3.744	3.722	3.701	3.683	3.667
10	3.665	3.621	3.583	3.550	3.522	3.496	3.474	3.453	3.435	3.419
11	3.474	3.430	3.392	3.359	3.330	3.304	3.282	3.261	3.243	3.226
12	3.321	3.277	3.239	3.206	3.177	3.152	3.129	3.108	3.090	3.073
13	3.197	3.153	3.115	3.082	3.053	3.027	3.004	2.983	2.965	2.948
14	3.095	3.050	3.012	2.979	2.949	2.923	2.900	2.879	2.861	2.844
15	3.008	2.963	2.925	2.891	2.862	2.836	2.813	2.792	2.773	2.756
16	2.934	2.889	2.851	2.817	2.788	2.761	2.738	2.717	2.698	2.681
17	2.870	2.825	2.786	2.753	2.723	2.697	2.673	2.652	2.633	2.616
18	2.814	2.769	2.730	2.696	2.667	2.640	2.617	2.596	2.576	2.559
19	2.765	2.720	2.681	2.647	2.617	2.591	2.567	2.546	2.526	2.509
20	2.721	2.676	2.637	2.603	2.573	2.547	2.523	2.501	2.482	2.464
21	2.682	2.637	2.598	2.564	2.534	2.507	2.483	2.462	2.442	2.425
22	2.647	2.602	2.563	2.528	2.498	2.472	2.448	2.426	2.407	2.389
23	2.615	2.570	2.531	2.497	2.466	2.440	2.416	2.394	2.374	2.357
24	2.586	2.541	2.502	2.468	2.437	2.411	2.386	2.365	2.345	2.327
25	2.560	2.515	2.476	2.441	2.411	2.384	2.360	2.338	2.318	2.300
26	2.536	2.491	2.451	2.417	2.387	2.360	2.335	2.314	2.294	2.276
27	2.514	2.469	2.429	2.395	2.364	2.337	2.313	2.291	2.271	2.253
28	2.494	2.448	2.409	2.374	2.344	2.317	2.292	2.270	2.251	2.232
29	2.475	2.430	2.390	2.355	2.325	2.298	2.273	2.251	2.231	2.213
30	2.458	2.412	2.372	2.338	2.307	2.280	2.255	2.233	2.213	2.195

References

Barrett, G. C. and Elmore, D. T. (1998). *Amino Acids and Peptides*. London: Cambridge University Press.

Bertalanffy, L. von (1938). A quantitative theory of organic growth (Inquiries on growth laws II). *Human Biology*, 181–213.

Bethea, R. M. and Duran, B. S. (1985). *Statistical Methods for Engineers and Scientists. Second Edition, Revised and Expanded*. New York: Marcel Dekker.

Bickel, P. J. and Doksum, K. A. (2007). *Mathematical Statistics: Basic Ideas and Selected Topics. Volume 1, Second Edition*. New Jersey: Prentice Hall.

Bowler, P. J. (1989). *The Mendelian Revolution: The Emeregence of Hereditarian Concepts in Modern Science and Society*. Baltimore: Johns Hopkins University Press.

Canadian Longitudinal Study on Aging (CLSA).

Casella, G. and Berger, R. L. (2002). *Statistical Inference. Second Edition*. California: Duxbury.

Clopper, C. and Pearson, E. S. (1934). The use of confidence or fiducial limits illustrated in the case of the binomial. *Biometrika* **26**: 404–413.

Curtis, C. F. and Townson, H. (1998). Malaria: existing methods of vector control and molecular entomology. *British Medical Bulletin* **54** (2), 311–325.

Dennis, C. and Gallager, R. (Editors) (2001). *The Human Genome*. New York: Nature Palgrave.

Freeman, S. (2006). *Biological Science. Second Edition*. New Jersey: Prentice Hall.

Hartl, D. L. and Jones, E. W. (2009). *Genetics: Analysis of Genes and Genomes*. Ontario: Jones and Bartlett Publishing.

Hogg, R. V., McKean, J. W. and Craig, A. T. (2005). *Introduction to Mathematical Statistics. Sixth Edition*. Pearson Education.

Johnson, N. L. and Kotz, S. (1970). *Continuous Univariate Distributions*. Vols. 1 and 2. Boston: Houghton Mifflin.

Keeling, M. J. and Grenfell, B. T. (1997). Disease Extinction and Community Size: Modeling the Persistence of Measles. *Science* 275, 65–67.

Kendall, M., Stuart, A. and Ord, J. K. (1987). *Kendall's Advanced Theory of Statistics. Fifth Edition. Volume 1:* Distribution Theory. Oxford University Press.

Kendall, M. and Stuart, A. (1979). *The Advanced Theory of Statistics. Fourth Edition. Volume 2:* Inference and Relationship. Charles Griffin.

Kuehl, R. O. (1994). *Statistical Principles of Research Design and Analysis*. California: Duxbury.

Larsen, R. J. and Marx, M. L. (2001). *An Introduction to Mathematical Statistics and Its Applications. Third Edition*. New Jersey: Prentice Hall.

Milos, D. A. and Freyer, G. A. (1990). *DNA Science: A First Course in Recombinant DNA Techniques*. New York: Cold Spring Harbor Laboratory Press.

Montgomery, D. C. (1991). *Design and Analysis of Experiments. Third Edition*. New York: John Wiley.

Royston, R. (2001). The Lognormal Distribution as a Model of Survival Time in Cancer, with an Emphasis on Prognostic Factors. *Statistica Neerlandica* 55, 89–104.

Runyon, D. and Bentley, E. C. (2008). *More than Somewhat*. Willard Press.

Samuels, M. L. and Witmer, J. A. (2003). *Statistics for the Life Sciences. Third Edition*. New Jersey: Pearson.

References

Seber, G. A. F. (2002). *The Estimation of Animal Abundance and Related Parameters. Second Edition.* Caldwell, New Jersey: Blackburn Press.

Stewart, J. (2011). *Calculus, Early Transcendentals. Seventh Edition.* California: Brooks/Cole.

Streiner, D. L. and Norman, G. R. (2008). *Health Measurement Scales: a practical guide to their development and use. Fourth Edition.* New York: Oxford University Press.

U. S. National Library of Medicine.

Wilks, D. S. (2006). *Statistical methods in the Atmospheric Sciences. Second Edition.* Burlington, MA: Academic Press.

Yates, F. (1981). *Sampling Methods for Censuses and Surveys. Fourth Edition.* London: Charles Griffin.

Index

2×2 contingency table, 180–184
$r \times c$ contingency table, 185–186
$2 \times c$ contingency table, 186–187

A

absolute frequency, 135
allele, 3, 9
alternate hypothesis, 154–158
analysis of variance, 191–220
ANOVA, 191–220
ANOVA table, 194, 195, 198, 200
approximate confidence interval, 150, 161
approximate standard error, 116
asymptotically efficient, 108

B

balanced design, 192, 193, 196, 204, 206
bar chart, 7, 28, 132, 133
basic event, 10
Bayes' Theorem, 17, 112
Bernoulli
    distribution, 54–55
    experiment, 53
    random variable, 55
    trial, 55, 152–153
Beta
    distribution, 81–82
    random variable, 81
biased, 102
bimodal, 127
binomial
    distribution, 55–59
    experiment, 56
    random variable, 56
bins, 132, 135, 139
bivariate distribution, 35, 40, 211, 215
bivariate normal distribution, 94, 213, 215
block design, 206
blocking, 197
box-and-whisker plot, 135
box plot, 135, 137

C

cardinality, 19
categorical data, 7, 132
categories, 16, 53, 132
causal model, 212
cdf, 27, 31, 43
cells, 177, 179, 182, 185
censoring, 130
Central Limit Theorem, 179, 150, 152, 164
central moments, 73

centroid, 215
Chebyshev's Theorem/Inequality, 42, 130
chi-square
    distribution, 80
    random variable, 88, 179
classes, 27, 135
clinical trials, 5
Clopper-Pearson exact method, 153
Cochran's Theorem, 88, 89, 194, 198, 199, 214
coefficient of determination, 41, 213
coefficient of variation, 131
combination, 12
comparative study, 201
complement, 20
complete block experiment, 198
complete blocks, 206
completely randomized design, 124, 191
composite hypothesis, 154
conditional distribution, 25, 34
conditional probability, 16–18
confidence, 3, 80
confidence coefficient, 148
confidence interval: symmetric, 148
confidence level, 148
confidence lower bound, 149
confidence upper bound, 149
confounding, 207
consistent, 102
contingency table, 180–187
continuity correction factor, 86
continuous distribution, 29–33
continuum, 19, 29
correlation, 5
correlation analysis, 212, 214, 215
cost, 3
countable, 19, 29
counting principles, 12
covariance, 41
covariate, 212
CRD, 191, 194, 198, 206
critical region, 157, 158

D

data, 3
datum, 7
decile, 131
degrees of freedom, 90, 91
descriptive statistics, 7, 28, 115, 125–127
design variable, 192, 196
directional alternative, 183–184
discrete distribution, 27–29
distribution, 15
dot diagram, 7

dot plot, 135
double blind experiment, 201

E

efficient, 102
element, 14
elementary event, 10
empirical distribution, 29, 179
Empirical rule, 42–43
empty set, 14, 19
error sum of squares, 111
errors, 202
estimate a single mean, 161
estimate a single proportion, 162
estimate a single variance, 162–163
estimated residuals, 208
estimates for two proportions, 166
estimates for two variances, 166–167
estimator, 99
event, 10
excess kurtosis, 83
exchangeability, 101
exhaustive events, 12
expectation, 36–43
expectation operator, 37
expected value, 36, 128
experimental design, 3, 10, 125, 200–208
experimental unit, 7, 10, 203
explanatory variable, 202, 204
exponential
    distribution, 79–80
    random variable, 99, 110

F

F
    distribution, 92–93
    random variable, 92
factor, 203
factorial designs, 207
false negative, 5
false positive, 5
fixed effects, 191–196

G

gamma
    distribution, 76–80
    random variable, 76, 77
genomics, 3
geometric
    distribution, 61
    experiment, 61
    random variable, 61

Index

global F test, 191–197
goodness of fit test, 177–180
Graeco-Latin squares, 207

H

heterogeneous, 201
hierarchical design, 207–208
histogram, 28, 139
    frequency, 135
    relative frequency, 135
homogeneous, 183
hypergeometric
    distribution, 61–62
    experiment, 62
    random variable, 62

I

iid, 100, 126
incomplete block design, 197–198
independence, 5, 182, 202
independent populations, 6, 166
indicator function, 179
inference, 2, 117
inferencial statistics, 2
interaction, 197
interquartile range, 131
intersection, 14, 20
interval estimate, 146
    mean, 146–149
    proportion, 152–153
    variance, 150–152

J

joint cdf, 33, 34
joint distribution, 33, 34, 41, 67
jpdf, 33, 35

K

Kolmogorov-Smirnov Test, 179

L

Latin square, 200, 206–207
least squares estimation, 111
left tail of a distribution, 44, 151
likelihood function, 103, 104
line chart, 135, 136, 137
linear model, 204
linear operator, 41
linear regression, 200, 211–218
log likelihood function, 104
log normal
    distribution, 86–87
    random variable, 86
longitudinal study, 201

M

marginal distribution, 33–34
maximum likelihood estimation, 103–111
maximum likelihood estimator, 103
mean, 128
mean absolute deviation, 131
mean corrected moments, 37
mean square error, 194
mean square for treatments, 194
measure of kurtosis, 46
measure of scale, 129
measure of skewness, 46
median, 43–44, 127
metric data, 7
mgf, 36, 37
MLE, 103, 105
modal frequency, 131
mode, 43, 127
Model 1, 191–196
Model 2, 191, 196–197
moment estimation, 110
moment generating function, 36
multimodal, 43
multinomial distribution, 63–64
multiplication principle, 12–13
multivariate distribution, 36–37
multivariate hypergeometric distribution, 63–64
mutually exclusive events, 12

N

negative binomial
    distribution, 59–61
    experiment, 59
    random variable, 60
nested design, 207–208
non-central moments, 73
non-linear model, 204
non-linear regression, 41
non-parametric test, 167
normal approximation to binomial, 85–86
normal
    distribution, 82–87
    random variable, 87–93
nuisance parameter, 149
null hypothesis, 154

O

observational study, 201
Occam's Razor, 4
odds, 18
ogives, 135
one sided alternative, 161
order statistics, 126
outcome, 7
outlier, 85

P

paired comparison, 6, 165
parameter, 51, 52
parent distribution, 101
Pareto diagram, 7
Pascal distribution, 59
pdf, 27, 29
Pearson's correlation coefficient, 41
percentile, 44, 131
permutation, 12
pie chart, 133
pilot project, 205
pivot, 115
pivotal statistic, 92, 115
placebo, 6, 26
placebo effect, 201
point estimate, 99
Poisson distribution, 64–66
pooled estimate of the variance, 164
population, 2, 7, 160–161, 163–166
population variance, 129, 163–164
power, 2, 155, 156, 180, 202
precision, 202
prediction, 4–5, 211–218
principle of maximum likelihood, 103–104
probability, 3, 9–20
probability density function, 29–33, 73
probability mass function, 222
proteomics, 3–4
p-value, 158–159

Q

quantile, 44, 131
quartile, 44, 131

R

random effects, 191, 196–197, 203, 212
random experiment, 2, 7
random number table, 7
random variable, 7, 25–46, 87–90
randomization, 201
ranked data, 7
Rao-Cramer Inequality, 105
realization, 11, 101, 103
regression analysis, 211, 212, 214
regression coefficients, 211, 212
regressor variable, 212
relative frequency, 25–26, 132
repeated measures, 201
replacement, 7
replication, 202, 206
residuals, 192, 202
response variable, 203
right tail of a distribution, 44

S

sample, 7
sample variance, 89, 129
sampling distribution of the mean, 114
sampling distribution of the proportion, 114
sampling distribution of the variance, 114
scatter plot, 212
score, 7
semi-interquartile range, 131
sign test, 167
significance level, 148, 155

simple hypothesis, 154
simple random sample, 7, 10, 100
single blind experiment, 201
skew, 3
skewness, 46
solution set, 18
special continuous distributions, 74–82
    Beta, 81–82
    bivariate normal, 94
    chi-sqare, 80
    exponential, 79–80
    F, 92–93
    gamma, 76–79
    log normal, 86–87
    normal, 82–86
    standard normal, 84, 91
    Student's t, 90–92
    uniform, 74–76
    Weibull, 95
special discrete distributions, 52–66
    Bernoulli, 54–55
    binomial, 55–59
    geometric, 61
    hypergeometric, 61–67
    multinomial, 64–65, 66–67
    multivariate hypergeometric, 63–64, 223
    negative binomial, 55, 59–61
    Pascal, 61
    Poisson, 64–66
    uniform, 51, 53
standard deviation, 36–37

standard error, 114, 116, 147
standard normal
    distribution, 84, 91
    random variable, 88
standardization, 39, 84
statistical association, 5
stem-and-leaf display, 134
stochastic, 1
stratified random sampling, 100
Student's t
    distribution, 90–92
    random variable, 90
sufficient, 102
sum of squares for error, 194
sum of squares for treatments, 194
sum of squares total, 193
symmetric, 44
symmetric confidence interval, 148

T

test of hypothesis, 117, 155, 159–160
total probability, 17
tractable, 102
treatment effect, 193
treatment sum of squares, 194
two means
    dependent populations, 165
    independent populations, 163–165
two-sided alternative, 167, 183
two-way classification, 199–200
type 1 experiment, 183–184

type 2 experiment, 184
type 3 experiment, 184
type I error, 155
type II error, 155

U

unbiased, 102
uncorrected moments, 37
uncountable, 19, 29
uniform distribution: continuous
    distribution, 74–76
    random variable, 74
uniform distribution: discrete
    distribution, 53
    experiment, 53
    random variable, 53
union, 20
universal set, 20

V

variance, 36–37
variation ratio, 131
Venn diagram, 14

W

Weibull distribution, 95
Wilcoxon-Mann-Whitney test, 167
Wilcoxon rank test, 167
without replacement, 7